Current Developments in Agricultural Research

Current Developments in Agricultural Research

Edited by **Laura Vivian**

SYRAWOOD
PUBLISHING HOUSE

New York

Published by Syrawood Publishing House,
750 Third Avenue, 9th Floor,
New York, NY 10017, USA
www.syrawoodpublishinghouse.com

Current Developments in Agricultural Research
Edited by Laura Vivian

International Standard Book Number: 978-1-68286-080-9 (Hardback)

Contents

Preface

Agricultural research is aimed at improving the quality and production of crops, irrigation, plant protection, management of agricultural resources, etc. This book focuses on the state-of-the-art research and developments in the fields of genetic engineering, plant breeding, food science and technology, role of agriculture in regional economic development, sustainable techniques and practices in agriculture, etc. As this field is emerging at a rapid pace, the contents of this book will help the readers understand the modern concepts and applications of the recent developments in agricultural research.

Significant researches are present in this book. Intensive efforts have been employed by authors to make this book an outstanding discourse. This book contains the enlightening chapters which have been written on the basis of significant researches done by the experts.

Finally, I would also like to thank all the members involved in this book for being a team and meeting all the deadlines for the submission of their respective works. I would also like to thank my friends and family for being supportive in my efforts.

Editor

FEASIBILITY STUDY OF CASSAVA MEAL IN BROILER DIETS BY PARTIAL REPLACING ENERGY SOURCE (CORN) IN REGARD TO GROSS RESPONSE AND CARCASS TRAITS

M.A. Hossain[1*], J.R. Amin[2] and M.E. Hossain[3]

Abstract

Day-old unsexed broiler chicks (Cobb 500) were used to investigate the growth responses and meat yield traits by nourishing them under four dietary treatment group in inclusion of cassava meal with partial substitution of valuable energy source (maize or corn). A total of 260 birds were assigned randomly into four dietary treatments [T_1 (0%); T_2 (10%); T_3 (20%), and T_4 (30%)]; each replicated 5 times, 13 birds/replicate in a completely randomized design. Birds were reared with *ad libitum* feeding, similar housing, and environmental management condition from d1-35 days. Growth responses of broilers in terms of feed intake, body weight, feed conversion ratio (FCR), livability and meat yield traits were assessed in this study. Except for first week, the feed intake of birds on cassava treated diets up to 21 and 35 days was significantly ($p<0.001$) higher than the non-cassava or control diet (T_1). Feed intake of broilers on cassava treated diet (T_4) had the highest (2795.8 g/b) while the birds of control (T_1) diet group consumed the lowest feed (2771.9 g/b) at 35 day. Live weight decreased ($p<0.001$) with the inclusion of cassava meal diets up to 21 day only; and live weights up to 7 and 35 days were identical between treatment. FCR up to 21 day was improved ($p<0.001$) on control diet (T_1) and deteriorated on cassava treated diets. FCR values up to 7 and 35 days were unaffected by all the treatment groups. Neither the livability nor the meat yield traits (thigh, breast, drumstick, shank, neck, giblet, wing and dressing yield) of broiler chickens was influenced by the dietary treatments regardless of feeding broiler chickens with cassava or non-cassava diets in this study. It may be deduced based on the present findings that, cassava tuber meal might be a potential ingredient to replace the costly maize up to 300 g kg^{-1} without affecting growth performance of the broiler chickens.

Keywords: Cassava Tuber Meal, Performance, Meat Yield Characteristics

[1]Associate Professor, Department of Dairy and Poultry Science, Chittagong Veterinary and Animal Sciences University, Khulshi, Chittagong-4225, Bangladesh

[2]School of Environmental and Rural Science, Armidale, NSW 2351, Australia

[3]Department of Animal Science and Animal Nutrition, Chittagong Veterinary and Animal Sciences University, Khulshi, Chittagong-4225, Bangladesh

*Corresponding author's email: rony745@gmail.com (M.A. Hossain)

Introduction

In poultry production, feed costs are the principal cost encompassing more than 70% (McNab, 1999), of which dietary energy sources occupy the greatest portion (70 -75% of the diets) (Van der Klis *et al.*, 2010). Birds have a normal tendency to eat feeds mainly to gratify their energy requirements and once this is satisfied, they won't show any trend to consume any more feeds, even if the requirements of other nutrients like protein, vitamins or minerals have not been met (Singh and Panda, 1992). For this reason, the energy ingredients of the diets play a pivotal role in diet formulation for poultry. Most of the energy ingredients come from plant sources in the form of starch from cereal grains. Corn/maize and wheat are the main cereal grains that are used predominantly as energy sources for manufacturing poultry diet. These cereal grains provide the energy component which accounts for 60 to 70% of the nutrient requirement for the poultry.

However, the considerable increase in the cost of poultry diets round the world has been driven a force to search for cheaper sources of dietary energy to be used to partially substitute the cereal grains traditionally used in broiler diets (Stevenson and Jackson, 1983). Cassava tuber meal might be as an alternative energy source to replace the costly cereal grains (corn/wheat), and to minimize the feed cost for poultry production. Its use in animal feeds not new, and its cultivation in the tropical countries are predominant. Africa alone is produced around 50% cassava of the world. Besides this, Nigeria, Brazil and Thailand are the most important cassava producing countries, with Thailand being the world's leading cassava exporter (Garcia and Dale, 1999).

Under tropical conditions cassava (*Manihot esculenta*), a tuber, is the most productive crop in terms of stability of production and high energy yield per unit of land area (Oke, 1978; Okezie *et al.*, 1982; Ravindran and Blair, 1991). Cassava root meal is mainly a source of energy, with a high starch content (about 60 -70%), and the yield may be counted less than 12.5 tonnes acre[-1]. However, the level of protein is very low, being approximately 2.5% of dry matter (Garcia and Dale, 1999).

The preconditions of profitable poultry business depend on the availability of quality feed ingredients at reasonable price. As major energy components (50-70%), maize and wheat are being used traditionally in poultry and other mono-gastric animal's diet formulations for long since (PAN, 1995). Apart from this, these ingredients (corn/wheat) are also being used to meet the ever increasing food demand of human beings. Owing to this food-feed competition between animals and human beings, the cost of cereal grains is going up day by day. High feed cost hardly permits remunerative and profitable poultry farming. Therefore, the use of cereal products as livestock feed is being increasingly unjustified in terms of economic stand point. Besides, the low production and high demand of cereal grains for both men and animals have been creating constant force to explore other potential energy sources (Raihan *et al.*, 2008). Therefore, there is a need to explore cheaper energy sources, to replace expensive cereals for livestock production, and to ease the food-feed competition in the future.

However, using inexpensive, unconventional feeds, locally available ingredients in poultry diets by replacing the costly grains, can become an effective way to minimize the feed cost and maximize the profitability of broiler chickens suggested by the Nutritionists (Hossain *et al.*, 1989). For this reason, in the past few decades, many efforts have been made to replace cereals with other carbohydrate sources, for example, sorghum, lentils, or cassava tuber meal (Garcia and Dale, 1999), Leucaena root meal (Bhatnagar *et al.*, 1996) in poultry feed. Cassava (*M. esculenta*) can be given more emphasis for using in poultry diets amongst others, because of its better yield, and rich in starch (60 to 70%) content. Furthermore, cassava is an important food crops grown in the tropical areas (Hahn, 1989; Phillips *et al.*, 2004) like Africa, where it provides a significant amount of energy for 500 million people (Mroso, 2003). If we can process properly the anti-nutritive factor (HCN) found in cassava meal can be minimized to non-toxic level (Padmaja, 1995). With the light of above advantages, the study was undertaken to investigate the growth responses and meat yield characteristics of broiler chickens fed diets incorporating with cassava tuber meal by partial replacement with corn.

Materials and Methods

Animal husbandry and bird management

A total of 260 day-old (Cobb 500) unsexed commercial broiler chicks were used to conduct this experiment. The chicks were divided randomly into 4 dietary treatment groups [T_1, T_2, T_3 and T_4]; each treatment replicated 5 times, 13 birds per replicate in a completely randomized design. The birds were reared under floor-pen system in a wire-netted house from d1 to 35 days. A total of 20 wire-netted pens of equal size (140 × 90 cm) were set up in a shed type house to accommodate the birds. Each pen was designed for 13 birds, and was equipped with a feeder and a drinker to supply diets and water *ad libitum* for the birds throughout the trial period. Rice husk litter materials to a depth of 2.5 cm were spread over the floor of each pen to maintain comfortable environment.

The birds were brooded at 33°C for the first two days, and then the temperature was reduced gradually to 24°C at 19 days of age, and maintained at this level to the end of the trial. Continuous lighting (23 hours light and 1 hour darkness) including the natural light and electrical bulb was provided the birds throughout the trial period. All the birds received the necessary vaccines against Newcastle disease, Infectious bursal diseases and Infectious Bronchitis disease and some medication for disease protection and immunity.

Dietary treatment

Processing cassava tuber

Collected cassava tubers were cleaned by removing dust, sand or any other foreign materials attached with these ingredients, then they were sliced, chopped, crushed and dried in the sunlight for 3 days. After drying tubers, meal was prepared by grinding sun-dried cassava tuber slices in a Mortar and Pestle. After that, these ground cassava meal (CM) was used in formulating diets at various dietary levels of 0, 100, 200, and 300 g kg[-1] CM by replacing maize (shown below in Tables 1 and 2).

Formulation of diets

Four experimental diets, identified as T_1, T_2, T_3 and T_4 were formulated with maize, cassava meal, and vegetable oil as main energy sources, along with soybean meal and fishmeal as protein ingredients (Tables 1 and 2). Except for diet T_1, cassava meal was included at the rate of 10%, 20% and 30% in T_2, T_3 and T_4 diets, respectively, by replacing the corn. All the diets were prepared to meet or exceed NRC (1994) recommendations, and supplied the birds in mash form in both starter and finisher period. All diets were iso-caloric and iso-nitrogenous and supplemented with exogenous microbial enzymes (Avizyme 0.5g and Phyzyme 0.1 g kg[-1] of each diet, Danisco Animal Nutrition, UK). Starter diets were fed the birds for the first three weeks, and finisher diets were used for rest of the trial period (22 to 35 d).

Table 1. Starter broiler diets

Ingredients (%)	Diets			
	T_1	T_2	T_3	T_4
Corn	60.00	54.00	48.00	42.00
Cassava meal	0.00	6.00	12.00	18.00
Soybean meal	22.80	23.00	24.00	24.20
Fishmeal	8.30	8.60	8.60	9.00
Limestone	3.00	3.00	3.00	2.40
Dicalcium phosphate	4.50	4.10	3.10	3.10
DL-methionine	0.21	0.21	0.21	0.21
L-Lysine	0.19	0.19	0.19	0.19
Table salt(NaCl)	0.50	0.50	0.50	0.50
[1]Vit- min-premix	0.25	0.25	0.25	0.25
Choline Chloride	0.06	0.06	0.06	0.06
Avizyme 1502	0.05	0.05	0.05	0.05
Phyzyme XP	0.01	0.01	0.01	0.01
Nutrient content				
ME (MJ/kg)	12.70	12.80	12.70	12.70
CP (%)	21.00	21.00	21.00	21.00
CF (%)	3.40	2.60	2.60	2.70
EE (%)	1.70	2.10	2.03	2.00
Ca (%)	2.00	2.00	1.80	1.64
Av.P (%)	0.98	0.91	0.73	0.73
Na (%)	0.24	0.23	0.24	0.23
Cl (%)	0.35	0.34	0.34	0.33
Lysine (%)	1.31	1.30	1.32	1.31
Arginine (%)	1.29	1.30	1.30	1.29
Methionine (%)	0.58	0.57	0.56	0.55

[1]Provided per kg of diet (mg): vitamin A (as all-trans retinol), 3.6 mg; cholecalciferol, 0.09 mg; vitamin E (as d-α-tocopherol), 44.7 mg; vitamin K_3, 2 mg; thiamine, 2 mg; riboflavin, 6 mg; pyridoxine hydrochloride, 5 mg; vitamin B_{12}, 0.2 mg; biotin, 0.1 mg; niacin, 50 mg; D- calcium pantothenate, 12 mg ; folic acid, 2 mg; Mn, 80mg; Fe, 60 mg; Cu, 8 mg; I, 1 mg; Co, 0.3 mg and Mo, 1 mg.

Table 2. Finisher broiler diets

Ingredients (%)	Diets			
	T_1	T_2	T_3	T_4
Corn	59.00	53.10	47.20	41.30
Cassava meal	0.00	5.90	11.80	17.70
Vegetable oil	0.70	0.70	1.70	2.10
Soybean meal	21.10	22.00	23.00	24.00
Fishmeal	8.20	8.20	8.20	8.20
Limestone	4.00	3.32	2.67	2.27
Dicalcium phosphate	5.60	5.50	4.10	3.10
DL-methionine	0.21	0.21	0.21	0.21
L-Lysine	0.19	0.19	0.19	0.19
Table salt (NaCl)	0.50	0.50	0.50	0.50
[1]Vit- min-premix	0.25	0.25	0.25	0.25
Choline Chloride	0.06	0.06	0.06	0.06
Avizyme 1502	0.05	0.05	0.05	0.05
Phyzyme XP	0.01	0.01	0.01	0.01
Nutrients content				
ME (MJ/kg)	12.90	12.90	12.90	12.90
CP (%)	20.00	20.00	20.00	20.00
CF (%)	3.30	3.40	3.40	3.50
EE (%)	1.60	1.60	1.50	1.40
Ca (%)	2.40	2.30	2.00	2.00
Av.P (%)	1.30	1.20	1.00	1.00
Na (%)	0.25	0.23	0.24	0.23
Cl (%)	0.35	0.34	0.34	0.33
Lysine (%)	1.25	1.26	1.27	1.28
Arginine (%)	1.29	1.30	1.30	1.29
Methinine (%)	0.56	0.55	0.55	0.54

[1]Provided per kg of diet (mg): vitamin A (as all-trans retinol), 3.6 mg; cholecalciferol, 0.09 mg; vitamin E (as d-α-tocopherol), 44.7 mg; vitamin K_3, 2 mg; thiamine, 2 mg; riboflavin, 6 mg; pyridoxine hydrochloride, 5 mg; vitamin B_{12}, 0.2 mg; biotin, 0.1 mg; niacin, 50 mg; D- calcium pantothenate, 12 mg ; folic acid, 2 mg; Mn, 80mg; Fe, 60 mg; Cu, 8 mg; I, 1 mg; Co, 0.3 mg and Mo, 1 mg.

Data collection

Body weight and feed intake data of the birds were recorded weekly. Mortality was recorded as it occurred. Feed conversion ratio was calculated weekly, and corrected for mortality. At the end of the trial period (35d), two birds from each pen were randomly selected, weighed and killed humanely to measure different body parts (thigh, shank, breast, giblet, neck, wing, drumstick and dressed yield) of the birds. These were described below.

Performance indices: Data were collected weekly on the parameters listed below:

Feed intake: Feed intakes of broiler chickens were assessed weekly basis. The amount of feeds served the birds and what were left unconsumed (collected daily) were weighed weekly. Then weekly feed intake was obtained from the difference of the amount of feed consumed by the birds and the left-over found in each replicate cage. To obtain the consumption per day, weekly feed intake was divided by 7 and the value further divided by the number of birds in each replicate to get consumption per day per bird.

Body weight and weight gain: Body weights of broiler chickens were also assessed weekly. Chicks were weighed initially in group (each replicate cage) before distributing them into each replicate cage. Weights of the birds in different replicates were taken weekly, and weights gained for the week were obtained by deducting the weekly average weight of each bird from the initial weight of the birds. The values were divided by the number of chicks to get gain per chick per week and from this gain per day was calculated as below:

Body weight gain= Weekly acquired body weight (g) – Initial body weight (g)

Feed/gain ratio or feed conversion ratio (FCR): The unit of feed needed per unit of production is called feed conversion ratio. This was calculated as the ratio of the feed consumed to the weight gained as follows:

$$FCR = \frac{Amount\ of\ feed\ consumed}{Body\ weight\ gain}$$

Livability (%): The percentage of livability was calculated by deducting mortality from hundred each replicate wise.

Organs and carcass yields studies: On the last day of the trial period (35d), two birds from each replicate cage (2 × 5 × 4=40 birds) representing the average body weight were randomly selected, weighed and slaughtered humanely. After slaughtering and bleeding, the birds were scalded and feathers were plucked. Carcasses were eviscerated, heads and shanks were separated, then the carcasses were chilled in a tap water for about 5 to 10 minutes. Eviscerated carcasses were individually weighted and dressing percentage was calculated (weight of carcass + giblet + abdominal fat/pre-slaughter weight × 100). The relative weights of other organs of carcass, for example, shank, neck, giblet (liver + heart + gizzard), thigh, drumstick, breast and wing weights were measured each replicate wise.

Statistical analysis

Statistical analyses were performed using Minitab software (Minitab version 15, 2000). The data were analyzed using one-way ANOVA with diet as factor. The significance of difference between means was determined by Fisher's least significant difference at $p \leq 0.05$.

Results

Gross responses

The results of gross responses of broilers in terms of feed intake, live weight and feed conversion ratio are shown in Table 3; and livability of birds is presented graphically below (Fig. 1).

Feed intake

The results of feed intake of broilers up to 7 day were identical between dietary treatment groups (Table 3). Except for first week, the feed intake up to 21 and 35 days differed significantly (p<0.001) between treatment. Birds on T_4 diet group had the highest feed intake while the birds on T_1 dietary group consumed the lowest feed during 21 and 35 days, respectively. The feed consumption of broilers on T_1 diet group was similar to the birds on T_2 diet group during 21 and 35 days, respectively.

Table 3. Feed intake (FI), live weight (LW), feed conversion ratio (FCR), of broiler chickens fed different diets from d1-35

	Age (day)	Dietary Treatments				Pooled SEM
		T_1	T_2	T_3	T_4	
FI (g/b)	1-7	135.8	132.6	133.0	129.3	1.23
	1-21	974.3[b]	983.7[b]	994.7[ab]	1000.6[a]	1.63
	1-35	2771.9[c]	2779.9[c]	2787.3[b]	2795.8[a]	1.29
LW (g/b)	1-7	134.1	132.3	131.0	128.1	0.73
	1-21	641.4[a]	632.8[b]	622.4[c]	614.9[d]	1.05
	1-35	1437.7	1424.8	1430.1	1415.7	9.08
FCR	1-7	1.55	1.55	1.59	1.59	0.023
	1-21	1.64[d]	1.68[c]	1.73[b]	1.76[a]	0.004
	1-35	2.00	2.10	2.12	2.13	0.013

Data represent means of five replicates consisting of 13 birds per replicate during d1-35 days; [a,b,c,d]Means bearing uncommon superscript within a row are significantly different at ***p<0.001; SEM= Standard error of mean.

Live weight

Live weight was similar between the treatment during first and fifth weeks. Apart from this, live weight up to 21 (p<0.001) days differed significantly between the treatment. Bird on control (T_1) diet had the highest body weight while the birds on T_4 diet being the least at 21day.

Feed conversion ratio (FCR)

FCR did not affect significantly (p>0.001) during first and fifth weeks only. After that, FCR up to 21 days significantly improved (p<0.001) between the treatment. Birds on T_1 diet group had the superior FCR while the birds on T_4 diet group being the poorest during 21day.

Livability (%)

The Livability of broilers of different dietary group was 96.9%, 95.4%, 95.4%, and 96.9% in T_1, T_2, T_3 and T_4 treatment groups, respectively, but did not differ significantly at 35 day (Fig. 1).

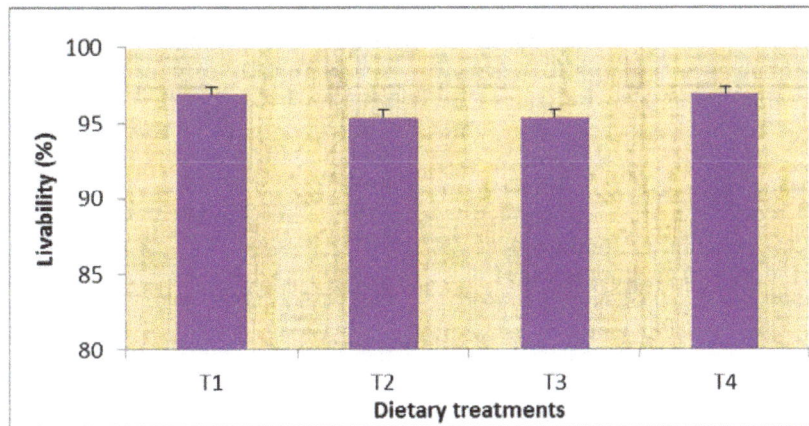

Fig. 1. Livability (%) of broilers of different dietary groups at 35 day (values mean ± SE)

Meat yield characteristics

The results of meat yield of different body parts (thigh, breast, drumstick, shank, neck, giblet, wing and dressing yield) of broilers fed different diets are shown in Table 4. All the body parts measured in this study were found to be similar between the treatment without affecting significantly. Apart from this, thigh, breast, drumstick, neck and wing weights (g/b) of broilers of different dietary treatments are tended to be significant (p=0.064 to 0.099).

Table 4. Meat yield (g/b) of various body parts of broilers fed different diets at age of 35d

Parameters	Dietary Treatments				Pooled SEM
	T_1	T_2	T_3	T_4	
Live weight (g/b)	1437.7	1424.8	1430.1	1415.7	9.08
Shank weight	52.2	48.1	46.3	46.1	1.19
Neck weight	50.7	46.7	48.6	42.2	1.12
Giblet weight	87.7	87.3	78.7	82.3	1.36
Thigh weight	134.0	131.1	132.4	131.5	0.40
Drumstick weight	131.5	124.7	126.2	122.6	1.14
Wing weight	129.3	125.8	125.2	121.5	0.89
Breast weight	237.1	227.4	222.5	223.0	2.13
Dressing (%)	62.3	61.1	60.5	60.2	0.35

Data represent mean values of two chicken of five replicate groups at age of 35d; SEM-Standard error of mean

Discussion

Feed intake

Feed intake of broilers on cassava treated diets was significantly increased throughout the trial period with having no effect on first week only. It would appear from the feed consumption of broilers on cassava treated diets that, cassava meals does not affect palatability of the feeds. So the increased feed intake of broilers on cassava meal diets in this study might be resulted from the palatability of feeds and some other factors might stimulate birds to ingest higher feeds (Tewe, 1993; Maust et al., 2000; Onjoro et al., 2001; Onyimoyi and Ugwu, 2007). However, feed intake was improved by inclusion of cassava meal (CM). It implied   that CM did not hamper the palatability of the formulated diets used in this study. On the other hand, the impaired feed intake of broilers offered on non-cassava meal diet may be possibly due to reduced palatability of the diets. Lower palatability of the diets might be another reason for reduced feed consumption by the broiler chickens (Mahmoudnia *et al.*, 2011). Other deleterious substances which may be present in the diets are protease inhibitors, lectins, polyphenolic compounds, saponins, HCN, tannin and non-starch polysaccharides (Reed *et al.*, 1982; Piva, 1987; Hughes and Choct, 1999), can also affect the feed consumption.

Live weight

Except for 21 day, the live weight of birds offered with cassava treated diets or non-cassava meal diet was found to be similar in this study. However, live weight of broiler was slightly decreased on cassava treated diets during mid growth point (21 day). This result was agreed with the Vogt (1966) who also found depressed performance of broilers by feeding cassava diets. Reduction of live weight may be due to the high level of fiber and poor protein contents in cassava meal diets which could reduce the digestibility of diets and subsequently affected the growth of broilers (Akintala *et al.*, 2002) at this stage. Apart from this, live weight of broilers on either diet groups was unaffected during first week and 35 days, respectively. These results are agreed with the findings of Stevenson and Jackson (1983) who found similar results either treating cassava meal diets or non-cassava meal diets to the broiler chickens.

Feed Conversion Ratio (FCR)

FCR values were unaffected between the treatments up to 7 and 35 days except for 21 day. FCR values of broilers were impaired by feeding cassava treated diets during 21 day. This would imply that birds fed diets with cassava meals demonstrated poorer feed efficiency during this period than the birds on non-cassava diets. The reason may be explained that cassava treated diets might have lower digestibility and lower protein and amino acids which might influence the feed efficiency of the broiler chickens. Moreover, some deleterious factors (HCN, tannin) are found in cassava meal (Reed *et al.*, 1982; Stevenson and Jackson, 1983; Ravindran *et al.*,1987; Garcia and Dale,1999), which might also affect the FCR of broiler chickens.

Livability (%)

In this present study, percentage of livability was insignificant among the dietary treatments suggesting that, CM did not cause any fatal effect to boiler chicken (Akintala *et al.*, 2002), and can be used safely in broiler diets instead of using cereal grains (i.e maize and wheat).

Meat yield characteristics

The meat yield characteristics of broiler chickens regardless of dietary treatments were preserved the similar characteristics and insignificant in terms of statistical analysis, suggesting that dietary supplementation of CM and feeding time did not affect notably on the meat yield characteristics and different organs of broiler chickens (Awojobi and Adekunmi, 2002).

Conclusion

It would appear from the performance of the broiler chickens that cassava meal could be used to replace corn in broiler diets at levels up to 300 g cassava per kg diet. This view is supported by the fact that dried cassava root meal had no adverse effect on growth responses in terms of live weight and FCR, livability, and meat yield characteristics. However, although the feed intake of broilers fed on cassava treated diets was a little bit higher than the non-cassava or control diet in this study, but it would no longer affect the profitability of rearing broiler chickens, because the price of cassava is about half than that of the corn/maize. From the results of this experiment, it can be assumed that, the maximum level of cassava meal of the type used in the experiment, which could be included in broiler diets is about 30%.

References

Akintala, E.O., Aderibigbe, A.O. and Matanmi, O. 2002. Evaluation of the nutritive value of cassava tuber meal as replacement for maize in the starter diets for broiler chicken. Cali, Colombia: CIPAV Foundation. 14: 1-6.

Awojobi, H.A. and Adekunmi, A. A. 2002. Performance of broilers fed graded levels of cassava tuber meal. New Delhi, India: Indian Council of Agricultural Research. 72: 1169-1172.

Bhatnagar, R., Kataria, M. and Verna, S.V.S. 1996. Effect of dietary Leucaena root meal on the performance and egg characteristics in white leghorn hens. *Indian J. Animal Sci.* 66 (12): 1291-1294.

Garcia, M. and Dale, N. 1999. Cassava root meal for poultry. *Appl. Poultry Sci.* 8: 132-137.

Hahn, S.K. 1989. An overview of African traditional cassava processing and utilization. *Outlook on Agric.* 18: 110-118.

Hossain, M.D., Bulbul, S.M. and Howlider, M.A.R. 1989. The composition of some unconventional feeds. *Poultry Adviser.* 22: 37-40.

Hughes, R.J. and Choct, M. 1999. Chemical and physical characteristics of grains related to variability in energy and amino acid availability in poultry. *Australian J. Agril. Res.* 50: 689-701.

Mahmoudnia, N., Boldaji, F., Dastar, B. and Zereharan, S. 2011. Nutritional evaluation of poultry by-product meal in broiler chickens. *Animal Biol. & Animal Husbandry Int. J. Bioflux Soc.* 3 (1) : 55-64.

Maust, L.E., Pond, W. G. and Choun, M. L. 2000. Energy value of cassava maize based diet with and without supplemental zinc for growing poultry. *J. Animal Sci.* 35: 935-957.

McNab, J. 1999. Advance in poultry nutrition in the world. Proceedings of the seminar and international poultry show, WPSA, Bnagladesh Branch. p. 52.

Minitab. 2000. Minitab Statistical Software User's Guide 2: Data Analysis and Quality Tools. Minitab Inc., State College, PA 16801-3008, USA.

Mroso, P.V. 2003. Cassava: An emerging food product, the consequence of its popularity. http://www.suite1101.com/article.Cfm/167 38/999964.

NRC. 1994. Nutrient Requirements of Poultry, 9th edn., National Research Council, National Academic Press, Washington, DC.

Oke, O.L. 1978. Problems in the use of cassava as animal feed. *Feed Sci. Tech.* 3 (4): 345-380.

Okezie, B.O. and Kosikowski, F.V. and Markakis, P. 1982. Cassava as food. *Critical Rev. Food Sci. & Nutri.* 17 (3): 59-275.

Onjoro, P.A., Bhattacharjee, M. and Ottaro, J.M. 2001. Bioconversion of cassava tuber by fermentation into broiler feed of enriched nutritional quality. Trivandrum, India: Indian Society for Root Crops, Central Tuber Crops Research Institute. 24: 105-110.

Onyimoyi, A.E. and Ugwu, S.O.C. 2007. Bioeconomic indices of broiler chicks fed varying ratios of cassava peel/bovine blood. *Int. J. Poultry Sci.* 6 (5): 318-321.

Padmaja, G. 1995. Cyanide detoxification in cassava for food and feed uses. *Critical Rev. Food Sci. & Nutri.* 35 (4): 299-339.

PAN. 1995. Annual Reports. Poultry Association of Nigeria. Lagos. Nigeria. 33: 34-37.

Phillips, T.P.,Taylor, D.S., Sanni, L. and Akoroda, M.O. 2004. A cassava industrial revolution in Nigeria: The potential for a new industrial crop. International Institute of Tropical Agriculture, Ibadan, Nigeria, International Fund for Agricultural Development, Food and Agriculture Organization of the United Nations, Rome, Italy. 43 p.

Piva, G. 1987. An evaluation of feeding stuffs: Alternatives for poultry diets. Feed International. July, 1987. pp. 26-30.

Raihan, S., Mahmud, N.T. and Linkages, P.A. 2008. Case study of the poultry industry in Bangladesh. Cuts International. pp. 1-15.

Ravindran, V. and Blair, R. 1991. Feed resources for poultry production for Asia and the Pacific regions. I Energy sources. *World's Poultry Sci. J.* 47: 213-231.

Ravindran, V., Kornegay, E.T. and Rajaguru, A.S. B. 1987. Influence of processing methods and storage time on the cyanide potential of cassava leaf meal. *Animal Feed Sci. & Tech.* 17: 227-234.

Reed, J. D., McDowell, R. E., Van Soest, P. J. and Horvath, P. J. 1982.Condensed tannins: a factor limiting the use of cassava foliage. *J. Sci. Food & Agric.* 41: 45-53.

Singh, K. S. and Panda, B. 1992. Poultry Nutrition. Kalyani Publisher, New Delhi, India. p. 61.

Stevenson, M.H. and Jackson, N. 1983. The nutritional value of dried cassava root meal in broiler diets. *J. Sci. Food & Agric.* 34: 1361-1367.

Tewe, O. 1993. Thyroid cassava toxicity in animals. pp. 114-118. *In:* Cassava toxicity end thyroid: research and public health issues, Proceedings, International Workshop on Cassava Toxicity, edited by F. Delange and R. Ahluwalia, 31 May-2 June 1993, Ottawa, Canada. IDRC-207e : Ottawa.

Van der Klis, J.D., Kwakernaak, C., Jansman, A. and Blok, M. 2010. Energy in poultry diets: Adjusted AME or Net Energy. Proceedings of Australian Poultry Science Symposium. 21: 44-49.

Vogt, H. 1966. The use of tapioca meal in poultry rations. *World's Poultry Sci. J.* 22: 113-125.

2

THE CONTRIBUTION OF TURMERIC RESEARCH AND DEVELOPMENT IN THE ECONOMY OF BANGLADESH: AN EX-POST ANALYSIS

M.K. Hasan[1*] and M.A.A. Mahmud[2]

Abstract

The study estimated the benefit and rates of returns to investment on turmeric research and development in Bangladesh. The Economic Surplus Model with ex-post analysis was used to determine the returns to investment and their distribution between the production and consumption. Several discounting techniques were also used to assess the efficiency of turmeric research. The adoption rate was found increasing trend over the period. The yield of BARI developed modern varieties of turmeric was 41 to 73% higher than those of the local variety. Society got net benefit Tk. 9333.88 million from the investment of turmeric research and extension. The net present value (NPV) and present value of research cost (PVRC) were estimated at Tk. 1200.84 and 157.88, respectively. The internal rate of return (IRR) and benefit cost ratio (BCR) were estimated to be 68% and 10.45, respectively indicated investment on turmeric research and development was a good and profitable investment. Seed production programme of turmeric should be taken largely to increase production by increasing area adoption.

Keywords: Turmeric, Contribution, Adoption, Yield Advantage, Benefit and Rate of Return

[1]Senior Scientific Officer (Agricultural Economics), Spices Research Centre, Bangladesh Agricultural Research Institute, Shibgonj, Bogra, Bangladesh
[2]Assistant Inspector General (Crime-East), Bangladesh Police, Police Headquarter, Dhaka, Bangladesh

*Corresponding author's email: kamrulspc@yahoo.com (M.K. Hasan)

Introduction

Turmeric is a spice derived from the rhizome of *Curcuma longa* L. which is used as condiment, flavouring and colouring agent, drug and cosmetic in addition to its use in social and religious ceremonies. It is a principal ingredient in the Bangladeshi kitchen as curry powder and paste. It adds flavour and colour to curries and has medicinal values also. In 1984-85, just before development and introducing of improved varieties of turmeric in Bangladesh yield per hectare was very low and it was only 1.92 tonnes per hectare. Yield per hectare of turmeric started to increase since 1985-86 and it stands 5.16 tonnes per hectare in 2007-08 production period, due to develop and dissemination of improved varieties of turmeric (BBS, 2012).

Due to increasing population, demand for cereal food increased significantly. To mitigate this demand, the land of turmeric crops is being diverted to cereal food crop cultivation. On the other hand, now-a-days many spice-processing industries such as Square, BD Foods, Pran, Archu, Advanced Chemical Industries (ACI), Amrita, Dekko etc. have been established in Bangladesh. They are exporting turmeric as a finished product outside the country. Due to that, demands for turmeric as raw material of these industries are increasing with the extension of their production. That is why; the total demands for turmeric are increasing at incremental rate. Resulting that, a big gap was observed between production and demand now. To meet up this gap the country has to spend a huge amount of foreign currency in every year for importing spices from abroad.

Realizing the importance of turmeric, Bangladesh government started turmeric research through Horticulture Division of BARI since 1980-81 for increasing turmeric production. But the fund was not sufficient to do research vastly. To keep the continuity of research of turmeric and other spices, Bangladesh government established Spices Research Centre (SRC) in 1994 under Bangladesh Agricultural Research Institute (BARI) for increasing the production of turmeric throughout the country. BARI has already released three improved turmeric varieties namely BARI Turmeric- 1, 2, and 3. These varieties are cultivated in the farmers' fields since the release of these varieties. BARI, BARC (Bangladesh Agricultural Research Council) and DAE (Department of Agricultural Extension), to some extent have strengthened their works to turmeric production. However, for the research work of turmeric and its development, the contribution of BARC and DAE are greatly associated with BARI.

The present analysis thus took into the benefits from past turmeric research and its farm level development in the country. However, this study provided information for the policy makers, donors, researchers, extension people and the public on the contribution and the rate of return to investment in turmeric research in Bangladesh.

Methodology

Sources of data

For the present study, data were collected from different sources like published and unpublished reports, and informal scientist's interview. The area, production and yield of BDMVs (BARI Developed Modern Varieties) of turmeric were collected from SRC (Spices Research Centre); adoption rates were collected through informal scientist's interview; and harvest price and consumer price index (CPI) were collected from various issues of Statistical Yearbooks (1985 to 2012) published by the Bangladesh Bureau of Statistics. The supply elasticity was taken from the study conducted by Dey and Norton (1993). Since SRC of BARI is the main organization for turmeric research, the research cost included mainly from SRC of BARI. The extension and promotion activities were done by DAE and the related costs were collected from this organization. BARC mainly provided the administrative costs. The on-farm yield data of BDMVs turmeric varieties were collected from the SRC, Bogra. Data on the input cost change was calculated by the researcher through analyzing increased production claimed higher labour costs for harvesting and transporting, expensive of seeds, and used slightly more fertilizers per hectare for improved variety than for traditional varieties.

Analytical procedure

The collected data were analyzed using the following statistical techniques.

Estimation of returns to investment

The Economic Surplus Model (ESM) with Ex-Post analysis was used to estimate the rate of returns to investment in turmeric research and extension. The analysis was done under small open-economy market situation. The theoretical concept of ESM has been illustrated below.

Theoretical concept of Economic Surplus Model (ESM): The concept of economic surplus was used to measure economic welfare and the changes in economic welfare from policy and other interventions (Alston *et al.*, 1995; Currie *et al.*, 1971). Usually the economic surplus concept is adopted to estimate the benefits from the adoption of improved varieties. The components of economic surplus are consumer surplus and producer surplus. Given the initial condition (i.e., pre-research supply curve S_1 and demand curve D_1), consumer surplus is depicted as Area P_0P_nb in Fig. 1. This is the surplus or benefit to consumers because of a functioning market. Consumer surplus is that area beneath the demand curve less the cost of consumption. The cost of consumption is the area below the price line P_n. Producer surplus is defined by area P_nbO in Fig. 1. Area P_nbO in the surplus left to the farmers after they have paid for the total costs of production, area ObQ_n (Alston *et al.*, 1995).

The adoption of an intervention by farmers, such as an improved variety usually means one of two things: (i) A farmer can supply more of the commodity using the same level of resources (i.e, same land area and other inputs), or (ii) A farmer can supply the same level of commodity output but do it with fewer resources.

In either case, this is depicted by a shift to the right of the supply curve as shown in Fig. 1 (the shift is from S_1 to S_2). The shift is the supply curve from the adoption of an intervention changes the initial equilibrium price and quantity of the commodity. This new price quantity equilibrium increases economic surplus. The change in economic surplus (economic benefits) is measured by comparing the difference in economic surplus between the pre-adoption period and the post-adoption period.

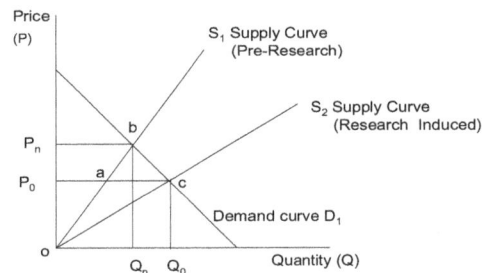

Change in Consumer Surplus	= Area abc + Area P_nbaP_0
Change in Producer Surplus	= Area Oac-Area P_nbaP_0
Change in Total Economic Surplus	= Area abc+ Area Oac

Fig. 1. Economic Surplus Model (Closed Economy)

Given a shift in the supply curve S_1 to S_2, the change in consumer surplus is depicted in Fig. 1 as Area abc + Area P_nbaP_0. The shift in the supply curve (due to the adoption of an intervention) has decreased the price consumers now have to pay for the commodity.

Given a shift in the supply curve S_1 to S_2, the change in producer surplus is depicted in Fig.1 as Area Oac-Area P_nbaP_0. Area Oac represents the decrease in the cost of production the same unit of the commodity that farmers now enjoy because they are using the intervention. This represents the benefits to the farmers from adopting the intervention and can be measured and quantified in monetary terms. The adoption of the intervention, however, has increased the quantity produced thereby decreasing the price of the commodity (P_n to P_0 in Fig. 1) and is a loss to farmers income. Farmers can recover some of this loss since they can sell more quantity (Q_n to Q_0 in Fig. 1) of the commodity.

The total social benefits to society from the adoption of an intervention is the summation of the change in consumer surplus plus the change in producer surplus (Area abc + Area Oac) minus the input cost change from adopting the new interventions.

For a closed economy model, the estimated price elasticity of demand is used in the above formulas. For small open-economy model where the elasticity of demand is perfectly elastic, use a sufficiently large number of η (Nagy and Alam,

2000). A small open economy market is one where the amount of exports or imports is small relative to total world trade in the commodity. Thus, there is little or no effect on the world price of the commodity (the small country assumption). In this case, the price of commodity does not change with the shift in the supply curve. For this study, the Bangladesh turmeric market is modelled as a small open economy market.

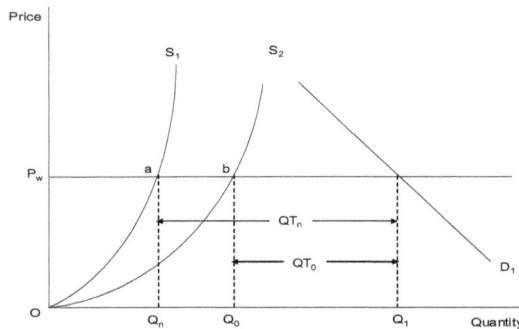

Fig. 2. Small Open-economy Importer Economic Surplus Model

The change in economic surplus for a small open-economy that is domestically produced but allows imports to cover shortfall (i.e., the Bangladesh turmeric market) is depicted Fig. 2. The world price P_w and quantity demanded by Bangladeshi consumers Q_1 defines the initial equilibrium. At price P_w, producers supply Q_n amount of turmeric when faced by the pre-research supply curve S_1. Turmeric imports are equal to QT_n. When faced by the research induced supply curve S_2 (the supply curve that exist because farmers have adopted new high yielding varieties). Turmeric producers increased production to quantity Q_n and increase Q_nQ_0. Spices imports are decreased by the same amount as the increase in production Q_nQ_0 and are now at QT_0. Since P_w does not change (small economy assumption), there is no change in consumer surplus- consumers are neither better off nor worse off. The enter change in economic surplus from the adoption of new turmeric varieties is thus a change in producer surplus only and is identified by area oab in Fig. 2 (corresponds to area oac in Fig. 1). The amount of foreign exchange saved by the adoption of improved varieties is equal to $P_w x (Q_nQ_0)$.

Empirical approach: The Akino and Hayami (1975) approximation formulas for calculating changes to producer and consumer economic surplus are described below and these are used in this study. The Akino and Hayami (1975) approximation formulas for calculating the change in economic surplus for a closed economy analysis (Fig. 1) is as follows:

Area A (abc)	=	$0.5 P_0Q_0 ((k(1+\gamma))^2/(\gamma+\eta))$	(1)
Area B (Oac)	=	kP_0Q_0	(2)
Area C (P_nbaP_0)	=	$((P_0Q_0k (1+\gamma))/(\gamma+\eta)) x (1- ((0.5k(1+\gamma)\eta)/(\gamma+\eta))-0.5k (1+\gamma))$	(3)

Where,

P_0	=	Price of turmeric (Tk./ton) (Existing market price)
Q_0	=	Production of BDMVs turmeric (ton) (Existing production)
P_n	=	Quantity price that would exist in absence of research
Q_n	=	Quantity of the turmeric produced that would exist in absence of research
k	=	Horizontal supply shifter
γ	=	Price elasticity of turmeric supply
η	=	Absolute price elasticity of the demand for the commodity.

The supply shifter (k): The supply shifter 'k' is the overall yield advantage of improved varieties of turmeric over the local variety weighed by the area sown to the improved varieties of turmeric. In the case of the Akino and Hayami (1975) approximation formulas, k is the horizontal shift The supply shifter k is calculated as follows: from the equilibrium price P_n given S_1 to the equilibrium price P_0 given S_2 which corresponds to a distance equal to Q_nQ_0 in Fig. 1 (Gardiner *et al.*, 1986; Nagy and Furtan, 1978).

$$k_t = \sum_{i=1}^{n} [1 - \frac{Y_t}{Y_{it}})] \times A_{it}$$

Where,

Yit	=	Yield of the improve varieties of turmeric in year t
Yt	=	The yield of a base (or average yield of local variety turmeric) that has been gown in the past and that would still be grown if no new varieties had been developed
Ait	=	The proportion of the total area sown to improved varieties of turmeric in year t
n	=	The number of improved turmeric varieties

Estimation of net present value (NPV)

The amount of total funds returned from the investment in research is called NPV. The benefits were calculated by using the following formula of NPV:

$$NPV = \left[\sum_{i=1}^{n} (TSB_t - C_t)(1+r)^{-t} \right]$$

Where,

C_t	=	The cost of research and extension investment in year t
r	=	The discount rate
n	=	The time horizon over which the benefits of the research investments are realized

Internal rate of return (IRR)

The IRR was calculated relating to the total social benefit (TSB) minus an input cost change, if any, in each year to the research expenditure (C) in each year and is the discount rate that results in a zero net present value of the benefits. The IRR is calculated as

$$O = \left[\sum_{t=1}^{n} (TSB_t - C_t)(1 + IRR)^{-t} \right]$$

The IRR can be defined as the rate of interest that makes the accumulated present value of the flow of costs equal to the discounted present value of the flow of returns, at a given point in time (Peterson, 1971).

Results and Discussion

Adoption status and yield advantages of BARI developed modern varieties of turmeric

The adoption of improved variety is very important factors by which the volume of change in economic surplus is determined. The more the adoption of improved varieties over traditional one, higher the change in surplus will be. Apart

from this, it gives us feedback as to why and how well a technology is being accepted by the farmers. There was no turmeric varietal adoption survey conducted in Bangladesh. The existing variety survey information along with the considerable field experience of the spices scientists is used to sketch out the per centage area sown by variety grouping which are presented in Table 1.

Three modern varieties of turmeric named BARI Turmeric-1, BARI Turmeric-2 and BARI Turmeric-3 were released from BARI. BARI Turmeric-1, BARI Turmeric-2 were released in the same year 1984-85 but BARI Turmeric-3 was released in 1998-99. The overall area coverage of BARI Turmeric-1 increased from 0.50 per cent in 1985-86 to 22 per cent in 2007-08 but BARI Hodud-2 increased from 0.50 per cent to 21 per cent over the years. On the contrary of another modern variety BARI Turmeric-3, the area coverage increased from 1 per cent in 1999-2000 to 13 per cent in 2007-08. Area covered by all BDMVs of turmeric occupied about 56 per cent of the area sown to turmeric (Table 1).

Table 1. Area of traditional variety replaced by BMVs of turmeric

Year	Total turmeric area		Area of LVs of turmeric		Area covered by BARI Turmeric-1		Area covered by BARI Turmeric-2		Area covered by BARI Turmeric-3		Area covered by all BMVs of turmeric	
	ha	%	ha	%	ha	%	ha	%	ha	%	ha	%
1979-80	13899	100	13899	100	0	0.00	0	0.00	0	0.00	0	0.00
1980-81	14251	100	14251	100	0	0.00	0	0.00	0	0.00	0	0.00
1981-82	14599	100	14599	100	0	0.00	0	0.00	0	0.00	0	0.00
1982-83	14080	100	14080	100	0	0.00	0	0.00	0	0.00	0	0.00
1983-84	14162	100	14162	100	0	0.00	0	0.00	0	0.00	0	0.00
1984-85	15049	100	15049	100	0	0.00	0	0.00	0	0.00	0	0.00
1985-86	15202	100	15050	99	76	0.50	76	0.50	0	0.00	152	1.00
1986-87	16160	100	15837	98	162	1.00	162	1.00	0	0.00	324	2.00
1987-88	16051	100	15409	96	321	2.00	321	2.00	0	0.00	642	4.00
1988-89	15773	100	14827	94	315	2.00	631	4.00	0	0.00	946	6.00
1989-90	16372	100	15226	93	491	3.00	655	4.00	0	0.00	1146	7.00
1990-91	16332	100	14699	90	817	5.00	817	5.00	0	0.00	1634	10.00
1991-92	16285	100	14168	87	977	6.00	1140	7.00	0	0.00	2117	13.00
1992-93	16089	100	13515	84	1287	8.00	1287	8.00	0	0.00	2574	16.00
1993-94	16132	100	12906	80	1613	10.00	1613	10.00	0	0.00	3226	20.00
1994-95	15986	100	12789	80	1758	11.00	1439	9.00	0	0.00	3197	20.00
1995-96	16026	100	12500	78	1923	12.00	1603	10.00	0	0.00	3526	22.00
1996-97	16132	100	12099	75	2097	13.00	1936	12.00	0	0.00	4033	25.00
1997-98	16063	100	11887	74	2088	13.00	2088	13.00	0	0.00	4176	26.00
1998-99	15846	100	11409	72	2218	14.00	2218	14.00	0	0.00	4436	28.00
1999-00	15575	100	10591	68	2336	15.00	2492	16.00	156	1.00	4984	32.00
2000-01	15800	100	9954	63	2686	17.00	2844	18.00	316	2.00	5846	37.00
2001-02	16018	100	9611	60	3043	19.00	2723	17.00	641	4.00	6407	40.00
2002-03	15992	100	9275	58	2879	18.00	3038	19.00	800	5.00	6717	42.00
2003-04	18441	100	9589	52	3688	20.00	3873	21.00	1291	7.00	8852	48.00
2004-05	18939	100	9470	50	3977	21.00	3788	20.00	1705	9.00	9470	50.00
2005-06	20413	100	9594	47	4083	20.00	4287	21.00	2450	12.00	10820	53.00
2006-07	21028	100	9673	46	4416	21.00	4626	22.00	2313	11.00	11355	54.00
2007-08	22008	100	9684	44	4842	22.00	4622	21.00	2861	13.00	12325	56.00

Source: Several issues of BBS

Note: BDMVs- BARI Developed Modern Varieties, LVs- Traditional Varieties and shaded area indicates no improved varieties were released.

Table 2. Adoption rate of BARI developed modern varieties of turmeric

Name of crop	Varietals Name	Adoption rate (%)
Turmeric	BARI Turmeric-1 (1985-86 to 2007-08)	1.06
	BARI Turmeric-2 (1985-86 to 2007-08)	1.05
	BARI Turmeric-3 (1999-00 to 2007-08)	1.58
	Total BDMVs of turmeric (1985-86 to 2007-08)	2.64

The annual adoption rates were estimated to 1.06 per cent for BARI Turmeric-1 (1985-86 to 2007-08), 1.05 per cent for BARI Turmeric-2 (1985-86 to 2007-08) and 1.58 per cent for BARI Turmeric-3 (1999-2000 to 2007-08) as depicted from Table 2. The adoption rate of BARI Turmeric-3 was higher than those of BARI Turmeric-1 & 2 due to attractive colour and higher yield. The table also showed that the adoption rate of all BDMVs of turmeric was 2.64 per cent.

The supply shifter k identifies the amount of production that can be attributed to the varietals improvement research in each year (i.e., the shift in the supply curve). The more the value of supply shifter the more is the shift in the supply curve, resulting higher benefit to the society. The supply shifter is the outcome of the simultaneous force of adoption per centage and yield advantage. It was calculated using the formula discussed in methodology.

Supply shifter k

Table 3. Calculation of the supply shifter (K) of BARI Turmeric over traditional variety

Year	% Area BARI Turmeric-1 replacing LVs	% Area BARI Turmeric-2 replacing LVs	% Area BARI Turmeric-3 replacing LVs	Supply shifter k
1983-84	0	0	0	0
1984-85	0	0	0	0
1985-86	0.5	0.5	0	0.0049
1986-87	1	1	0	0.0099
1987-88	2	2	0	0.0197
1988-89	2	4	0	0.0279
1989-90	3	4	0	0.0337
1990-91	5	5	0	0.0494
1991-92	6	7	0	0.0633
1992-93	8	8	0	0.0790
1993-94	10	10	0	0.0987
1994-95	11	9	0	0.1004
1995-96	12	10	0	0.1103
1996-97	13	12	0	0.1243
1997-98	13	13	0	0.1283
1998-99	14	14	0	0.1382
1999-00	15	16	1	0.1595
2000-01	17	18	2	0.1866
2001-02	19	17	4	0.2087
2002-03	18	19	5	0.2184
2003-04	20	21	7	0.2528
2004-05	21	20	9	0.2691
2005-06	20	21	12	0.2894
2006-07	21	22	11	0.2919
2007-08	22	21	13	0.3082

Table 3 shows each year adoption per centage and supply shifter of turmeric. It was found that the rate of shift gradually increased. The shifter accounted for the yield advantage of BARI developed turmeric varieties over the traditional varieties. The supply shifter of turmeric was found to be 0.308 for the year 2007-08, meaning that 31% more turmeric production was made available during 2007-08 because of farmers' adoption of BARI developed turmeric variety.

Yield advantages: This is very important factor to determine the economic surplus. The higher yield advantage always ensures higher level of economic surplus. Two types of data exist in most

of the less developed countries for good estimation of yield advantage (YA) as well as the aggregate production function shifter. They are on-station yield trial data and on-farm yield data. The on-station yield data is readily available and most often the only reliable source. One of the arguments against using on-station yield trial data is that of superior management practices and techniques are used and therefore, the results may not reflect on the on-farm situation. Another argument placed by different author (Hertford and Schmits, 1971; Ayer and Schuh, 1972; Akino and Hayami, 1975; Scobie and Posada, 1977; Nagy and Alam, 2000) and showed that the yield advantage estimation from the on-station yield trial data would be biased upward because the estimation might also include the contribution made by inputs such as fertilizer and water. To account for this problem, the estimated yield advantage of new varieties by estimating production functions of yield as a function of new varieties and other inputs. This process requires a substantial data which is not readily available in Bangladesh.

For the present study, on-farm yield trial data were considered as a more reliable source for the

calculation of yield advantage rather than the on-station yield data in Bangladesh. The yield advantages have been calculated for this study following Gardiner *et al.* (1986), Nagy and Furtan (1978) and Nagy (1991).

Developed varieties of Spices Research Centre (SRC) of BARI have replaced the traditional varieties starting in 1985-86 for BARI Turmeric-1 and BARI Turmeric-2 and 1999-00 for BARI Turmeric-3.

Table 4. Yield advantages of improved varieties of turmeric over traditional varieties

Name of turmeric crop	Average (weighted) yield of Improved variety (t/ha)	Average yield of traditional variety (t/ha)	Yield difference (t/ha)	Yield advantage
BARI Turmeric-1	4.123	1.74	2.38	0.578
BARI Turmeric-2	2.946	1.74	1.21	0.409
BARI Turmeric-3	6.476	1.74	4.74	0.731

The weighted yields were calculated by taking the average of the irrigated optimum, late irrigated and non irrigated yield multiplied by the mean of irrigated, late irrigated and non irrigated area of turmeric. Per hectare average yield of high yielding varieties of turmeric i.e. BARI Turmeric-1, 2 and 3 were found to be 4.12, 2.95 and 6.48 tonnes, respectively. In case of traditional variety it was only 1.74 tonnes. Therefore, the yield advantages of BARI Turmeric-1, 2 and 3 over traditional variety were found to be 58, 41 and 73%, respectively (Table 4).

Estimating benefits from turmeric research and extension

This section deals with the estimation of returns to investment in turmeric research and extension using the economic surplus approach. This approach estimates the benefits to agricultural research by measuring the change in consumers' surplus (CS) and producers' surplus (PS) from a rightward shift in the supply curve that is brought about through technological change. It should be mentioned here that aggregate consumers' surplus, producers' surplus and total surplus were calculated by summing up corresponding surpluses of all turmeric rather than summing up from the areas of the model. In order to calculate the net benefits (NB) research and extension expenditures are subtracted from total surplus. All these estimates of benefits are expressed in real term by using 2007-08 constant prices. The rates of return and NB are then discounted using 10% interest rate for obtaining the efficiency of investment. First, the yearly total social benefits are estimated using the small-open economy model (Fig. 2).

This is done by assigning a very high number to the demand elasticity parameter (ŋ) since in a small open-economy model, ŋ is perfectly elastic. The analysis is undertaken for each year 1980-81 to 2007-08 for turmeric.

Turmeric research and extension in Bangladesh are seemed to be continued by three different organizations. The Organizations are Bangladesh Agricultural Research Institute (BARI), Bangladesh Agricultural Research Council (BARC) and Department of Agricultural Extension (DAE). The turmeric research and extension expenditure comprised the expenditure of three organizations are furnished in the following sequence.

The year wise expenditures behind variety development and dissemination for the new varieties to the farmers of turmeric are shown in Table 5. The expenditures of BARI/SRC and BARC were estimated from 1980-81 to 2007-08. The accumulated expenditures over the years of BARI/SRC and BARC were estimated at Tk. 95.88 and 17.71 million, respectively. Extension expenditures and input cost change were estimated after development of improved variety and they were started since 1985-86. The cumulative expenditures of extension and input cost changes were respectively amounted at Tk. 67.02 and 530.04 million. Over the years, expenditures accruing for BARI/SRC, BARC, DAE and input cost change were Tk. 710.65 million. For the analysis, the current total expenditures were converted to 2007-08 constant prices using the Bangladesh Middle Income Group CPI Index and it was Tk. 987.70.

Table 5. Turmeric research and extension expenditures by sources from 1980-81 to 2007-08

Year	Total BARI/ SRC Research Expenditures (current Taka)	BARC Administrative Expenditures (current Taka)	Total Extension Expenditures (current Taka)	Input cost change (current Taka)	Total Expenditures (current Taka)	Total Expenditures (2007-08 Tk.)
1980-81	125000	124550	0	0	249550	1475139
1981-82	425000	133000	0	0	558000	2854002
1982-83	235000	141900	0	0	376900	1753902
1983-84	200000	885500	0	0	1085500	4599356
1984-85	112500	818050	0	0	930550	3567626
1985-86	164250	634100	220309	281200	1299859	4485540
1986-87	174350	956650	1026029	635364	2792393	8630861
1987-88	187300	539250	1134000	1334500	3195050	9064153
1988-89	177950	562700	1284993	2084397	4110040	10811807
1989-90	284450	819500	1291320	2676577	5071847	12305861
1990-91	635100	746050	1684024	4045322	7110496	15892219
1991-92	755700	1107250	1767249	5555559	9185758	19550597
1992-93	1630450	1346100	1828555	7160137	11965242	25244078
1993-94	2741150	798650	2199600	9512245	15251645	31216936
1994-95	660000	225500	2317525	9992339	13195364	25648035
1995-96	1980000	171400	2238367	11681879	16071645	29921794
1996-97	7788000	261250	2408661	14163300	24621211	45035932
1997-98	8699064	675850	2660407	15545425	27580746	48295222
1998-99	2882220	1181750	2806472	17504089	24374531	40648556
1999-00	1728804	1221000	3312387	20846439	27108630	43055350
2000-01	2132064	1114500	3354363	25919017	32519944	48889658
2001-02	1985280	645000	3424078	30110662	36165019	51292050
2002-03	3334320	173500	3592067	33461606	40561493	54271199
2003-04	3859020	271300	4216270	46743229	55089818	69537802
2004-05	14129808	301300	4854348	53006990	72292446	86086822
2005-06	13091496	442200	5433602	64197231	83164529	93427779
2006-07	11586960	724950	6909080	71413783	90634773	96056526
2007-08	14171520	690350	7058884	82165158	104085911	104085911
Total	95876756	17713100	67022590	530036448	710648890	987704713

Note: $1.00 =Tk. 80.00

Table 6. Estimation of surplus from turmeric research and extension investments

Year	Change in consumer surplus (Tk.)	Change in producer surplus (Tk.)	Change in total surplus (Tk.)	Total expenditure (Based on 2007-08 Tk.)	Net Benefit (Tk.)
A	B	C	D=B+C	E	F=D-E
1980-81	0	0	0	1475139	-1475139
1981-82	0	0	0	2854002	-2854002
1982-83	0	0	0	1753902	-1753902
1983-84	0	0	0	4599356	-4599356
1984-85	0	0	0	3567626	-3567626
1985-86	0.14	12567323.59	12567323.72	4485540	8081784
1986-87	0.20	18699643.14	18699643.35	8630861	10068782
1987-88	0.39	35892454.42	35892454.81	9064153	26828302
1988-89	0.59	54539682.00	54539682.59	10811807	43727876
1989-90	0.83	76900261.89	76900262.72	12305861	64594402
1990-91	1.49	139643807.17	139643808.67	15892219	123751590
1991-92	1.87	176522118.24	176522120.11	19550597	156971524
1992-93	2.31	219729031.58	219729033.89	25244078	194484956
1993-94	2.63	252430504.29	252430506.92	31216936	221213571
1994-95	1.93	185271298.96	185271300.88	25648035	159623266
1995-96	2.06	198973912.42	198973914.48	29921794	169052120
1996-97	3.49	340896087.72	340896091.21	45035932	295860160
1997-98	4.70	460210029.27	460210033.97	48295222	411914812
1998-99	5.00	492267151.11	492267156.11	40648556	451618600
1999-00	5.40	538506185.99	538506191.39	43055350	495450841
2000-01	4.29	434278640.51	434278644.80	48889658	385388987
2001-02	3.49	358382061.88	358382065.37	51292050	307090016
2002-03	3.59	370955714.75	370955718.34	54271199	316684519
2003-04	8.45	892087989.68	892087998.12	69537802	822550196
2004-05	10.59	1130017034.46	1130017045.05	86086822	1043930223
2005-06	12.18	1316432587.41	1316432599.59	93427779	1223004821
2006-07	10.52	1138926169.57	1138926180.09	96056526	1042869654
2007-08	13.50	1477455192.84	1477455206.34	104085911	1373369295
Total	99.64	10321584882.89	10321584982.52	987704713	9333880272

Note: $1.00 =Tk. 80.00

The total over years changes in consumers' and producer' surplus were estimated Tk. 99.64 and Tk. 10321.58 million respectively from turmeric research and extension. Consumers' surplus was very much lower compared to producers' surplus due to perfectly elasticity of demand for turmeric in the small-open economy market. The estimated total surplus/total benefits ranged from Tk. 12.57 million in 1985-86 to 1477.46 million in 2007-08 and the total surplus accrued as Tk. 10321.58 million from the turmeric research and extension in Bangladesh. Besides, the total net benefits obtained from turmeric research and extension was Tk. 9333.88 million for the year 1980-81 to 2007-08 (Table 6).

Rate of return to turmeric research and extension

The rates of returns are the indicators which help to estimate the investment efficiency of the research programme. There are many types of measures that can be used to estimate the rates of return. Among them, Net Present Value (NPV) of benefit, External Rare of Return (ERR) and Internal Rate of Return (IRR) was considered as the rates of return to turmeric research and extension investments in Bangladesh. For comparing the net benefits with the total research costs, Present Value of Research Costs (PVRC) was also calculated. All the estimates were calculated at constant (2007-08) prices with 10% discount rate. Table 7 was used to calculate the NPV, PVRC, ERR, IRR and BCR under small-open economy condition. Under open economy, the producers' benefits were found much higher compared to consumers' benefits since the elasticity of demand for turmeric were very high.

Table 7 revealed that the society was benefited substantially from the investment in spices research and extension in Bangladesh. The NPV of benefit indicates the total social benefit for a country and it was found negative up to 1984-85 and then it was positive. It means that the country did not receive any benefit from turmeric research up to 1984-85 (Table 6). After 1984-85, the country as a whole benefited with a big amount and found increasing trend up to 2007-08. The NPV was found to be Tk. 1200.84 million while PVRC over the period was Tk. 157.88 million for turmeric research and extension investment. The ERR was found to be 1049.50%. This means that the average taka spent on research and extension in turmeric earn return 10% annually from the start of the initial investment (1980-81) and is now paying off at the rate of 1049.50% annually into perpetuity. In the benefit/cost mode, using 10% external interest rate, a one taka investment returned 104.95 taka over the period. The IRR of 68% means that on the average, each taka invested in turmeric research and extension returned 68% annually from the date of the initial investment. It implies that the expenditure on turmeric research and extension (Tk. 157.88 million) could have been borrowed at 68% real rate of interest without incurring loss (Table 7).

Table 7. Estimated rates of returns to turmeric research and extension

Name of crop	Net present value (NPV)	Present value of research cost (PVRC)	External Rate of Return (ERR)	Internal Rate of Return (IRR)	Benefit Cost Ratio (BCR)
	Million taka in 2007-08 constant prices		%		
Turmeric	1200.84	157.88	1049.50	68	10.45

Note: $1.00 =Tk. 80.00

The benefit cost ratios were found to be 10.45 for turmeric. The value of the parameter clearly indicated that the investment in research and extension of turmeric in Bangladesh is a good investment and highly profitable.

Foreign exchange savings

The yearly increase in production due to research save the country's foreign exchange to a remarkable extends. First, the research induced productions for turmeric for the past years were calculated by multiplying the country's total turmeric production by their respective production function shifter k. Multiplying the results by world turmeric price, foreign exchange savings was obtained.

Considerable amounts of turmeric are imported in Bangladesh every year to meet the internal demand for increasing population. In 2006/07, the imported value of turmeric was Tk. 148.70 million (BBS 2008). In reality, the amount imported is higher due to the illegal border trade of spices from neighbouring countries. Thus, the increased production attributed to turmeric improvement saved foreign exchange amounting to Tk. 9084.63 million from turmeric research and extension (Table 8).

Table 8. Foreign exchange savings from investment in turmeric research

Year	Import (cif) Price (Taka)	Supply Shifter K (per cent)	Turmeric Production (ton)	Increase in Production from Research (ton)	Foreign Exchange Savings (Taka)
	1	2	3	4=2x3	5=1x4
1986-87	152500	0.0099	34665	343.13	52326879
1987-88	98540	0.0197	34740	686.00	67598076
1988-89	58998	0.0279	35928	1003.14	59183308
1989-90	19455	0.0337	37925	1278.57	24874717
1990-91	20564	0.0494	38780	1915.49	39390074
1991-92	36713	0.0633	41115	2604.02	95602759
1992-93	39686	0.0790	41170	3251.77	129049292
1993-94	41083	0.0987	41990	4145.54	170311411
1994-95	41115	0.1004	39550	3971.42	163285631
1995-96	42093	0.1103	40580	4476.18	188416125
1996-97	36584	0.1243	42140	5236.54	191571573
1997-98	36771	0.1283	41500	5326.49	195858544
1998-99	35063	0.1382	40850	5645.77	197960223
1999-00	35027	0.1595	40875	6520.30	228388340
2000-01	33075	0.1866	42900	8004.13	264734501
2001-02	40688	0.2087	42640	8897.47	362019408
2002-03	37915	0.2184	43820	9570.80	362878682
2003-04	41639	0.2528	70730	17878.51	744444648
2004-05	48141	0.2691	79175	21305.47	1025660930
2005-06	49548	0.2894	92395	26735.56	1324687145
2006-07	51819	0.2919	110000	32107.60	1663787020
2007-08	42561	0.3082	116832	36009.14	1532602974
Total Foreign Exchange Savings:				Tk.	9084632260

Note: $1.00 =Tk. 80.00

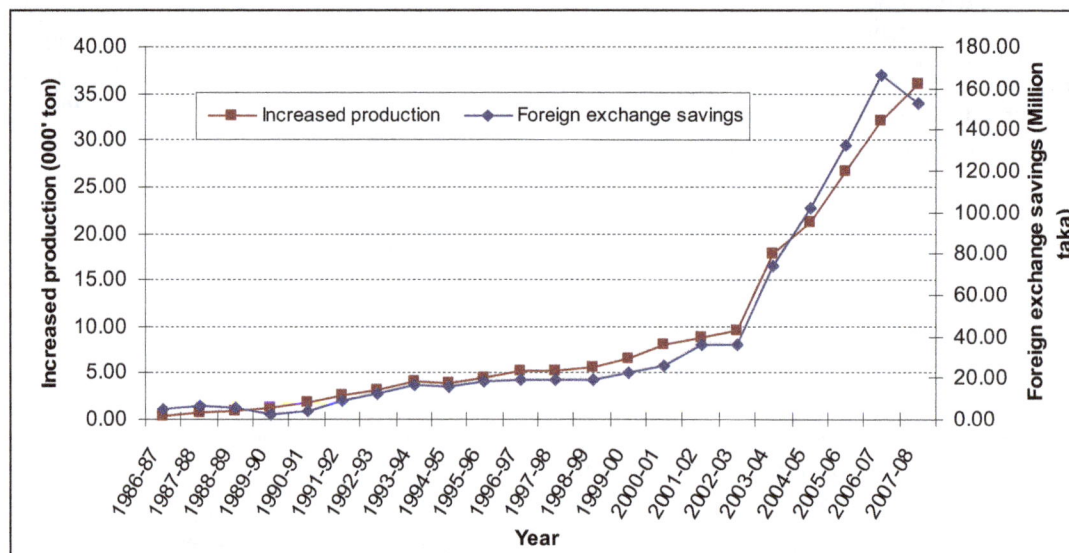

Fig. 3. Foreign exchange savings due to turmeric research and development over time

Policy implications

The empirical results indicate that the expenditure on turmeric research and development paid a favourable rate of returns and the society were also benefited enormously out of it. The IRR to turmeric research and development expenditure was found to be 68%. A 68% IRR on investment in turmeric research and development is a good rate of return. The consumer's surplus is found to be very few only due to small open economy. But this situation might not be the good sign for the economic prosperity. For the survival of the consumer, price support should be given by government.

The annual adoption rates of BARI Turmeric-1, 2 and 3 are not good because of non availability of seed. Seed production programme should be taken largely by the government and nongovernmental organization so that the farmers can get quality seed easily with a reasonable price.

References

Akino, M. and Hayami, Y. 1975. Efficiency and Equity in Public Research: Rice Breeding in Japan's Economic Development. *American J. Agril. Econ.* 57: 1-10.

Alston, J.M., Norton, G.W. and Pardey, P.G. 1995. Since Under Scarcity: Principles and Practice for Agricultural Research. Evaluation and Priority Setting. Cornell University Press Ithaka . p. 237.

Ayer, H.W. and Schuh, G.E. 1972. Social Rates of Return and Other Aspects of Agricultural Research: The Case of Cotton Research in Sao Paulo, Brazil. *American J. Agril. Econ.* 54: 557-569.

BBS. 2008. Statistical Year Book of Bangladesh Bureau of Statistics Ministry of Planning Government of the People's Republic of Bangladesh Dhaka.

BBS. 2012. Statistical Year Book of Bangladesh Bureau of Statistics Ministry of Planning Government of the People's Republic of Bangladesh Dhaka.

Currie, J.M., Murphy, J.A. and Schmitz, A. 1971. The Concept of Economic Surplus and Its Use in Economic Analysis. *Econ. J.* 18: 741-798.

Dey, M. and Norton, G. 1993. Analysis of Agricultural Research Priorities in Bangladesh. BARC, ISNAR. p. 300.

Gardiner, J.C., Sanders, J.H. and Barker, T.G. 1986. An Economic Evaluation of the Prude Soft Red Winter Wheat Programme. Department of Agricultural Economics, Agricultural Experimental Station, Prude University, West Lfayette. p. 258.

Hertford, R. and Schmitz, A. 1971. Measuring Economic Returns to Agricultural Research. Resource Allocation and Productivity in National and International Agricultural Research. Minneapolis. University of Minnesota Press. pp. 148-167.

Nagy, J.G. 1991. Returns from Agricultural Research and Extension in Wheat and Maize in Pakistan. Research and Productivity in Asian Agriculture. Cornell University Press. Ithaca. New York.

Nagy, J.G. and Alam, M.F. 2000. The Impact of Agricultural Research in Bangladesh: Estimating Returns to Agricultural Research. Project Report, BARC, Dhaka and International Fertilizer Development Centre, Muscle Shoals, Alabama.

Nagy, J.G. and Furtan, W.H. 1978. Economic Cost and Returns from Crop Development Research: The Case of Rapeseed Breeding in Canada. *Canadian J. Agril. Econ.* 26 (1): 1-14.

Peterson, W.L. 1971. The Returns to Investment in Agricultural Research in the United States. *In:* W.L. Fishel, ed., Resources Allocation in Agricultural Research.

Scobie, G.M. and Posada, R.T. 1977. The Impact of High Yielding Rice Varieties in Latin America-with Special Emphasis on Colombia. Series JE-01. CIAT. Calli. Colombia.

3

WOMEN'S PARTICIPATION IN RICE MILLS IN SHERPUR SADAR UPAZILA OF BANGLADESH

Naima Sultana[1] and M.S.I. Afrad[2]*

Abstract

The main purpose of this study was to determine women's participation in rice mills of Sherpur sadar upazila of Bangladesh. Ninety respondents were selected using cluster random sampling method. The researchers personally collected data from August to September 2012. Aimed at explaining the assessment of the present status of women worker of rice mills and explored the relationship between their selected characteristics and extent of participation. Majority of the respondents had long experience in working in the rice mill and almost everybody had participation around the year. Women were mostly involved in drying and on an average spent time for helping in husking machine operating (3.02 hours) followed by final drying of paddy (2.05 hours), drying the paddy before boiling (1.25 hours), piling the paddy (1.12 hours), cleaning the threshing floor (1.05 hours), packing the rice (1.02 hours) per day. They worked about 8-10 hours per day. Results show that the women worker's performance decreased with the increase of their age.

Keywords: Rice Mills, Women's Participation

[1]Former MS student, Dept. of Agril. Ext. and Rural Development, BSMRAU, Gazipur, Bangladesh
[2]Professor, Dept. of Agril. Ext. and Rural Development, BSMRAU, Gazipur, Bangladesh

*Corresponding author's email: afrad69@gmail.com (M.S.I. Afrad)

Introduction

Rice is the most important staple food in Bangladesh. More than 70 per cent of the total calorie intake comes from this food item. The per capita consumption (about 400g/day) of rice in Bangladesh is one of the highest in the world (IFRI, 2013). Rice cultivation in Bangladesh covers more than 70 per cent of the total cropped area (Begum, 1997). Three types of rice, namely *aus* (summer), *aman* (winter) and *boro* (spring) are grown in Bangladesh. When the paddy is harvested from fields, it needs to be processed for consumption. Parboiling, drying and milling are the different stages of getting rice from paddy. These are done both at home (small scale) and at the rice mills (large scale). Processing at home, the non–commercial sector of milling is perhaps the oldest and almost obsolete form of rice processing method in Bangladesh. In this method, paddy is processed in *dheki*[1] after it is parboiled and dried. Paddy processing and milling in Bangladesh is mostly performed at the rice mills. These mills are called commercial milling center. There are three kinds of commercial mills available in Bangladesh. They are husky, major and automatic. Increasing trend in mechanical process unit in the market has gradually replaced the traditional method of rice processing. These commercial- processing units have been playing a significant role in the rice marketing system (Zaman *et al.*, 2003).

In Bangladesh, there are approximately 40,000 rice mills. Rice mills are very much dependent on human labor, and almost 5 millions of unorganized workers are working in different rice mills and more than 60 per cent of them are female worker (Anonymous, 2003). Women workers are responsible for drying and husking paddy and packing the husked rice into sacks. Male workers are responsible for doing the less laborious works (Siddika, 2012).

The above-mentioned discussion implies the importance of women's participation in Bangladeshi rice milling sector. Therefore, keeping the importance in mind the present study was undertaken to explore the extent of participation by the respondents in the rice mills in the selected area. The relationship between the selected characteristics of the respondents and their participation in the rice mills will be also identified.

Methodology

Sherpur sadar upazila was purposively selected for the study. Accordingly, a list of rice mills was collected from the rice mill owner association of Sherpur. List shows that there are 391 rice mills in the study area. From the list, four rice mills namely, M/S Bangla Rice Mill, M/S Ruman Suman Rice Mill, M/S Jisan Rice Mill and M/S

Aditi Rice Mill were selected following cluster sampling technique. Then the researcher collected list of women workers from the managers or the owners of the selected rice mills. In each mill, 20 to 30 women have been working daily or permanent basis. Thus, all women workers of these four rice mills were the sample of this study, which was 90 in number (from M/S Bangla Rice Mill 27, from M/S Ruman Suman Rice Mill 22, from M/S Jisan Rice Mill 21 and from M/S Aditi Rice Mill 20 respondents). Data were collected through face-to-face interview method using pre-design and pre-tested interview schedules.

Selected characteristics of the respondents such as age, education, family size, properties, earning member in the family, marital status, family type, physical fitness, working experience, relation with superiors or mill-owner, relation with colleague/ fellow worker, contribution to the family were considered as the independent variables of the

study and computing appropriate scores. The characteristics were measured either raw scores or through rated scores. Amount of time spent by women rice mill worker was the dependent variable of the study. It was measured by calculating the actual time spent (in hour) in the rice mill by the respondent women. Total time spent by the respondents in different activities of rice mill was identified in hours per day.

Results and Discussion

Personal characteristics of the respondents

The results indicate that most of the respondents are middle aged (65%), illiterate (54%), married (66%), medium family size (59%) and respondent and their husbands are the main earning member in their family (49%).

Table 1. Distribution of the respondents according to their personal characteristics

Characteristics	Categories	Number	Per cent	Mean	SD
Age	Young (up to 35 year)	1	1	48.31	6.14
	Middle (36 to 50year)	58	65		
	Old (>50 year)	31	34		
Education	Illiterate	49	54	1.51	0.06
	Signature ability	36	40		
	Primary education	5	6		
Family size	Small (up to 4members)	35	39	4.96	1.11
	Medium (5-7 members)	53	59		
	Large (above 7 members)	2	2		
Properties	Landless	26	29	1.77	0.55
	Only house	58	65		
	House with a piece of land	6	6		
Earning member	Only respondents	28	31	1.88	0.71
	Respondents and her husband	44	49		
	Respondents and other members of family	18	20		
Marital status	Married	59	65	3.47	0.82
	Divorced	17	19		
	Widow	14	16		
Family type	Nuclear family	81	90	1.1	0.30
	Joint family	9	10		
Physical fitness	Physically well	83	92	1.07	0.26
	Physically sick	7	8		
Experience	Less than12years experience	34	38	13.47	6.43
	12-18 years experience	36	40		
	More than 18 years experience	20	22		
Relation with superiors or mill owner	Amiable	4	5	2.08	0.46
	Cooperative	76	84		
	Respectable	8	9		
	Not good	2	2		
Relation with colleague or fellow worker	Amiable	8	9	2.08	0.62
	Cooperative	72	80		
	Respectable	4	4		
	Not good	6	7		
Financial contribution to the family	Low(up to 16)	14	15	18.45	2.05
	Medium (17-18)	42	47		
	High(Above 18)	34	38		

Majority of the respondents (65%) have only house and of them 92 per cent are physically well. It is also found that 40 per cent of the respondents have long experience (12 to 18 years) in working in the rice mill and almost everybody (93%) had participation around the year. The mill owners or superiors are cooperative to respondents (84%) and they have good relationship with their colleagues (80%). About half of the respondents (47%) have medium contribution to their family (Table 1).

Gender wise involvement in different activities in rice mill

Most of the activities performed within the rice mill jointly by men and women and, therefore, it is very difficult to find out the exact works performed only by the women. Results presented in the Table 2 indicate that women are mostly involved in drying (94.44% in drying before boiling and 96.66% in final drying).

Table 2. Gender wise involvement in different activities in rice mill

Sl. No.	Activities	Female only	Male only	Jointly
1	Carrying the sack of rice	0 (0)	86 (95.55)	4 (4.44)
2	Drying the rice before boiling	87(96.66)	0(0)	3(3.33)
3	Helping in boiling	0(0)	88(97.77)	2(2.22)
4	Firing the boiler machine	0(0)	90(100)	0(0)
5	Boiler operating	0(0)	90(100)	0(0)
6	Piling the rice	74(82.22)	7(7.77)	9(10)
7	Helping in husking machine operating	64(71.11)	19(21.11)	7(7.77)
8	Cleaning the threshing floor	83(92.22)	3(3.33)	4(4.44)
9	Final drying /drying after boiling	85(94.44)	0(0)	5(5.56)
10	Packing	14(15.55)	65(72.23)	11(12.22)
11	Others (covering incomplete dried paddy, winnowing etc.)	33(36.60)	15(16.66)	42(46.64)

(Figures in parenthesis indicate per cent)

There are less involvement of the women respondent workers in helping in boiling and operating the boiler machine. Rice packing and piling are done jointly (10 and 12.22%, respectively). Carrying the sack (95.55%), helping in boiling (97.77%) and boiler operation is almost done by their male counter parts (Table 2).

Extent of participation in the rice mills by women worker

Results shown in Table 3 reveal that respondent women worker on an average spend most of their time in rice mills for helping in husking machine operating (3.02 hours) followed by final drying of paddy (2.05 hours), drying the paddy before boiling (1.25 hours), piling the paddy (1.12 hours), cleaning the threshing floor (1.05 hours) and packing the rice (1.02 hours) per day. Muhibbullah (2010) reported similar results.

Table 3. Extent of respondent women's participation in the rice mills (hour/day)

Sl. No.	Activities	Time spent (in hour/day)	
		Mean	SD
1	Carrying the sacks of rice	0.30	0.06
2	Drying the paddy before boiling	1.25	0.38
3	Helping in boiling	0.38	0.27
4	Piling the rice	1.12	0.46
5	Helping in husking machine operating	3.02	0.65
6	Final drying/drying after boiling	2.05	0.34
7	Packing	1.02	0.35
8	Cleaning the threshing floor	1.05	0.34
9	Others (covering incomplete dried paddy, winnowing)	0.25	0.05
	Total	10.44	

Relationship between the extent of participation and the selected characteristics of the respondents

Based on the correlation analysis it is observed that out of seven selected characteristics of the respondents only age have significant negative relation with their extent of participation in rice mill (Table 4). That is if the age of respondents increased then their extent of participation in different activities of rice mill would also be increased.

Table 4. Relationship between extent of participation of the respondents in rice mill and their selected characteristics

Independent variables	Dependent variable (value of 'r')
1. Age	-0.224*
2. Education	0.035[NS]
3. Family size	0.151[NS]
4. Earning member in the family	0.073[NS]
5. Marital status	-0.087[NS]
6. Experience	0.120[NS]
7. Contribution to the family	0.137[NS]

(* Significant at 0.05 level of probability, ** Significant at 0.01 level of probability, NS= Not significant)

This might be because with the increase of age, an individual's capacity tends to decrease. Begum (2001) observed similar findings but Tuli (2011) observed different findings in her study.

Conclusion

From the findings and their logical interpretation, it can be concluded that most of the respondents of the study were middle aged, illiterate, physically sound, married, had a medium family size, respondent and her husband are earning member in their family, possessed only house. Most of the respondents have long experience of working in the rice mill and almost everybody has round the year participation. The mill owner or superiors are cooperative to respondents and good relation with colleagues and medium contribution to the family. Women are mostly involved in drying the paddy and on an average spent time for helping in husking machine operating (3.02 hours) followed by final drying of paddy (2.05 hours), drying the paddy before boiling (1.25 hours), piling the paddy (1.12 hours), cleaning the threshing floor (1.05 hours), packing the rice (1.02 hours) per day. Therefore, women in rice mill work about 8-10 hours per day.

References

Anonymous. 2003. Development project in AGRO-BASED industry. Bangladesh Occupational Safety, Health and Environment Foundation, OSHE. Available at: http://oshebd.org/grain.html

Begum, F. 2001. Contribution of farmwomen in post-harvest activities of boro rice, pulses and oilseed crops in Narsingdi district of Bangladesh. PhD Thesis. Department of Agricultural Extension and Rural Development. Banghabandhu Sheikh Mujibur Rahman Agricultural University, Gazipur.

Begum, S. 1997. Commodity Report on Food Grain, Research and Planning, Department of Agricultural Marketing, Dhaka.

IFRI. 2013. Zinc Rice. Available at: http://www.unscn.org/layout/modules/resources/files/HarvestPlus_Rice_Strategy_EN.pdf

Muhibbullah, M. 2010. Women's Participation in Homestead Agriculture as a Rural Culture: a micro level study on Barisal district in Bangladesh. Proceedings of International conference on "The Roles of Humanities and Social Sciences in Engineering" held on 12-14 November 2010, Penang, Malaysia. 88p.

Siddika, A. 2012. Women in informal sector. The News Today. Newscorp Publications Limited, Dhaka, Bangladesh. July 18, 2012.

Tuli, A.N. 2011. Women's Participation in Household and Agricultural Activities in Selected Area of Gazipur District. MS Thesis, Department of Agricultural Extension and Rural Development, Banghabandhu Sheikh Mujibur Rahman Agricultural University, Gazipur.

Zaman, Z., Tokuzo, M. and Shuji, H. 2003. The role of rice processing industries in Bangladesh: a case study of the Sherpur district. *The Review Agril. Econ.* 57: 121-133.

4

PARTICIPATORY RURAL APPRAISAL APPROACHES: AN OVERVIEW AND AN EXEMPLARY APPLICATION OF FOCUS GROUP DISCUSSION IN CLIMATE CHANGE ADAPTATION AND MITIGATION STRATEGIES

M.N. Uddin[1*] and N. Anjuman[2]

Abstract

Different tools and techniques of participatory approaches are the basic way of conducting qualitative research especially in the field of applied social science. Focus Group Discussion (FGD) is one of the main Participatory Rural Appraisal (PRA) technique often used in combination with others to achieve desired goals. Considering this concept, this paper attempts to review the PRA approach and then application of FGD, in combination with matrix scoring and ranking to identify problems and causes of climate change along with possible mitigation and adaptation strategies. A group of 20 students at post graduate level under the faculty of Agriculture and Horticulture at Humboldt University of Berlin, Germany those from different corner of the world was considered as target people of the study. The results concluded that "unpredictable weather events" was ranked as the present outstanding visible climate change problem caused by "human activities". However, it was noted that if alternative renewable energy sources are exploited, this could contribute to solving the present climate change problem. This finding might have the good reference for the policy makers in the same line not only for developing countries but also for developed countries.

Keywords: PRA, FGD, Climate Change, Adaptation and Mitigation

[1]Assistant Professor, Department of Agricultural Extension Education, Bangladesh Agricultural University (BAU), Mymensingh-2202, Bangladesh
[2]Ex- MS Student, Department of Rural Sociology, BAU, Mymensingh-2202, Bangladesh

*Corresponding author's email: nasirbau@gmail.com (M.N. Uddin)

Introduction

PRA is a process which extends into analysis, planning and action. The World Bank defines PRA as a 'family of participatory approaches and methods which emphasize local knowledge and enable local people to do their own appraisal, analysis and planning. 'PRA uses group animation and exercises to facilitate information sharing, analysis and action among stakeholders' (World Bank, 1995).

It originated in the early 1990's, deriving its basic principles from activist participatory research, agro-ecosystem analysis, applied anthropology, field research on farming systems and most significantly Rapid Rural Appraisal (RRA). While there is no concrete definition, RRA can be defined as a series of techniques for research that are claimed to generate results of less apparent precision, but greater evidential value, than classic quantitative survey techniques[1]. From parallel research work in different parts of the world, RRA emerged as an idea in the 1970's. Later in the 1980's the word "Participatory" found footing in RRA. At the 1985 Khon Kaen International Conference a typology of seven types of RRA were generated, (KKU, 1987) of which "Participatory Rapid Rural Appraisal" (PRRA) was one. From here RRA further evolved to PRA in 1988 – 1990 mainly in Kenya & India at NGO's and various government bodies. This was then promoted by bodies like International Institute for Environment and Development (IIED), Ford Foundation (FF) and Swedish International Development Cooperation Agency (SIDA). In RRA, information is more elicited and extracted by outsiders while in PRA it is more shared and owned by the locals (Chambers, 1994). Alam and Ishan (2012) explained that PRA is the most suitable and appropriate method to indentify the existing situation of the community. Recently, PRA has come to mean Participatory Reflection and Action (Chambers, 2007) while Participatory Learning and Action (PLA) method, which is much broader and includes other related or similar approaches is sometimes equally used in the place of PRA (Chambers, 2007).

In the three decades from its origin PRA witnessed a period of constant evolution. At the core of these changes was the goal to address two primary concerns, which were not tackled by the pre-existing research methods. First was to integrate local perspective in the development process by becoming more responsive to local people and local situations. Second was to develop an adaptive methodology that would provide timely and cost effective information.

Today the principles of PRA are: 1) 'handing over the stick' which means surrendering authority to local people in the learning processes, 2) ability to conduct critical examination by and of facilitators of their own roles, personal responsibility i.e. 'using one´s own best judgment at all times', 3) multi way sharing of ideas and information and 4) stimulation of 'community awareness' (Chambers, 1992; Chambers, 1997; Weber and Ison, 1995).

Its applications include but are not limited to general analysis of a specific topic, question, or problem; needs assessment; feasibility studies; identification and establishment of priorities for development or research activities, monitoring and evaluation of development or research activities and identification of conflicting interests between groups. This is usually achieved by use of one or more tools of PRA. This paper aims at: 1) briefly clarify the theoretical concepts of some of the existing PRA tools including Focus Group Discussion (FGD) as one of the PRA tool and 3) practically illustrating how a FGD can be use to determine problems and causes of climate change as well as mitigation and adaptation strategies.

Methodology

Different literatures review and FGD were practiced to explain objectives of the study. About 20 post graduate students those coming from different part of the globe whose study under the faculty of Agriculture and Horticulture at Humboldt University of Berlin, Germany were considered as target people of the study. The details of the FGD conduction along with result is explained at the second part of the paper while first part of the paper revealed with PRA approaches.

Overview of PRA tools

There are several tools and techniques that belong to the PRA family. Some of these are briefly described along their main goal below:

Some existing PRA tools

Historical Timelines involves the analysis of past events such as conflicts, natural disasters (floods, froughts, cyclones etc.), changes in the natural, social, political or economic environment and the ways in which community members have dealt with them (Callens *et al.*, 1999). The goal of this tool is to understand the history of the

community and identify trends and their influences throughout history (SEPP, 2007). Village resource maps are a compilation of the perception of resources in a given community and are usually drawn by the community members on large pieces of paper or on the ground to indicate spatial representations of resources such as infrastructure, water sources, agricultural landscapes, agro-ecological zones, forest and grazing areas (AFN, 2002). Moreover, it helps to the researchers or policy makers to assess & evaluate the resources of the community (Carey and Etling, 1997).

Another improtant tool of PRA is seasonal calendar which is used to explore seasonal changes in a given community. Changes such as the distribution of rainfall patterns, income, agricultural and non-agricultural labour, food consumption, animal fodder, gender-specific workload and migration can be shown on such calendars (Chambers, 1994). The main objective of the seasonal calendar is to learn about changes in livelihoods over the year and to show food availability, gender-specific workload, water availability, credit availability throughout the year, as well as holidays available within the community. The use of open ended questions is important in obtaining more detailed information. Some key questions to ask the community could be: How does credit availability vary over the year? What are the busiest months of the year? How does rainfall vary over the year? (Sontheimer *et al.*, 1999).

Wealth ranking is a sensitive PRA tool aimed at investigating perceptions of wealth differences and inequalities in a community. It involves placing people on the different steps of the social ladder according to their own criteria, chiefly to discover which community members belong to the richest, middle-income and poorest categories (Lekshmi *et al.*, 2008). During the interviews, Callen *et al.* (1999) suggests the use of questions such as: What socio-economic groupings are there in a community and who belongs to what group? ; What are the local perceptions of wealth, well-being and inequality in the community? Questions such as these are aimed at understanding local indicators and criteria of well-being and wealth in a community. The responses got can also be used to address livelihood concerns for different wealth groups in the community especially the poor (Vietnam, 2003).

Transect Walks involve members of the outside investigating team (researchers and facilitators) walking through the community with local people to record significant social and physical features of the region (Maarten *et al.*, 2008). Observations and discussions involving asking open-ended questions and listening are carried out, as different zones, soil types; land uses, vegetation, crops and livestock are identified. Problems as well as possible solutions are also sought as the different zones and resources are mapped and diagrammed (Chambers, 1994). Maarten *et al.* (2008) add that the use of transect walks helps researchers gain the confidence of the local people and can be used to identify circumstances under which climate change may have an impact on a given village. Diagrams are a pictorial representation of information, used to illustrate flows, causal relationships and other connections as well as the analysis of spatial data (Adepo, 2000). Examples are Venn diagrams, Flow diagrams, Transect diagrams, Causal-linkage diagrams and Systems diagrams. Conroy (2002) indicates that diagrams encourage participants to get involved in the research process and express the information in a way that is best understood by them, while at the same time openly discussing options of correcting and refining the information.

Schwedes and Werner (2010) describe matrix scoring and ranking as an exercise that involves placing something in order to determine what is important and what is less important or less appropriate, that is; different options or solutions are ranked according to criteria. A matrix is a dual entry network that can be applied to evaluate two sets of variables (Conroy, 2002). The major objective of this tool is to identify the common problems within the community, rank and score them in order of importance, then scrutinize them and brainstorm for possible solutions. Problem-Cause-Effect-Solution Trees (Problem Trees) involves collectively identifying, listing and prioritizing problems within a community, their causes and possible solutions. This tool helps recognize linkages between causes and effects of problems as well as their solutions. It can be used to plan activities within a given community, pertaining to issues such as health, nutrition, education and gender issues (SEPP, 2007). Interview such as semi-structured interviews, key informant interviews and expert interviews can be used with individuals, key informants, experts, interest groups or other small groups of villagers (Cavestro, 2003).

Focus Group Discussion (FGD)

Campbell (2008) defines a FGD as *"a planned, facilitated discussion among a small group of stakeholders designed to obtain perceptions in a defined area of interest in a permissive, non threatening environment"*. It is the method of rapid assessment and data gathering in which participants congregate to talk about the specific issues and concern based on a list of key themes drawn up by the researcher/facilitator (Kumar,

1987). The main objective of focus group discussion is to acquire knowledge regarding the particular issue. It can be used to collectively assemble and analyse information for many purposes such as the adoption of a particular innovation (Ndah *et al.*, 2011), needs assessment (Tipping, 1998), program evaluation (Packer *et al.*, 1994) etc. For conducting a focus group discussion, a facilitator and assistant to facilitator are needed. The facilitator leads the group discussion and encourages the participants. The assistant to the facilitator is to take notes, run the tape recorder, respond to the unexpected interruptions, and is always ready to follow the facilitator's mode of action. Knowledgeable, pleasing personality, politeness, ability to speak local language, respect to local norms and behaviour, ethics, patience etc. are the main criteria of a good facilitator.

Exemplary application of a FGD in climate change

The main purpose of this issue was to practically demonstrate how a FGD as a PRA tool alongside others can be used to identify and analyse some of the adverse climate change problems, causes as well some possible mitigation and adaptation strategies. The target group for this exercise was a group of students under the faculty of agriculture and horticulture at Humboldt University of Berlin. Therefore, represents the statements and conclusions arrived at by this group of students within a FGD session.

Specific objectives of the FGD exercise

This exercise was meant specifically to:

1) Understand the group's perceptions of climate change by identifying and ranking some of the main climate change problems presently under debate.
2) Identify and understand the major cause or triggers of the identified problems
3) Identify and understand some of the possible mitigation and adaptation strategy to Climate change.

All these were meant to in effect expose the individual as well as groups perception of the present climate change issue under debate with the use of a FGD.

Organisation of the FGD

Participants included about 20 students from different cultural as well as disciplinary backgrounds. The purpose of choosing such a heterogeneous group for this purpose was meant to ideally bring to a common platform the differences in perceptions with regards to the present climate change issue. Two PRA-team members were in charge of the organisation and running of the exercise. While one was in charge of facilitating the discussion, one was taking notes and assisting with the compilation of the results for the feedback session.

Guided questions and methods used

To effectively achieve the desired objectives, the focus group exercise was combined with other PRA tools such as "matrix scoring and ranking". The key questions which guided the discussion with the corresponding steps or activities that were followed through in chronological order included:

1) According to you, what are the major climate change problems that people have faced during the 10 past years? (exercise 1)
2) In your view, what are the possible causes for the problems of climate change you have identified? (exercise 2)
3) According to you, what could be possible mitigation and adaptation strategies to the climate change problems you have identified? (exercise 3)

Activities in chronological order during the FGD (Exercise 1: Problems)

- Participants were asked to write down three problems each on cards which were collected and pasted on the pin board
- The pasted cards were then grouped with the help of participants according to categories and boldly printed numbers printed against each category (e.g. 1, 2, 3...)
- Each participant was then asked to individually rank these categories according to his/her perception by casting three votes for the three most important (severe) problems. This was by writing down three numbers selected from the represented (preferred) categories on cards.
- The votes were then counted and the three top ranked categories with the highest frequency of votes (selection) were then singled out as the most severe climate change problems perceived by the group.

The same exercise and activities was repeated for exercise 2 (causes) and exercise 3 (solutions) respectively to obtain results which answered the three questions and met the three objectives (see results).

Results and Discussion

After the three exercises, three major problems of climate change were identified, three main causes as well as three main mitigation and adaptation strategies as presented in Figures 1, 2 and 3.

Problems of climate change

Amongst the identified problems of climate change, those that fell under the category "unpredictable weather events" were ranked the most severe with 50% severity rate. This was closely followed by the "water scarcity and desertification category" with 33% severity rate while the "Air and water pollution" category occupied the third position with 17% severity according to the perception of the group (Fig. 1). Semilar finding has been explained by the several researchers and institutes.

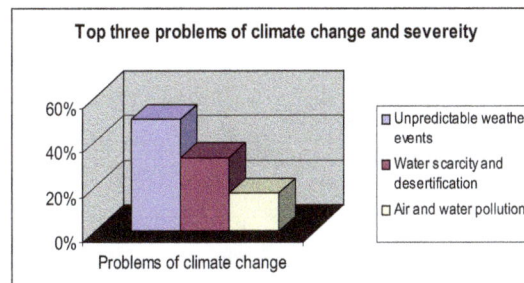

Fig.1. Top three problems of climate change according to the FGD

IPCC (2001) found the weather variability in the most places in the world due to cliamte change while Pickup (1998) decribed desertification is trigged by climate variability. Water scacity is the result of climate change as explained by Morrison *et al.* (1998). Besides, Kinney (2008) and Delpla *et al.* (2009) reaveled that air and water qulaity is affected by the cliamte change. So, the top three problems of cliamte change that have identified by this study are also justified by the others research. Therefore, policy makers, researchers, developers etc. might be considered while taking strategies for future context.

Causes of climate change

After categorising and ranking the identified possible causes of climate change by the group (exercise 2), the "human activities and use of fossil fuel category emerged as that with the most severe effect on climate (70% magnitude). Second on the list was the category "increasing temperatures" with 20% while "increase deforestation" occupied the third position with 10% (Fig. 2). Semilar result has been found by the several researches while Hamilton and Stampone (2013) decribed that anthropogenic activities is the cause of cliamte change and fossil fule is also the responsible for the same (IPCC, 2007). Morevoer, emission of methane that plays an important role in global warming has been increased due to higher temperatures (Science News, 2010). Bloom *et al.* (2010) found about 7% methane has been increased during 2003-2007 due to warming of mid-latitude and wet arctic region. Now a days, higher temperatures are not consequence of climate change rather it can also worsen cause of it. Besides, Nobre *et al.* (2009) described that deforestation is one of the cause of climate change. Now a days, deforestation, Green House Gas (GHG) emission are the crucial issue for altering the claimte (Nordhaus, 1991). Responsible authority might have interests to give emphasize of these indentified problems that makes climate change adaptations effective.

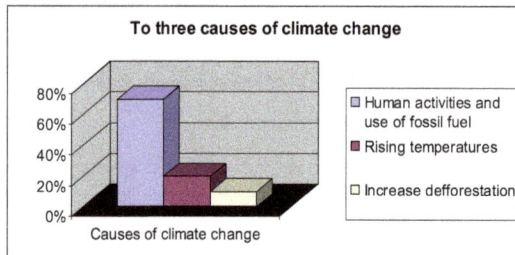

Fig. 2. Top three causes of climate change and magnitude

Mitigation and adaptation strategies

The last exercise (exercise 3), focused on analysing possible adaptation and mitigation strategies ended as well with three main categories: "Use of renewable energy plus a positive change in attitude and behaviour" towards the climate was identified as the category of solutions, which could have a significant positive effect on climate change (60% chance). This was followed by that which gave suggestions on favourable policies with regards to climate issues with a 30% chance (Fig. 3).

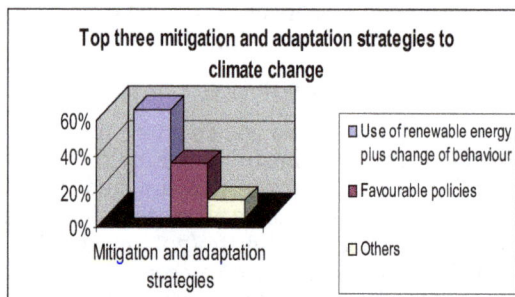

Fig. 2. Top three mitigation and adaptation strategies of climate change strategies

The last category which remained un-classified made mentioned of change in farming practices, and other suggestions which were deemed irrelevant and could have no influence on climate e.g. one suggestion was "to kill people", which we found to be out of context. Ziuku and Meyer (2012) explains that renewable energy reduces the 30% GHG by 2030 while McKibbin and Wilcoxen (2003) mentioned optimal policy can be mitigated as well as adaptation to climate change at low cost. They also specified the policies on land use change, water use property rights that migh have good effect on cliamte

change adaptations and mitigation as well. In the developing countries, agriculture sector is highly affected by climate change (FAO, 2009) resulting food shortage. Therefore, it is necessary to adapt with climate change quickly to produce more food by adopting different adaptations strategies such as increased use of irrigation, practicing crop diversification, integrated farming system, use of drought & salinity tolerant varieties etc (Uddin, 2012).

From the three categories of problems, causes and solutions, it could by concluded that unpredictable weather events as the most outstandingly identified category of climate change problems, is possibly caused by human activities especially through increase used of fossil fuels. The group then agreed that this problems could be possibly remedied to a certain extend if there is an increase search of alternative renewable energy sources followed by general awareness towards positively influencing peoples attitude and behaviour with regards to the present climate change issue.

Conclusion

In conclusion PRA is an ever changing trans-disciplinary process which uses adaptive methodology as and when problems arise. It is important to remember that in this research process one must always try to reduce the big questions to specific queries, immediate gratification is a rare event, local development is a two way street (feedback is critical) and that the behaviour and attitude of outsiders matter as much as the methods and their correct performance. One should be an active learner rather than claim to be an expert. Focus Group Discussion is one of the important PRA technique often used in combination with others to achieve desired goals. This paper tried to explain this technique, in combination with matrix scoring and ranking with a group of 20 students to identify causes of climate change as well as possible mitigation and adaptation strategies. The results concluded that "unpredictable weather events" was ranked as the present outstanding visible climate change problem caused by "human activities". However it was noted that if "alternative renewable energy sources are exploited, this could contribute to solving the present climate change problem. This finding might have the good reference for the policy makers in the same line not only for developing countries but also for developed countries.

Acknowledgement

The authors are grateful to the participants who gives their opinion regarding the purposes of the study.

References

Adepo, S. 2000. Training Manual on Participatory Rural Appraisal. Online available: http://www.myfirecommunity.net/discussionimages/NPost8220Attach1.pdf and accessed on 21st January 2013.

AFN. 2002. Participatory Rural Appraisal for Community Forest Management: Tools and Techniques. Asia Forest Network, California USA. pp. 18-19.

Alam, A. and Ishan, S. 2012. Role of Participatory Rural Appraisal in Community Dvelopment (A Case Study of Barani Area Development Project in Agriculture, Live Stock and Forestry Development in Kohat). *Int. J. Acad. Res. Busi. & Soc. Sci.* 2 (8): 25-38.

Bloom, A.A., Palmer, I.P., Fraser, A. Reay, D.S. and Frankenberg, C. 2010. Large-Scale Controls of Methanogenesis Inferred from Methane and Gravity Spaceborne Data. *Science.* 327 (5963): 322. doi:10.1126/sciencce.1175176

Callens, K., Seiffert, B. and Sontheimer, S. 1999. The PRA Tool Box, Technical Backstopping to the Preparatory Phase of GCP/ETH/056/BEL. Online available: http://www.fao.org/docrep/003/x5996e/x5996e06.htm#6.2.%20Modified%20PRA%20Tools and accessed on 20th August, 2013.

Campbell, R. 2008. Guide to Focus Group Discussion, micro report no. 138. USAID. p. 1.

Carey, H.A. and Etling, A.W. 1997. Constructing and Conducting Rural Appraisal. *J. Int. Agric. Extn. Edn.* 4 (3): 27-37.

Cavestro, L. 2003. PRA - Participatory Rural Appraisal Concepts Methodologies and Techniques. MS Thesis. University of Padova, Italy. pp. 16 -26.

Chambers, R. 1992. Rural Appraisal: Rapid, Relaxed and Participatory. Discussion Paper #311, Institute of Development Studies, Sussex, UK. pp. 15-16.

Chambers, R. 1994. The Origins and Practice of Participatory Rural Appraisal. *World Dev.* 22 (7): 953-969.

Chambers, R. 1997. Whose Reality Counts? Putting the First Last. London: ITDG Publishing. pp. 157-158.

Chambers, R. 2007. From PRA to PLA and Pluralism: Practice and Theory. Working Paper #286, Institute of Development Studies, University of Sussex, Sussex, UK. pp. 7-12.

Conroy, C. 2002. PRA Tools used for Research into Common Pool Resources: Socio-economic Methodologies for Natural Resources Research, Best Practice Guidelines. Natural Resources Institute, University of Greenwich, UK. pp. 7-8.

Delpla, I., Jung, A.V., Baures, E., Clement, O. and Thomas. 2009. Impacts of climate change on surface water quality in relation to drinking water production. *Env. Int.* 35: 1225–1233.

FAO. 2009. Food Security and Agricultural Mitigation in Developing Countries: Options for Capturing Synergies. Rome: FAO. 11 p. Hamilton, L.C. and Stampone, M.D. 2013. Blowin' in the Wind: Short-Term Weather and Belief in Anthropogenic Climate Change. *Wea. Climate Soc.* 5: 112–119.

IPCC. 2001. Impacts, Adaptation and Vulnerability. Third Assessment Report. Cambridge University Press, Cambridge, UK. 92 p.

IPCC. 2007. Climate change 2007. Fourth Assessment Report. Cambridge University Press, Cambridge, UK. 134 p.

Kinney, P.L. 2008. Cliamte change, Air Quality, and Human Health. *American J. Prev. Med.* 35 (5): 459-467.

KKU. 1987. Rapid Rural Appraisal. Proceedings of the 1985 International Conference on Rapid Rural Appraisal, Rural Systems Research and Farming Systems Research Projects, University of Khon Kaen, Thailand. pp. 14-18.

Kumar, K. 1987. Conducting focus group interviews in developing countries. AID Program Design and Evaluation Methodology Report No. 8. Washington, DC: USAID. pp. 3-5.

Lekshmi, P.S.S., Venugopalan, R. and Padmini, K. 2008. Livelihood Analysis using Wealth Ranking Tool of PRA. *Indian Res. J. Extn. Edn.* 8 (2&3): 75-77.

Maarten, K.A., Cannon, T. and Burton, I. 2008. Community Level Adaptation to Climate Change: The Potential Role of Participatory Community Risk Assessment. *Global Env. Change.* 18: 165-179.

McKibbin, W.J. and Wilcoxen, P.J. 2003. Climate Policy and Uncertainty: The Roles of Adaptation versus Mitigation. Online available: http://een.anu.edu.au/download_files/een0306.pdf and accessed on 10th October, 2013.

Morrison, J., Morikawa, M., Murphy, M. and Schulte, P. 2009. Water Scarcity & climate change: Growing Risks for Businesses & Investors. A ceres report. Pacific Institute. Oakland. Online available: http://www.pacinst.org/wp-content/uploads/2013/02/full_report30.pdf and accessed on 20th September, 2013.

Ndah, H.T., Knierim, A. and Ndambi, O.A. 2011. Fish Pond Aquaculture in Cameroon: A Field Survey of Determinants for Farmers' Adoption Behavior. *J. Agril. Edn. & Extn.* 17 (4): 309-323.

Nobre, Paulo, Malagutti, M., Domingos, F.U., Roberto A.F., De Almeida. and Giarolla, E. 2009. Amazon Deforestation and Climate Change in a Coupled Model Simulation. *J. Climate.* 22: 5686–5697.

Nordhaus, W.D. 1991. Economic approaches to greenhouse warming. pp. 33-68. *In:* Global warming: Economic policy approaches, *ed.* R.D. Dornbush and J.M. Poterba, Cambridge, MA: MIT Press.

Packer, T., Race, E.K. and Hotch, F.D. 1994. Focus groups: a tool for consumer-based program evaluation in rehabilitation agency settings. *J. Rehabn.* 60 (3):30-33.

Pickup, G. 1998. Desertification and climate change—the Australian perspective. *Climate Res.* 11: 51–63.

Schwedes, S. and Werner, W. 2010. Manual for participatory land use planning facilitators. Ministry of Lands and Settlement and German Technical Cooperation (GTZ). pp. 139-141.

Science News. 2010. Higher Temperatures Can Worsen Climate Change, Methane Measurements from Space Revea. Online available: http://www.sciencedaily.com/releases/201 0/01/100115204416.htm and accessed on 10th September, 2013.

SEPP. 2007. Socio-Economic Planning Process. Participatory Rural Appraisal Manual. Quang Ngai Province, Vietnam. Online available: http://www.rdsikkim.org/Files/3%20PRA %20Facilitators%20Manual.pdf and accessed on 19th September, 2013.

Sontheimer, S., Callens, K. and Seiffert, B. 1999. Conducting a PRA Training and Modifying PRA Tools to Your Needs. An Example from a Participatory Household Food Security and Nutrition Project in Ethiopia. Food and Agricultural Organization of the United Nation. Online available: http://www.fao.org/Participation/english_ web_new/content_en/Sector_doc/PRA_nu trition.pdf and accessed on 18th September, 2013.

Tipping, J. 1998. Focus groups: A method of Needs Assessment. *J. Cont. Edu. Health Prof.* 18: 150–154. Online available: http://www.iisd.org/casl/caslguide/rapidru ralappraisal.htm and accessed on 18th September, 2013.

Uddin, M.N. 2012. An Analysis of Farmers' Perception and Adaptation Strategies of Climate Change in Bangladesh. MS Thesis. Humboldt University of Berlin, Germany. pp. 46-49.

Vietnam, H. 2003. PRA tools for identifying activities for the mid-term and yearly socio-economic Village Development Plan (VDP). Extension and Training Support Project for Forestry and Agriculture in the Uplands. pp. 25-26.

Weber, L. and Ison, R. 1995. Participatory Rural Appraisal design: Conceptual and process issues. *Agril. Sysm.* 47: 107–31.

World Bank. 1995. The Participation Sourcebook, Washington DC, World Bank. p. 175.

Ziuku, S. and Meyer, E.L. 2012. Mitigating climate change through renewable energy and energy efficiency in the residential sector in South Africa. *Int. J. Renewable Energy Tech. Res.* 2 (1): 33 – 43.

5

IMPROVEMENT OF FINISHING ANTIFRICTION TREATMENT WITHOUT ABRASIVE OF THE RUBBING PARTS SURFACES OF AGRICULTURAL MACHINERIES

I.V. Shepelenko[1], V.V. Cherkun[2], A. Warouma[3]*

Abstract

The wear of machines and mechanisms after rubbing is a major concern. The costs of manufacturing and restoration parts for agricultural machinery are enormous such as the trunnions pinions of hydraulic pumps. Finishing treatment antifriction without abrasive (FTAA) is the existing method of manufacturing and restoration, but it has disadvantages like low work efficiency and the frequent replacement of the instrument. That is why a new method of FTAA parts type "tree" has been developed called vibratory finishing treatment antifriction without abrasive, (VFTAA) method. The study was conducted at the laboratory of the State Technical University of Kirovograd (Ukraine) where turn 16K20, the profilograph-profilometer "Talysurf-5", the scanning electron microscope REM-106I, friction machine MI-1M and the stand KI-28097-02M were used respectively for machining, study of micro relief before and after the VFTAA, microstructure, wear resistance and the determination of the break-in period of parts. The results showed that the VFTAA helped reduce the roughness Ra of the samples studied by half compared to the samples processed by polishing and 1.3 times compared to those treated with the FTAA, the break-in period has been reduced four times , this leads to an increase in the life of the hydraulic pump. This technology can be recommended for manufacturing and repair of hydraulic units of agricultural machineries.

Keywords: Finishing Treatment, Abrasive, Trunnions Pinions, Wear Resistance, Rotation Speed

[1]Department of Agricultural Machinery Operating and Repair, National Technical University of Kirovograd, Ukraine
[2]Department of Mechanic and Technology, Agrotechnological University of State Tavri, Ukraine
[3]Department of Rural Engineering & Water and Forests, University of Maradi, Niger

*Corresponding author's email: warouma@yahoo.com (A. Warouma)

Introduction

The wear of machines and mechanisms after friction is one of the most pressing issues of our time. Manufacturing and restoration costs of worn parts of agricultural machineries are enormous due to rubbing. They are even higher for the parts such as pins sprockets hydraulic pumps that require high precision machining. The existing technologies for manufacturing and repairing of high precision parts require a long period of machining in order to increase their lifespan (Garkunov, 2009; Fedorenko et al., 2011). These costs can be reduced significantly by using rational methods based on tribotechnic. These methods must include the FTAA which is a finishing treatment of surfaces parts whose essence consists that the parts of the sliding surfaces should be covered with a thin layer of brass, bronze or copper by using the phenomenon of metal transfer during the rubbing (Kuzharov and Kuzharov, 2011). A coating of ductile metal with a thickness of 50 to 500 nm is applied to the rubbing surfaces of different machine parts. The FTAA is applied in order to reduce the intensity of wear, increasing the crack resistance of the rubbing surfaces and the intensification of the training process of films protection during the break-in period, after manufacturing or restoration of parts (Balabanov et al., 2010; Katkov, 2009).

However, these traditional methods of FTAA have poor performance (Bystrov, 2011); this stops their widespread use as a finishing treatment of parts in industrial conditions. Among the disadvantages of existing process of FTAA, in addition to inadequate work performance, we can mention the frequent replacement of the instrument because of its irregular wear.

The irregular wear of the instrument can only be removed when it rotates during the process of FTAA. Performance of FTAA can be improved by increasing the speed of the relative movement of the couple "tool-parts" not by increasing the speed of rotation of the part, but by the creation of the longitudinal oscillations of the instrument when machining. This will accelerate the FTAA without compromising the quality of the process.

By considering these lacks of performances, the machining tool must have a movement of rotation and an oscillation. For that reason, it was developed for the first time a method of FTAA of parts type "tree" with the use of vibration called vibratory finishing treatment antifriction without abrasive (VFTAA) method.

Thus, the objective of this work is to study the influence of VFTAA on the quality of the machined surfaces, the structure, physico - mechanical properties and antifriction of the surface layer journals gears hydraulic pumps agricultural machineries.

Materials and Methods

The study was conducted at the laboratory of the State Technical University of Kirovograd, Ukraine between 2010 and 2012.

The VFTAA parts took place on a turn 16K20 by using the schema developed (Fig. 1), the device (Fig. 2) and the surface active medium with the following plans: instrument pressure P = 6 MPa; speed sliding of the instrument, Vsl = 1, 82 m / s; ratio of the number of strokes of the instrument and the speed of the instrument, $n_{\partial 6.x.}/ n_{\partial} = 57$;

the value of oscillation of the instrument $\ell_0 =$ 7.12 mm; longitudinal movement of feed, S = 1 mm / rotation.

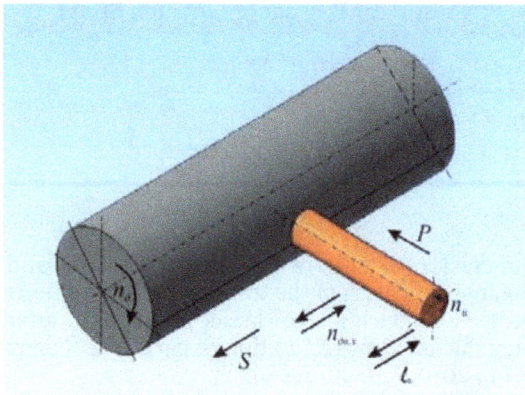

Fig.1. Diagram of the implementation of vibratory finishing treatment antifriction without abrasive: P - pressure of the instrument; n_{∂} - speed of rotation of the part, n_{u} - speed of rotation of the instrument; $n_{\partial 6.x.}$ - Number of strokes of the instrument, S - longitudinal feed movement of the instrument, l_{o} - oscillation of the instrument.

Fig. 2. Vibratory finishing treatment antifriction without abrasive of trunnions pinions of hydraulic pumps of agricultural machineries

The study of the surface of the micro relief before and after the VFTAA was realized by using a profilograph-profilometer "Talysurf-5", the microstructure of the carburized 18CrMnTi samples were investigated by the scanning electron microscope REM-106I, the wear resistance with machine friction MI-1M. The stand KI-28097-02M was used for doing the speed tests of hydraulic pumps.

In order to determine the break-in period of the main components of pumps NSH50A-3 and NSH100A-3, the trunnions pinions treated with existing and proposed technologies have been investigated on the stand. The power of electric motor transmitted on the training of the hydraulic pump has been analyzed in terms of different types of treatment in order to assess the duration of running.

Results

The results showed that to obtain a high-quality coating, it is necessary that the initial surface roughness of the steel is not less than 1.25 Ra. During the machining process, the VFTAA reduced the roughness Ra of the samples studied twice compared to the samples treated by polishing, and 1.3 times relative to FTAA (Fig.3).

Ra, μm

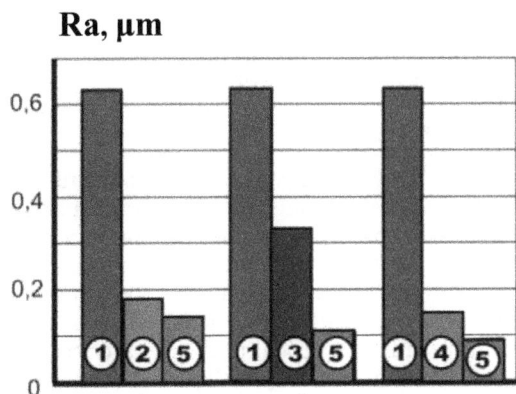

Fig. 3. Variation of roughness parameter of samples during machining in different types of finishing treatment: A - finish turning (1) polish (2) and break-in (5); B - finish turning (1) FTAA (3) and running (5); C - finish turning (1) VFTAA (4) and running (5).

The VFTAA reduced machining time, improved the break-in process of the rubbing couple "steel - bronze" and reduced considerably the intensity of initial wear of the surface layers (Fig. 4) compared to samples treated with a finish turning and a polishing by obtaining a uniform and dense coating.

J, x 10^{-4}, g/mn

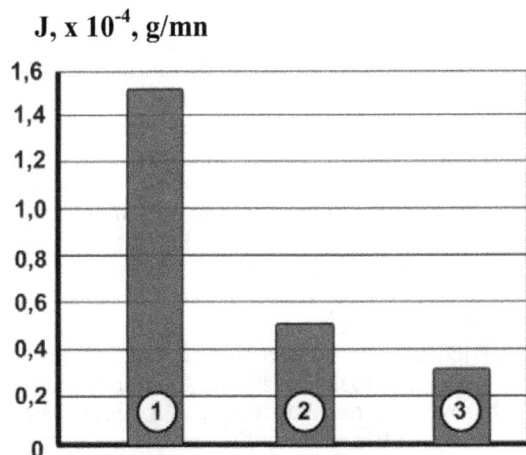

Fig. 4. Diagram intensity of initial wear J in different types of finish: 1 - finish by turning and polishing; 2 - finishing turning and FTAA; 3 - finishing turning and VFTAA.

Use of VFTAA reduced the total wear of the joint "trunnion-hub" (Fig. 5) and the friction coefficient (Fig. 6).

U_{1-2}, **μm**

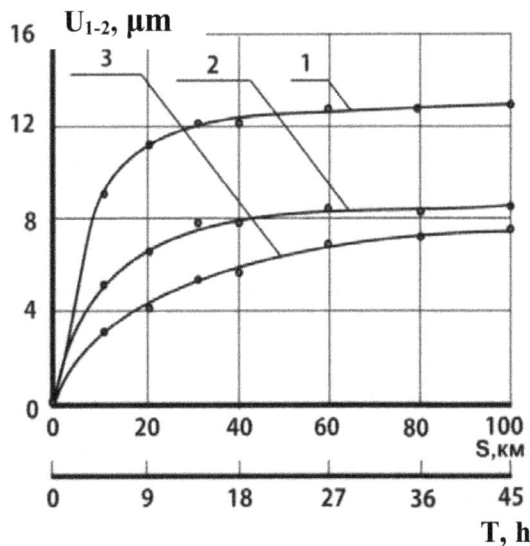

Fig. 5. Dependence total wear of the joint U_{1-2} depending on the path S and time T of rubbing during different types of finishing treatment: 1 - polishing, 2 - FTAA, 3 - VFTAA.

f

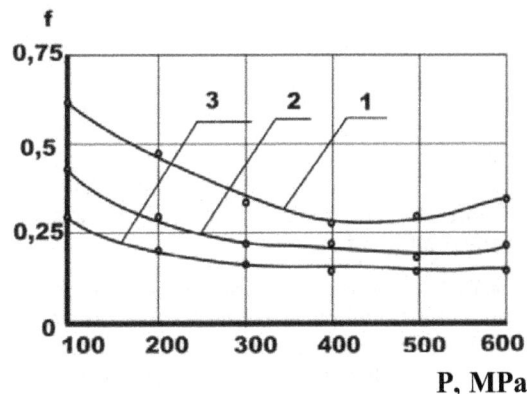

Fig. 6. Dependence of the rubbing coefficient (f) according to the loads (P) according to the different processing methods: 1 - polishing, 2 - FTAA, 3 - VFTAA.

The dependence of the break-in period of trunnions pinions of hydraulic pumps treated with different technologies according to the energy function transmitted by the electric motor to the hydraulic pump shown in Fig. 7.

N, Kw

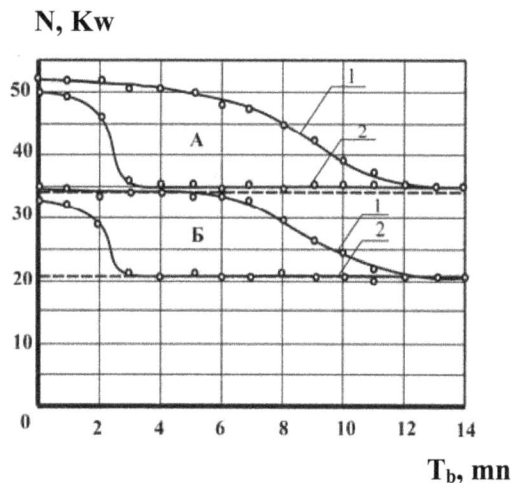

Fig. 7. Dependence of the break-in period (T_b) of trunnions of hydraulic pumps (NSH100A-3 and NSH50A-3) according to the force transmitted and by using different technologies: 1 - existing technology; 2 - the proposed technology with the adoption of the VFTAA.

Testing stand revealed that the break-in period, characterized by the stabilization of energy consumption by the electric motor of the stand during the application of existing technology was 12 minutes, while the technology by using the VFTAA of trunnions pinions of hydraulic pumps was 3 minutes.

Discussion

The FTAA has a low yield compared to the proposed technology: VFTAA. This low yield is explained by the fact that the speed of the relative movement of the tool and part is performed only by the rotation of the part.

The VFTAA reduced roughness Ra of the samples studied twice compared to the samples treated by polishing, and 1.3 times according to FTAA. This reduction is due to the more intensive rubbing brass in cavities of micro ledges and flattening the tops of micro ledges.

With the use of VFTAA, the break-in period of trunnions pinions of hydraulic pumps was 3 minutes against 12 minutes with FTAA. Therefore, this leads to the reduction of wear of the friction couple during the break-in period, to the stabilization of play in the joint, and consequently to an increase in the life span of the hydraulic pump during operation.

The studies conducted on the FTAA by Garkunov (2009) and Katkov (2009) gave successively the periods of running parts: 9 and 11 minutes, which are significantly lower than which are obtained by the developed technology (3 min).

Conclusion

The technology developed of VFTAA of trunnions pinions reduced significantly the period running of hydraulic pumps, the cost of recovery, and improved the life span of hydraulic pumps and can be recommended like at the time of manufacturing, as at the time of repair of hydraulic units of agricultural machineries. The use of coating technology with TVFASA allows in perspective the replacement of parts of non-ferrous alloys, steel and cast iron with anti-friction coating.

References

Balabanov, I.V., Bolgov, V.Y. and Ichenko, S.A. 2010. Application by friction nanoscale of antifriction coatings on parts nanotechnology, ecology, production. 1 (3): 104-107.

Bystrov, V. 2011. Use of devices for mechanical input by friction of wear resistant coatings in the repair mass conditions /Invention. 11 (3): 29-34.

Fedorenko, V.F., Erohin, M.N., Balabanov, V.I., Buklagin, D.S., Golubev, G.I. and Ishchenko, S.A. 2011. Nanotechnologies and nanomaterials in agriculture: ed.sci. / M.: FNGU "Rosinformagroteh". 312p.

Garkunov, D.N. 2009. Finishing treatment antifriction without abrasive (FTAA) parts of the friction surfaces of the parts PRM (Preparation. Restoration. Modernization). 3: 36-41.

Katkov, D.S. 2009. Increase of longevity of friction parts for agricultural machineries by using tribological methods / Saratov, 251p.

Kuzharov, A.S. and Kuzharov, A.A. 2011. More and a little differently on the plating of metals, the FTAA and the nonwear. *In:* Proc. Samara Scientific Center of the Russian Academy of Sciences. 4 (3): 772-775.

6

GENETIC DIVERSITY ANALYSIS OF RICE (*Oryza sativa* L.) LANDRACES THROUGH RAPD MARKERS

M.S. Alam[1]*, S.N. Begum[2], R. Gupta[3] and S.N. Islam[1]

Abstract

The molecular marker is a useful tool for assessing genetic variations and resolving cultivar identities. Information on genetic diversity and relationships among rice landraces from Bangladesh is currently very limited. Thirty-five rice genotypes including 33 landraces and 01 HYV of Bangladesh and 1 Indian landrace of particular interest to breeding programs were evaluated by means of random amplified polymorphic DNA (RAPD) technique. For molecular characterization, RAPD markers *viz.,* OPC 03, OPC 04 and OPA 01 gave reproducible and distinct polymorphic amplified products. A total of 20 RAPD bands were scored of which 15 polymorphic amplification products were obtained by using these arbitrary primers. The size of amplified fragments were ranged from 550 to 1775 bp. Based on analysis performed on a similarity matrix using UPGMA, 35 genotypes were grouped into 2 main clusters. Landrace Sylhet balam and Mota aman was totally different from other genotypes. The information will facilitate selection of genotypes to serve as parents for effective rice breeding programs in Bangladesh.

Keywords: Crop Diversity, Characterization, Genetic Distance, Genetic Identity, Polymorphic Loci

[1]Department of Genetics and Plant Breeding, Bangladesh Agricultural University, Mymensingh, Bangladesh.
[2]Senior Scientific Officer, Plant Breeding Division, Bangladesh Institute of Nuclear Agriculture, Mymensingh, Bangladesh.
[3]Scientific Officer, Bangladesh Institute of Nuclear Agriculture, Sub Station, Khagrachari, Bangladesh.
*Corresponding author's email: agt.shahed@gmail.com, m.s.alam@irri.org (M.S. Alam)

Introduction

Rice (*Oryza sativa* L.) diversity consists of landraces, improved cultivars, hybrids and closely related wild relatives. Landraces are local crop varieties developed in primitive agricultural system, rather than being deliberately bred, selected by the farmers over many generations. Landraces of rice played a very important role in the local food security and sustainable development of agriculture, in addition to their significance as genetic resource for rice genetic improvement. To maintain crop diversity, collection, characterization and conservation of traditional landraces are vital. Bangladesh is a good source of landraces of rice. To formulate a sustainable breeding program precise knowledge about genetic divergence for yield components is a crucial one as varietal improvement depends mainly on the selection of parents with high genetic divergence in hybridization that is supposed to increase the chance of obtaining maximum heterosis and give broad spectrum of variability in segregating generations. PCR-based molecular marker technique RAPD analysis is a reliable tool for assessing genetic diversity. This technique use a single short oligonucleotide primer (9-10 bp) of arbitrary DNA sequence and polymerase chain reaction (PCR) mediated amplification of random fragment from genomic DNA. So, among the available DNA molecular techniques, RAPD has many advantages over others such as ease and rapidity of analysis, a relatively low cost, availability of large numbers of primers and the requirement of a very small amount of DNA for analysis (Weising *et al.,* 1995). Advantages associated with RAPDs have made them a favorite marker technique in mapping, the determination of phylogenetic relationships, genetic diversity, and identification of cultivars and parents in a number of plant species. Since the RAPD technique involves enzymatic amplification of target DNA by PCR using arbitrary primers, it is also called Arbitrarily Primed Polymerase Chain Reaction (AP-PCR) or DNA Amplification Fingerprinting (DAF). RAPD markers tend to estimate intra or inter genetic distances more distantly related individuals. In addition, no prior knowledge of sequence is required. Since primers can be chosen arbitrarily, any organism can be mapped with the same set of primers. These advantages make RAPD markers far easier to work. The objective of this present study was to evaluate genetic divergence of 35 rice genotypes with three RAPD markers.

Materials and Methods

The experiment was carried out at the Biotechnology Laboratory, Biotechnology Division, Bangladesh Institute of Nuclear Agriculture, Mymensingh, during July 2012 to December 2012.

Plant materials

Thirty five rice genotypes including 33 landraces (i.e. Hati bajore, Malagoti, Kuchra, Enghi, Jamai naru, Hari, Dakh shail, Moina moti, Marish shail, Patnai, Bhute shallot, Kute patnai, Moghai balam, Sada gotal, Khak shail, Jota balam, Khainol, Hamai, Sylhet balam, Mota aman, Ghigoj, Piarjat, Lal biroi, Lalanamia, Golapi, Asam binni, Kakua binni, Ledra binni, Rotisail, Genggeng binni, Jolkumri, Mowbinni and Bogi) and 1 HYV of Bangladesh (i.e. Binadhan-8) and 1 Indian landrace (i.e. Nona bokhra) were used for genetic diversity analysis through RAPD markers.

DNA extraction

In order to carry out RAPD analysis, young, vigorously growing fresh leaf samples were collected from 25 days old seedling of each genotype and used as the source of genomic DNA. DNA was extracted from the leaves of each genotype using the Cetyl Trimethyl Ammonium Bromide (CTAB) mini-prep method. The simplified mini scale procedure for DNA isolation in PCR analysis developed at IRRI was followed. Confirmation of the DNA was done through electrophoresis on a 0.8% agarose gel and DNA quantification was through the spectrophotometer's (spectronic@ Genesisr™). Ten primers of random sequence were screened for amplification of the DNA sequences. Primers resulting in faint or irreproducible bands were excluded from subsequent analysis. A final subset of seven primers exhibiting good quality banding patterns and sufficient variability from where finally three primers were selected for further analysis.

PCR reaction and electrophoresis

PCR reactions were performed on each DNA sample in a 10µl reaction mix containing 1 µl Ampli *Taq* polymerase buffer (10X), 2.5 µl Primer (10 µM), 1 µl dNTPs (250 µM), 0.2 µl Ampli *Taq* DNA polymerase and 3.3 µl sterile deionized water. Two µl genomic DNA was added and finally, total volume was made 10 µl. DNA amplification was performed in an oil-free thermal cycler. The reaction mix was preheated at 94°C for 3 minutes followed by 40 cycles of 1 min denaturation at 94°C, 1 min annealing at 54°C and elongation or extension at 72°C for 2 minutes. After the last cycle, a final step for 7 minutes at 72°C to allow complete extension of all amplified fragments. After completion of cycling program, reactions were held at 4°C. Electrophoresis was carried out in 0.5 X TBE buffer on a 1.5 % agarose gel and amplified fragments were visualized by staining with ethidium bromide.

Data analysis

The amplified bands were visually scored as present (1) and absent (0) separately for each individual primer. The scores of bands were pooled to create a single data matrix. This was used to estimate polymorphic loci, Nei's (1972) gene diversity, population differentiation (Gst), gene flow (N_m). Genetic distance (GD) and to construct a UPGMA (Unweighted Pair Group Method of Arithmetic Means) dendrogram among populations using a computer program, POPGENE (Version 1.31); (Yeh *et al.*, 1997).

Results and Discussion

RAPD banding pattern

Ten primers were initially screened for their ability to produce polymorphic patterns and out of 10, three primers *viz.*, OPC 03, OPC 04 and OPA 01 gave reproducible and distinct polymorphic amplified products. DNA amplification from all the primers tested in this study was not consistently reproducible, is a very common feature of RAPD technique. A total of 20 RAPD bands were scored of which 15 (73.61%) polymorphic amplification products were obtained by using these arbitrary primers. This proportion of polymorphism was similar compared to previous RAPD analysis in rice genotypes by Skaria *et al.* (2011) who obtained 72.27% of polymorphic products. The size of the amplification products ranged from 550-1775 bp (Table 1). The selected 3 primers produced comparatively maximum number of high intensity band with minimal smearing, good technical resolution and sufficient variation among different cultivars. The dissimilar numbers of bands were generated by primer OPC 03, OPC 04 and OPA 01. Besides, the primer OPC-03 amplified maximum number of polymorphic bands (100%) while the primer OPA-01 generated the least (33.33%) polymorphic bands which were minimal in number. The banding patterns of 35 rice using primers OPC 03, OPC 04 and OPA 01 are shown in Figs. 1, 2 and 3, respectively. The DNA polymorphisms were detected according to the presence and absence of band. Absence of band may be caused by failure of primer to anneal a site in some individuals due to nucleotide, sequences difference or by insertions or deletions between primer sites (Clark and Lanigan, 1993). Frequencies of polymorphic loci (RAPD markers) in 35 rice genotypes were presented in Table 2.

Table 1. RAPD primers with corresponding bands score and their size range together with polymorphic bands observed in 35 rice genotypes

Primer code	Sequences (5´-3´)	Total number of bands scored	Size ranges (bp)	Number of polymorphic bands	Proportion of polymorphic loci (%)
OPC 03	GGGGGTCTTT	6	550-1650	6	100.00
OPC 04	CCGCATCTAC	8	600-1700	7	87.50
OPA 01	CAGGCCCTTC	6	600-1775	2	33.33
Total		20		15	220.83
Average		6.67		5.00	73.61

Genetic variation

The values of Nei's (1973) gene diversity and Shannon's information index for different accessions of rice across all loci are shown in Table 2. The estimate of Nei's (1973) genetic diversity for 35 genotypes of rice was 0.12 and Shannon's information index was 0.22. There was a high level of genetic variation among the studied genotypes of rice from the proportion of polymorphic loci point of view.

Table 2. Summary of Frequencies of polymorphic gene, genetic diversity statistics and Shanon information index for all loci in 35 rice genotypes

Loci	Gene frequency	Gene diversity (h)	Shanon information index (i)
OPC 03-1	0.8000	0.3200	0.5004
OPC 03-2	0.8857	0.2024	0.3554
OPC 03-3	0.9429	0.1078	0.2190
OPC 03-4	0.9429	0.1078	0.2190
OPC 03-5	0.9714	0.0555	0.1297
OPC 03-6	0.9143	0.1567	0.2925
OPC 04 -1	0.7714	0.3527	0.5375
OPC 04 -2	0.6000	0.4800	0.6730
OPC 04 -3	0.9714	0.0555	0.1297
OPC 04 -4	0.9714	0.0555	0.1297
OPC 04 -5	0.0000	0.0000	0.0000
OPC 04 -6	0.2286	0.3527	0.5375
OPC 04 -7	0.9714	0.0555	0.1297
OPC 04 -8	0.9714	0.0555	0.1297
OPA 01-1	0.0000	0.0000	0.0000
OPA 01-2	0.0857	0.1567	0.2925
OPA 01-3	0.0000	0.0000	0.0000
OPA 01-4	0.0000	0.0000	0.0000
OPA 01-5	0.0000	0.0000	0.0000
OPA 01-6	0.0286	0.0555	0.1297
Mean		0.1285	0.2203
St. Dev.		0.1420	0.2059

Genetic distance and genetic identity

Pair-wise comparisons of Nei's (1972) genetic distance (GD) between rice genotypes were computed from combined data for the three primers and the values ranged from 0.0000 to 0.9000 (Table 3). Comparatively higher genetic distance was observed between the genotypes Golapi vs. Bogi.

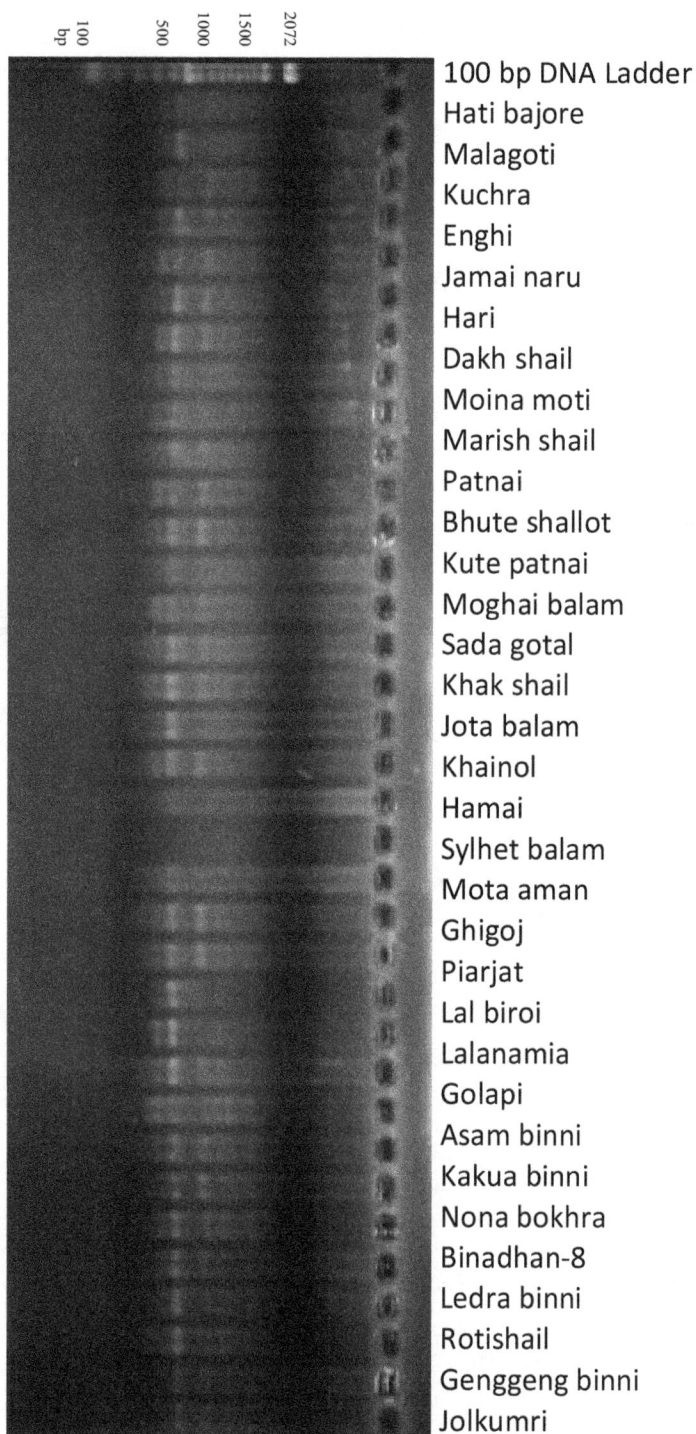

100 bp DNA Ladder
Hati bajore
Malagoti
Kuchra
Enghi
Jamai naru
Hari
Dakh shail
Moina moti
Marish shail
Patnai
Bhute shallot
Kute patnai
Moghai balam
Sada gotal
Khak shail
Jota balam
Khainol
Hamai
Sylhet balam
Mota aman
Ghigoj
Piarjat
Lal biroi
Lalanamia
Golapi
Asam binni
Kakua binni
Nona bokhra
Binadhan-8
Ledra binni
Rotishail
Genggeng binni
Jolkumri

Fig. 1. RAPD profiles of different 35 rice genotypes using primer OPC-03

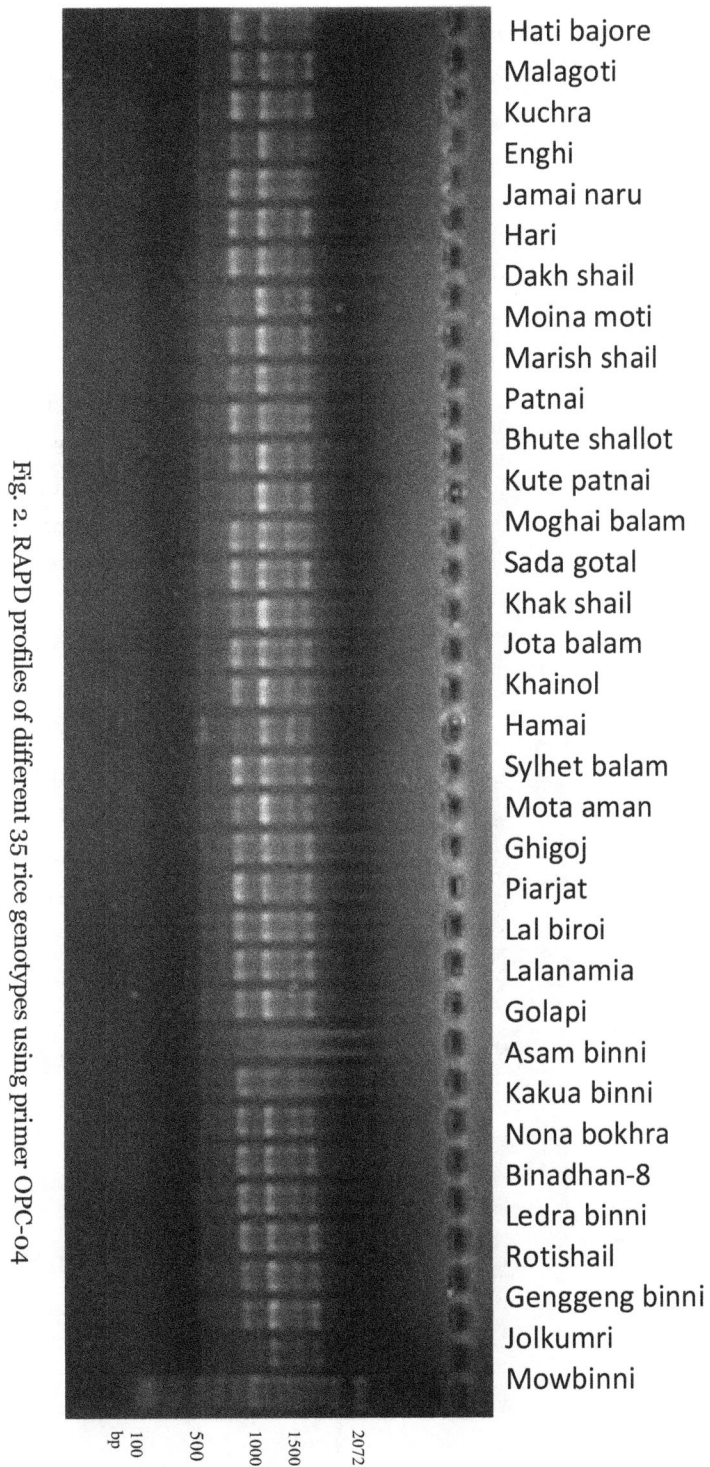

Fig. 2. RAPD profiles of different 35 rice genotypes using primer OPC-04

Hati bajore
Malagoti
Kuchra
Enghi
Jamai naru
Hari
Dakh shail
Moina moti
Marish shail
Patnai
Bhute shallot
Kute patnai
Moghai balam
Sada gotal
Khak shail
Jota balam
Khainol
Hamai
Sylhet balam
Mota aman
Ghigoj
Piarjat
Lal biroi
Lalanamia
Golapi
Asam binni
Kakua binni
Nona bokhra
Binadhan-8
Ledra binni
Rotishail
Genggeng binni
Jolkumri
Mowbinni

100 bp
500
1000
1500
2072

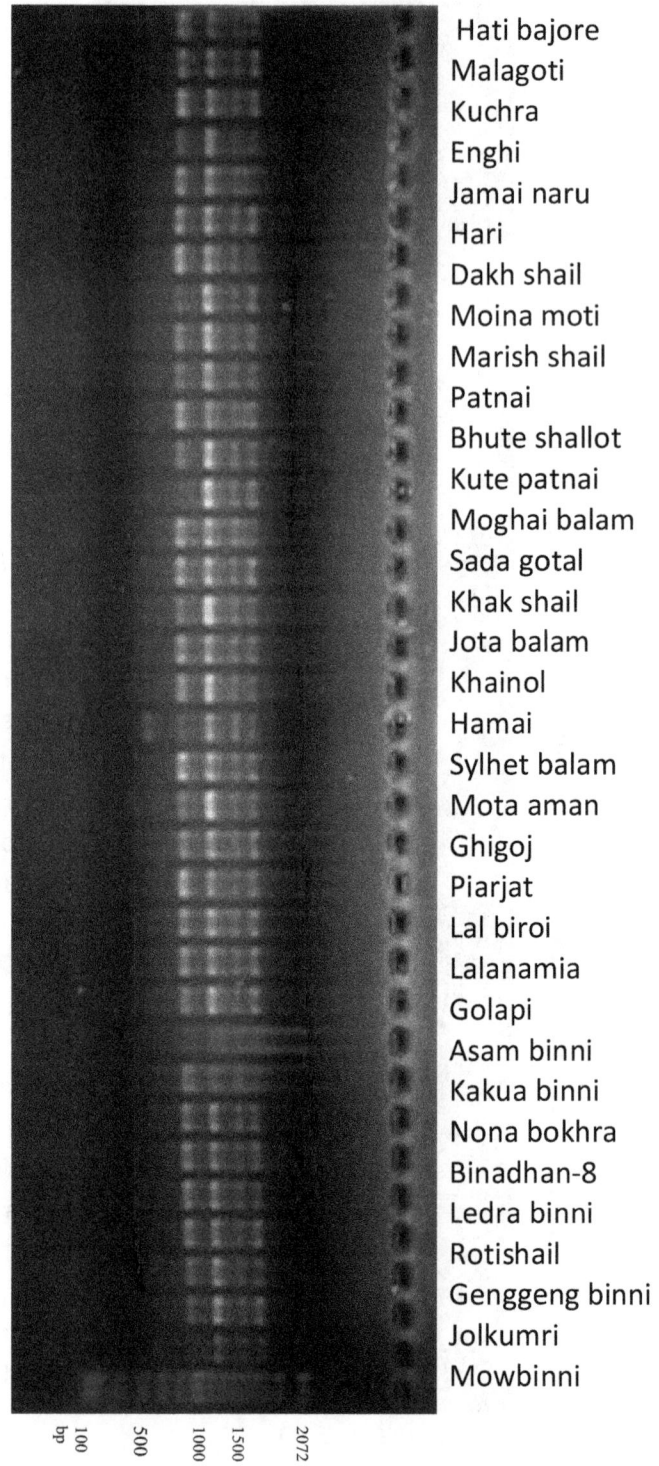

Fig. 3. RAPD profiles of different 35 rice genotypes using primer OPA-01

Table 3. Summary of Nei's genetic identity (above diagonal) and genetic distance (below diagonal) between 35 rice genotypes

Acce.	1	2	3	4	5	6	7	8	9	10	11	12	13	14	15	16	17	18
1	****	0.9500	1.0000	0.9500	0.9000	0.9500	0.9000	0.9000	0.9000	0.9500	0.8500	0.9000	0.9500	0.8500	0.9000	0.9000	0.9000	0.9500
2	0.0513	****	0.9500	0.9000	0.8500	1.0000	0.8500	0.8500	0.9500	0.9000	0.8000	0.8500	0.9000	0.8000	0.8500	0.8500	0.9500	0.9000
3	0.0000	0.0513	****	0.9500	0.9000	0.9500	0.9000	0.9000	0.9000	0.9500	0.8500	0.9000	0.9500	0.8500	0.9000	0.9000	0.9000	0.9500
4	0.0513	0.1054	0.0513	****	0.9500	0.9000	0.9500	0.9500	0.9000	0.9000	0.9000	0.9500	0.9000	0.9000	0.9500	0.9500	0.9000	0.9000
5	0.1054	0.1625	0.1054	0.0513	****	0.8500	0.9000	0.9000	0.8500	0.8500	0.9500	0.9000	0.8500	0.9500	0.9000	0.9000	0.8500	0.8500
6	0.0513	0.0000	0.0513	0.1054	0.1625	****	0.8500	0.8500	0.9500	0.9000	0.8000	0.8500	0.9000	0.8000	0.8500	0.8500	0.8000	0.9000
7	0.1054	0.1625	0.1054	0.0513	0.1054	0.1625	****	1.0000	0.9000	0.9500	0.9500	1.0000	0.9000	0.9500	1.0000	1.0000	0.9500	0.9500
8	0.1054	0.1625	0.1054	0.0513	0.1054	0.1625	0.0000	****	0.9000	0.9500	0.9500	1.0000	0.9000	0.9500	1.0000	1.0000	0.9000	0.9500
9	0.1054	0.0513	0.1054	0.0513	0.2231	0.0513	0.1054	0.1054	****	0.9500	0.8500	0.9000	0.9500	0.8500	0.9000	0.9000	1.0000	0.9500
10	0.0513	0.1054	0.0513	0.1054	0.1625	0.1054	0.0513	0.0513	0.0513	****	0.9000	0.9500	0.9500	0.9000	0.9500	0.9500	0.9500	1.0000
11	0.1625	0.2231	0.1625	0.1054	0.0513	0.2231	0.0513	0.0513	0.1625	0.1054	****	0.9500	0.9000	1.0000	0.9500	0.9500	0.8500	0.9000
12	0.1054	0.1054	0.1054	0.0513	0.1054	0.1054	0.0000	0.0000	0.1054	0.0513	0.0513	****	0.9500	0.9500	1.0000	1.0000	0.9000	1.0000
13	0.0513	0.1054	0.0513	0.1054	0.1625	0.1054	0.0513	0.0513	0.0513	0.0000	0.1054	0.0513	****	0.9000	0.9500	0.9500	0.9500	0.9000
14	0.1625	0.2231	0.1625	0.1054	0.0513	0.2231	0.0513	0.0513	0.1625	0.1054	0.0000	0.0513	0.1054	****	0.9500	1.0000	0.8500	0.9000
15	0.1054	0.1625	0.1054	0.0513	0.1054	0.1625	0.0000	0.0000	0.1054	0.0513	0.0513	0.0000	0.0513	0.0513	****	1.0000	0.9000	0.9500
16	0.1054	0.1625	0.1054	0.0513	0.1054	0.1625	0.0000	0.0000	0.1054	0.0513	0.0513	0.0000	0.0513	0.0513	0.0000	****	0.9000	0.9500
17	0.1054	0.0513	0.1054	0.1625	0.2231	0.0513	0.1054	0.1054	0.0000	0.0513	0.1625	0.1054	0.0513	0.1625	0.1054	0.1054	****	0.9500
18	0.0513	0.1054	0.0513	0.1054	0.1625	0.1054	0.0513	0.0513	0.0513	0.0000	0.1054	0.0513	0.0000	0.1054	0.0513	0.0513	0.0513	****
19	0.4308	0.5108	0.4308	0.3567	0.4308	0.5108	0.4308	0.4308	0.5978	0.5108	0.5108	0.4308	0.5108	0.5108	0.4308	0.4308	0.5978	0.5108
20	0.4308	0.5108	0.4308	0.5108	0.5978	0.5108	0.4308	0.4308	0.4308	0.3567	0.5108	0.4308	0.3567	0.5108	0.4308	0.4308	0.4308	0.3567
21	0.0513	0.1054	0.0513	0.1054	0.1625	0.1054	0.0513	0.0513	0.0513	0.0000	0.1054	0.0513	0.0000	0.1054	0.0513	0.0513	0.0513	0.0000
22	0.0513	0.1054	0.0513	0.1054	0.1625	0.1054	0.0513	0.0513	0.0513	0.0000	0.1054	0.0513	0.0000	0.1054	0.0513	0.0513	0.0513	0.0000
23	0.1625	0.1054	0.1625	0.2231	0.2877	0.1054	0.1625	0.1625	0.0513	0.1054	0.2231	0.1625	0.1054	0.2231	0.1625	0.1625	0.1625	0.1054
24	0.1625	0.2231	0.1625	0.2231	0.2877	0.2231	0.1625	0.1625	0.1625	0.1054	0.2231	0.1625	0.1054	0.2231	0.1625	0.1625	0.1625	0.1054
25	0.1054	0.1625	0.1054	0.1625	0.2231	0.1625	0.1054	0.1054	0.1054	0.0513	0.1625	0.1054	0.0513	0.1625	0.1054	0.1054	0.1054	0.0513
26	0.1054	0.0513	0.1054	0.1625	0.2231	0.0513	0.1054	0.1054	0.0000	0.0513	0.1625	0.1054	0.0513	0.1625	0.1054	0.1054	0.0000	0.0513
27	0.1054	0.1625	0.1054	0.0513	0.1054	0.1625	0.0000	0.0000	0.1054	0.0513	0.0513	0.0000	0.0513	0.0513	0.0000	0.0000	0.1054	0.0513
28	0.1054	0.1625	0.1054	0.0513	0.1054	0.1625	0.0000	0.0000	0.1054	0.0513	0.0513	0.0000	0.0000	0.0513	0.0000	0.0000	0.0513	0.1054
29	0.0513	0.1054	0.0513	0.1054	0.1625	0.1054	0.0513	0.0513	0.0513	0.0000	0.1054	0.0513	0.0000	0.1054	0.0513	0.0513	0.0513	0.0000
30	0.1054	0.1625	0.1054	0.0513	0.1054	0.1625	0.0000	0.0000	0.1054	0.0513	0.0513	0.0000	0.0513	0.0513	0.0000	0.0000	0.1054	0.0513
31	0.2877	0.3567	0.2877	0.2231	0.2877	0.3567	0.2877	0.2877	0.4308	0.3567	0.3567	0.2877	0.3567	0.3567	0.2877	0.2877	0.4308	0.3567
32	0.0513	0.1054	0.0513	0.1054	0.1625	0.1054	0.0513	0.0513	0.0513	0.0000	0.1054	0.0513	0.0000	0.1054	0.0513	0.0513	0.0513	0.0000
33	0.0513	0.1054	0.0513	0.1054	0.1625	0.1054	0.0513	0.0513	0.0513	0.0000	0.1054	0.0513	0.0000	0.1054	0.0000	0.0513	0.0513	0.0000
34	0.2231	0.1625	0.2231	0.2877	0.3567	0.1625	0.2231	0.2231	0.1054	0.1625	0.2877	0.2231	0.1625	0.2877	0.2231	0.2231	0.1054	0.1625
35	0.2231	0.1625	0.2231	0.2877	0.3567	0.1625	0.2231	0.2231	0.1054	0.1625	0.2877	0.2231	0.1625	0.2877	0.2231	0.2231	0.1054	0.1625

Table 3. Contd.

Acce.	19	20	21	22	23	24	25	26	27	28	29	30	31	32	33	34	35
1	0.6500	0.6500	0.9500	0.9500	0.8500	0.8500	0.9000	0.9000	0.9000	0.9000	0.9500	0.9000	0.7500	0.9500	0.9500	0.8000	0.8000
2	0.6000	0.6000	0.9000	0.9000	0.9000	0.8000	0.8500	0.9500	0.8500	0.8500	0.9000	0.8500	0.7000	0.9000	0.9000	0.8500	0.8500
3	0.6500	0.6000	0.9500	0.9500	0.8500	0.8500	0.9000	0.9000	0.9000	0.9500	0.9500	0.9000	0.7500	0.9500	0.9500	0.8000	0.8000
4	0.7000	0.6000	0.9000	0.9000	0.8000	0.8000	0.8500	0.8500	0.9500	0.9000	0.9000	0.9500	0.8000	0.9000	0.9000	0.8000	0.7500
5	0.6500	0.5500	0.8500	0.8500	0.7500	0.7500	0.8000	0.8000	0.9000	0.9000	0.8500	0.9000	0.7500	0.8500	0.8500	0.7000	0.7000
6	0.6000	0.6000	0.9000	0.9000	0.9000	0.8000	0.8500	0.9500	0.8500	0.8500	0.9000	0.8500	0.7000	0.9000	0.9000	0.8500	0.8500
7	0.6500	0.6500	0.9500	0.9500	0.8500	0.8500	0.9000	0.9000	1.0000	1.0000	0.9500	1.0000	0.7500	0.9500	0.9500	0.8000	0.8000
8	0.6500	0.6500	0.9500	0.9500	0.8500	0.8500	0.9000	0.9000	1.0000	1.0000	0.9500	1.0000	0.7500	0.9500	0.9500	0.8000	0.8000
9	0.5500	0.6500	0.9500	0.9500	0.9500	0.8500	0.9000	0.9000	0.9000	0.9000	0.9500	0.9000	0.6500	0.9500	0.9500	0.9000	0.9000
10	0.6000	0.7000	1.0000	1.0000	0.9000	0.9000	0.9500	1.0000	0.9500	0.9500	1.0000	0.9500	0.7000	1.0000	1.0000	0.8500	0.8500
11	0.6000	0.6000	0.9000	0.9000	0.8000	0.8000	0.8500	0.8500	0.9500	0.9500	0.9000	1.0000	0.7500	0.9000	0.9000	0.7500	0.7500
12	0.6500	0.6500	0.9500	0.9500	0.8500	0.8500	0.9000	0.9000	1.0000	1.0000	0.9500	1.0000	0.7500	0.9500	0.9500	0.8000	0.8000
13	0.6000	0.7000	1.0000	1.0000	0.9000	0.9000	0.9500	0.9500	0.9500	0.9500	1.0000	0.9500	0.7000	0.9000	1.0000	0.8500	0.8500
14	0.6000	0.6000	0.9000	0.9000	0.8000	0.8000	0.8500	0.8500	0.9500	0.9500	0.9000	0.9500	0.7000	0.9000	0.9000	0.7500	0.7500
15	0.6500	0.6500	0.9500	0.9500	0.8500	0.8500	0.9000	0.9000	1.0000	1.0000	0.9500	1.0000	0.7500	0.9500	0.9500	0.8000	0.8000
16	0.6500	0.6500	0.9500	0.9500	0.8500	0.8500	0.9000	0.9000	1.0000	1.0000	0.9500	1.0000	0.7500	0.9500	0.9500	0.8000	0.8000
17	0.5500	0.6500	0.9500	0.9500	0.9500	0.8500	0.9000	1.0000	0.9000	0.9000	0.9500	0.9000	0.6500	0.9500	0.9500	0.9000	0.9000
18	0.6000	0.7000	1.0000	1.0000	0.9000	0.9000	0.9500	0.9500	0.9500	0.9500	1.0000	0.9500	0.7000	1.0000	1.0000	0.8500	0.8500
19	****	0.7000	0.6000	0.6000	0.6000	0.7000	0.6500	0.5500	0.6500	0.6500	0.6000	0.6500	0.6000	0.6000	0.6000	0.5500	0.5500
20	0.3567	****	0.7000	0.7000	0.7000	0.8000	0.7500	0.6500	0.6500	0.6500	0.7000	0.6500	0.5000	0.7000	0.7000	0.7500	0.7500
21	0.5108	0.3567	****	1.0000	0.9000	0.9000	0.9500	0.9500	0.9500	0.9500	1.0000	0.9500	0.7000	1.0000	1.0000	0.8500	0.8500
22	0.5108	0.3567	0.0000	****	0.9000	0.9000	0.9500	0.9500	0.9500	0.9500	1.0000	0.9500	0.7000	1.0000	1.0000	0.8500	0.8500
23	0.5108	0.3567	0.1054	0.1054	****	0.9000	0.8500	0.9500	0.8500	0.8500	0.9000	0.8500	0.6000	0.9000	0.9000	0.7500	0.7500
24	0.3567	0.2231	0.1054	0.1054	0.1054	****	0.9500	0.9000	0.8500	0.8500	0.9000	0.8500	0.7000	0.9000	0.9000	0.8500	0.8500
25	0.4308	0.2877	0.0513	0.0513	0.1625	0.0513	****	0.9000	0.9000	0.9000	0.9500	0.9000	0.6500	0.9500	0.9500	0.9000	0.6500
26	0.5978	0.4308	0.0513	0.0513	0.0513	0.1625	0.1054	****	0.9000	0.6500	0.6000	0.6500	0.6000	0.9000	0.9000	0.9000	0.9000
27	0.4308	0.4308	0.0513	0.0513	0.1625	0.1625	0.1054	0.1054	****	1.0000	0.9000	1.0000	0.7500	0.9500	0.9500	0.8000	0.8000
28	0.4308	0.4308	0.0513	0.0513	0.1625	0.1625	0.1054	0.1054	0.0000	****	1.0000	1.0000	0.7500	1.0000	1.0000	0.8000	0.8000
29	0.5108	0.3567	0.0000	0.0513	0.1054	0.1625	0.0513	0.0513	0.0513	0.0513	****	0.9500	0.6000	0.9000	0.9000	0.8000	0.8000
30	0.4308	0.4308	0.0513	0.0513	0.1054	0.1625	0.1054	0.1054	0.0513	0.0000	0.0513	****	0.6500	0.9000	0.9000	0.8000	0.8000
31	0.5108	0.6931	0.3567	0.3567	0.5108	0.3567	0.2877	0.4308	0.2877	0.2877	0.3567	0.2877	****	0.9500	0.9500	0.6500	0.6500
32	0.5108	0.3567	0.0000	0.0000	0.1054	0.1054	0.0513	0.0513	0.0513	0.0513	0.0000	0.0513	0.3567	****	1.0000	0.8500	0.8500
33	0.5108	0.3567	0.0000	0.0000	0.1054	0.1054	0.0513	0.0513	0.0513	0.0513	0.0000	0.0513	0.3567	0.0000	****	0.8500	0.8500
34	0.5978	0.2877	0.1625	0.1625	0.1625	0.1625	0.1054	0.1054	0.2231	0.2231	0.1625	0.2231	0.4308	0.1625	0.1625	****	1.0000
35	0.5978	0.2877	0.1625	0.1625	0.1625	0.1625	0.9000	0.1054	0.2231	0.2231	0.1625	0.2231	0.4308	0.1625	0.1625	0.0000	****

Genetic differentiation and rate of migration among subdivided population

Nei's analysis of gene diversity in subdivided populations presented the gene flow (Nm+) value of 0.000 and the proportion of total genetic diversity (Gst) was 1.0000. Hardy-Weinberg expectation of average heterozygosity in sub- population (Ht) was 0.1285, whereas the heterozygosity (Hs) was 0.0000 (Table 4) (McDermott and McDonald, 1993).

Table 4. Summary of genetic variation statistics across all loci

Loci	Sample Size	Ht	Hs	Gst	Nm+
OPC 03-1	35	0.3200	0.0000	1.0000	0.0000
OPC 03-2	35	0.2024	0.0000	1.0000	0.0000
OPC 03-3	35	0.1078	0.0000	1.0000	0.0000
OPC 03-4	35	0.1078	0.0000	1.0000	0.0000
OPC 03-5	35	0.0555	0.0000	1.0000	0.0000
OPC 03-6	35	0.1567	0.0000	1.0000	0.0000
OPC 04 -1	35	0.3527	0.0000	1.0000	0.0000
OPC 04 -2	35	0.4800	0.0000	1.0000	0.0000
OPC 04 -3	35	0.0555	0.0000	1.0000	0.0000
OPC 04 -4	35	0.0555	0.0000	1.0000	0.0000
OPC 04 -5	35	0.0000	0.0000	****	****
OPC 04 -6	35	0.3527	0.0000	1.0000	0.0000
OPC 04 -7	35	0.0555	0.0000	1.0000	0.0000
OPC 04 -8	35	0.0555	0.0000	1.0000	0.0000
OPA 01-1	35	0.0000	0.0000	****	****
OPA 01-2	35	0.1567	0.0000	1.0000	0.0000
OPA 01-3	35	0.0000	0.0000	****	****
OPA 01-4	35	0.0000	0.0000	****	****
OPA 01-5	35	0.0000	0.0000	****	****
OPA 01-6	35	0.0555	0.0000	1.0000	0.0000
Mean	35	0.1285	0.0000	1.0000	0.0000
St. Dev.		0.0202	0.0000		

H_t = Hardy-Weinberg average heterozygosity expected in sub-population
H_s = Hardy-Weinberg average heterozygosity obtained in sub-population
G_{st} = Co-efficient of gene differentiation
N_m = Estimate of gene flow from Gst or Gcs. e.g., $N_m = 0.5(1-G_{st})/G_{st}$
**** = Infinity
The number of polymorphic loci is : 15
The percentage of polymorphic loci is : 75.00

UPGMA Dendrogram

A dendrogram was constructed based on Nei's (1972) genetic distance following the Unweighted Pair Group Method of Arithmetic Means (UPGMA). The 35 genotypes of rice were grouped into 2 main clusters namely cluster 1 and cluster 2 (Fig. 4).

Genotypes Sylhet balam and Mota aman was included in cluster 2. Cluster 2 i.e. Sylhet balam and Mota aman was totally different from other genotypes. So, genetic relationship was not present between cluster 1 genotypes with cluster 2.

Genotypes belong in cluster 1 were Hati bajore, Moghai balam, Golapi, Malagoti, Sada gotal, Asam binni, Kuchra, Khak shail, Kakua binni, Enghi, Jota balam, Nona bokhra, Jamai naru, Khainol, Binadhan-8, Hari, Hamai, Ledra binni, Dakh shail, Rotisail, Moina moti, Genggeng binni, Marish shail, Ghigoj, Jolkumri, Patnai, Piarjat, Mowbinni, Bhute shalot, Lal biroi, Bogi, Kute patnai and Lalanamia.

The genotypes of cluster 1 again divided into two sub-cluster 1 and sub-cluster 2. Sub-cluster 1 consisted of genotypes Hati bajore, Moghai balam, Golapi, Malagoti, Sada Gotal, Asam binni, Kuchra, Khak shail, Kakua binni, Enghi, Jota balam, Nona bokhra, Jamai Naru, Khainol, Binadhan-8, Hari, Hamai, Ledra binni, Dakh shail, Moina moti, Genggeng binni, Marish shail, Ghigoj, Jolkumri, Patnai, Piarjat, Mowbinni, Bhute shalot, Lal biroi, Bogi, Kute patnai and Lalanamia. Sub-cluster 2 formed by only one genotype Rotisail, genetic relationship was present between sub-clusters.

The genotypes of sub-cluster 1 again divided into two sub-sub-cluster 1 and sub-sub-cluster 2. Sub-sub-cluster 1 consisted of genotypes Hati bajore, Moghai balam, Golapi, Malagoti, Sada gotal, Asam binni, Kuchra, Khak shail, Kakua binni, Enghi, Jota balam, Nona bokhra, Jamai naru, Khainol, Binadhan-8, Hari, Hamai, Ledra binni, Dakh shail, Moina Moti, Genggeng binni, Marish shail, Ghigoj, Jolkumri, Patnai, Piarjat, Bhute shalot, Lal biroi, Kute patnai and Lalanamia. Sub-sub-cluster 2 formed by two genotypes Mowbinni and Bogi.

Genotypic variations based on molecular characterization indicated that genotypes belonging to different clusters depend on their genetic components itself, but not at geographical origin at all. Therefore, it could be concluded that for further research program, especially for hybridization, genotype could be selected from different clusters will be provided maximum heterosis regarding yield.

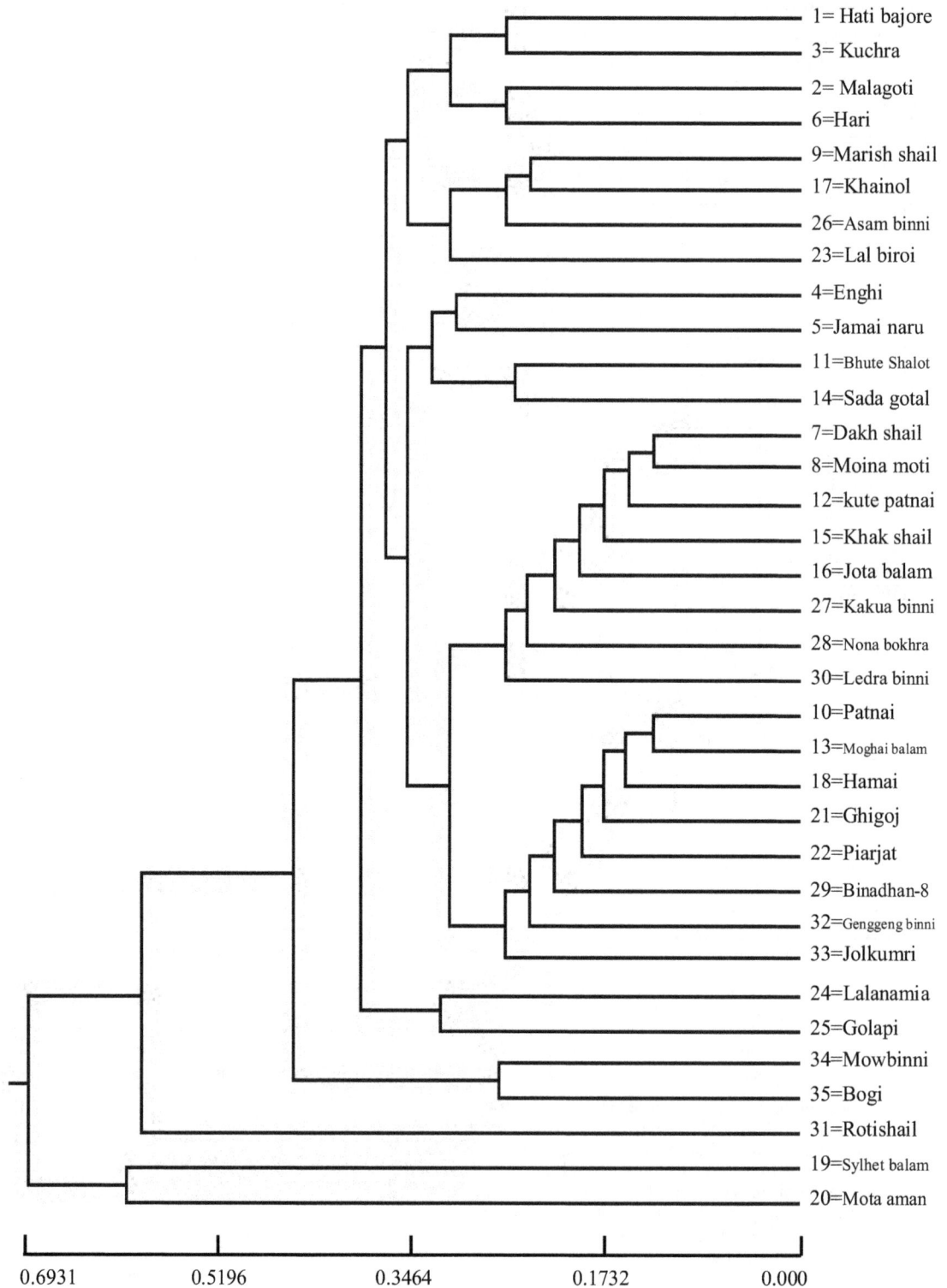

Fig. 4. Unweighted pair group method of arithmetic mean (UPGMA) dendrogram based on Neis's (1972) genetic distance, summarizing data on differentiation in 35 rice genotypes according to RAPD analysis.

References

Clarck, A.G. and Lanigan, C.M.S. 1993. Prospects for estimating nucleotide divergence with RAPDS. *Mol. Biol. Evol.* 10: 1069- 1111.

Mcdermott, J.M. and Mcdonald, B.A. 1993. Gene flow in plant pathosystems. *Ann. Rev. Phytopath.* 31: 353-373.

Nei, M. 1972. Genetic distance between populations. *American Naturalist* 106: 283-292.

Nei, M. 1973. Analysis of gene diversity in subdivided populations. *Proc. Natl. Acad. Sci.* 70: 3321-3323.

Skaria, R., Sen, S. and Muneer, P.M.A. 2011. Analysis of genetic variability in rice varieties (*Oryza sativa* L.) of Kerala using RAPD markers. *Genet. Eng. Biotech. J.* 24: 1-9.

Weising K., Atkinson, G. and Gardner, C. 1995. Genomic fingerprinting by microsatellite-primed PCR: a critical evaluation. *PCR Methods Applications* 4: 249-255.

Yeh, F.C., Yang, R.C., Boyle, T.B.J., Ye, Z.H. and Mao, J.X. 1997. POPGENE, the user-friendly shareware for population genetic analysis. Molecular Biology and Biotechnology Centre, University of Alberta, Canada.

7

CHANGES IN SEED WEIGHT IN RESPONSE TO DIFFERENT SOURCES: SINK RATIO IN OILSEED RAPE

Francisco M. Iglesias[1] and Daniel J. Miralles[1,2]*

Abstract

Little knowledge exists about the degree of source, sink and source: sink limitations on mean seed weight in oilseed rape (*Brassica napus* L.). The objective of this work was to analyze the nature and magnitude on seed weight response to assimilate availability during the effective seed-filling period in oilseed rape. Three Argentinean varieties, Eclipse, Impulse, and Master, were grown under field conditions, and at the beginning of the effective seed filling period, a broad range of source: sink manipulation combinations were produced. Source manipulations consisted of two incoming radiation (R) level reductions: 0% (Rn) and ~50% (Rs) combined with three different sources: sink treatments were applied: C, control; PR, ~50% pod removal, and D, 100% defoliation. Rs significantly reduced yield (15%) and MSW (12%) with respect to Rn, without significant effects on the rest of the sub yield components. Source:sink manipulation treatments significantly affected all yield components. PR diminished yield by 29%, reducing ca. 40% seeds pl^{-1} by reductions pods pl^{-1} (41%) with respect to Rn, whereas PR increased MSW by 19%, counterbalancing the reduction in seeds pl^{-1} and thereby in yield. When considering different seed positions along the main raceme, Rs reduced MSW by 12% independently of seed positions onto the raceme. On the contrary, PR increased MSW in average 17% with respect to C. Results reported here suggest that oilseed rape has source: sink co-limitation during the effective seed filling period, which is apparently higher than wheat and lower than maize.

Keywords: Seed Weight; Seed Number; Source: Sink Ratio; Oilseed Rape

[1]Cátedra de Cerealicultura, Dto. Producción Vegetal, Facultad de Agronomía UBA, Av San Martín 4453 (C1417DSE) Buenos Aires, Argentina; [2]CONICET and IFEVA

*Corresponding author's email: miralles@agro.uba.ar (Daniel J. Miralles)

Introduction

The process of yield production can be interpreted as the balance between the source and sink activity during the reproductive period. This simple view involves the production of assimilates by the photosynthetic organs (source) and the utilization of these assimilates by the seeds (sink) (Egli, 1998). Seed number per unit area, the yield component that accounts for most of the variations in yield of grain crops (Peltonen-Sainio and Jauhiainen, 2008), is directly limited by the availability of assimilates produced during flowering and pod set by canopy photosynthesis (Keiller and Morgan, 1988). As seed number is defined during that period, it could be considered as the critical time in terms of yield, since source limitations during the period immediately after flowering will determine reductions in the number of seed established by the crop (Diepenbrock, 2000; Berry and Spink, 2006).

Final mean seed weight (MSW) depends on the relationships between the sink capacity and the availability of assimilates to fill this sink (Jenner *et al.*, 1991). Variations in MSW contribute in a lower range to variations in yield, being the source:sink ratio the major factor of this variation (Jullien *et al.*, 2011). In oilseed rape, canopy photosynthesis activity declines rapidly during the effective seed-filling period (Gabrielle *et al.*, 1998), and current photosynthesis from pods (Müller and Diepenbrock, 2006) becomes more important than leaves from mid to late reproductive phase. Some evidence, obtained under controlled conditions, showed that when source:sink ratio was reduced, MSW decreased between 10 and 25% depending on the degree of source limitations (Pechan and Morgan, 1985). These findings support the statement that oilseed rape could be source limited during the effective seed-filling period. Nevertheless, these results may not be directly extrapolated to field situations. This is because gradual source limitations along main stem could occur and be more severe at the canopy scale than in isolated plants due to the reduction of the incoming radiation at lower sections of pod and leaf canopy with respect to those at upper sections (Yates and Steven, 1987; Leach *et al.*, 1989).

Although in the literature has been reported source limitation in various crops and its impact on MSW (Borrás et al., 2004). Little knowledge exists about the nature and magnitude on MSW response to assimilate availability during the effective seed-filling period in oilseed rape. Lack of information is especially evident about the effects of changes in source: sink ratio on seeds in different sections within the main raceme and primary branches in field-grown plants. The objective of this work was to evaluate the source-sink limitations during effective seed-filling period, analyzing seed growth and final seed weight responses into a broad range of assimilate supply per seed in three field grown canola varieties.

Materials and Methods

A field experiment was conducted at the experimental field of the Department of Plant Production (Faculty of Agronomy) of the University of Buenos Aires (34°35´S, 58°29´W) during the 2003 growing season. The soil was a silty clay loam type, classified as Vertic Argiudoll, according to the USDA taxonomy. Three oilseed rape Argentinean varieties (Eclipse, Impulse and Master) sown on 3rd June, were subjected to different radiation and source: sink manipulations. Thus, treatments consisted of a combination of varieties, two radiation levels and three levels of source: sink manipulation. Crops were grown without water and nutriment limitations and protected during the whole crop cycle against weeds, insects and diseases. Plant density was 100 pl/m² at emergence and the dimension of the plots was 2.5 m² (7 rows, 0.15 m apart and 2.5 m long).

Shadow treatment (Rs) was imposed by black nylon shade from flowering to physiological maturity. Source: sink manipulation treatments were: i) control (C); ii) sink reduction, in which 50% of pods were alternately removed from the main raceme and branches (PR) in order to produce 50% reduction in the number of seeds per plant and iii) total defoliation (D). The different source: sink balance combinations were imposed at the beginning of the effective seed-filling period (i.e. G₃ stage according to the scale of INRA-CETIOM, 1985). Thus, assimilate availability per seed was expected to be increased or reduced.

The experimental design used was a split-split-plot design with three replications, where the main factor was the genotype. Sub-plots consisted of two radiation levels: i) Control without restrictions of natural incident radiation (Rn) and ii) shadow treatment reducing 48% of incident radiation (Rs) and sub-sub-plots corresponding to the source: sink manipulation treatment combinations (Rn-C, Rn-PR, Rn-D, Rs-C, Rs-PR and Rs-D).

Weight of seeds during grain filling period was determined twice weekly from G₂ stage (INRA-CETIOM, 1985) to maturity for seeds from specific sections within the main raceme and for seeds of pooled pods from all branches. The dynamics of grain weight for different positions within each source:sink ratio treatment was followed in 20 uniform plants per replication per sampling date randomly selected and tagged at onset of flowering from the central rows of each sub-sub-plot. Pods were selected to represent different sections within the main raceme as follows: (i) lower, (ii) middle and (iii) upper pod positions. Seeds from branches were pooled sampled. In each sample, seeds from selected positions described above were oven-dried and weighed separately. Mean seed weight was determined by randomly weighting a sample of 100 seeds, three times, for each replication. These data were used to determine the physiological components of MSW (i.e. rate and duration of grain filling).

The dynamics of seed filling from G₂ onwards were fitted by a bilinear model (see equation 1) and the rate and duration (total and effective duration) calculated for seeds of different sections and treatments into the plant.

$$MSW = a + b*x \ (x \leq c) + b*c \ (x \geq c) \qquad [1]$$

In this model, b represents the seed growth rate during the effective seed-filling period (mg °Cd⁻¹) and c, the duration of the effective seed-filling period (°Cd). The bilinear model was fitted to the seed dry weight data using iterative optimization techniques applying Table Curve V3 (Jandel Scientific 1992). Daily thermal time (TT) values were calculated considering the mean daily temperature minus the base temperature of 4.5°C (Mendham et al., 1981).

At physiological maturity, 10 plants within each sub-sub-plot were harvested. The samples were oven-dried and yield per plant, seed number per plant, number of seeds per pod and average seed weight were determined.

Results

Radiation and source: sink alteration on yield and its components

The main effects (i.e. genotype-G, radiation-R (Rn and Rs) and source: sink ratio-SS) differentially modified yield and its components. Regarding the genotypic differences, Eclipse showed the highest yield (3.4 g/pl), followed by Master (3.1 g/pl) and Impulse (2.8 g/pl). A similar trend was shown by seed/pl and seed/pod. Contrary to what was observed in yield and seeds/pl, Eclipse registered the lowest MSW (2.8 mg/seed, with respect to Impulse and Master (3.4 mg/seed) (Table 1).

Table 1. Effects of genotype, source:sink ratio (control, C; pod removal, PR; and defoliation, D) and radiation levels (100%, Rn; and 48%, Rs, of incident radiation) on yield components

Genotype	Radiation	S:S	Yield/pl			Seeds/pl			MSW			Seeds/pod			Pods/pl		
Eclipse	Rn	C	4265	±	375	1525	±	121	2.79	±	0.06	17,6	±	0,41	86	±	5,0
		PR	2992	±	383	904	±	101	3.30	±	0.10	17,3	±	1,26	52	±	2,4
		D	3452	±	39	1164	±	25	2.97	±	0.04	14,5	±	0,58	80	±	1,7
	Rs	C	3503	±	280	1396	±	156	2.52	±	0.11	17,0	±	1,99	82	±	0,8
		PR	2816	±	194	952	±	79	2.96	±	0.05	17,9	±	2,17	53	±	2,2
		D	3079	±	485	1231	±	234	2.52	±	0.11	16,1	±	0,59	77	±	13,5
Impulse	Rn	C	3558	±	75	1058	±	25	3.37	±	0.12	15,9	±	0,05	66	±	1,6
		PR	2220	±	197	556	±	97	4.08	±	0.40	15,0	±	2,15	37	±	1,4
		D	3326	±	420	993	±	134	3.36	±	0.10	14,6	±	0,90	68	±	8,3
	Rs	C	2987	±	150	997	±	77	3.00	±	0.11	14,4	±	0,44	69	±	3,2
		PR	2121	±	81	585	±	32	3.64	±	0.25	15,9	±	1,03	37	±	0,5
		D	2636	±	100	905	±	24	2.91	±	0.04	14,7	±	0,32	62	±	2,9
Master	Rn	C	4030	±	521	1194	±	183	3.40	±	0.16	17,2	±	0,58	69	±	8,9
		PR	2867	±	40	713	±	16	4.02	±	0.06	17,9	±	0,56	40	±	0,5
		D	3293	±	247	1017	±	79	3.24	±	0.06	16,1	±	0,72	63	±	5,1
	Rs	C	3437	±	68	1157	±	32	2.98	±	0.14	16,6	±	1,23	70	±	3,3
		PR	2437	±	263	689	±	42	3.52	±	0.17	17,3	±	0,29	40	±	3,0
		D	2379	±	214	807	±	55	2.94	±	0.06	14,2	±	0,87	57	±	0,5
Genotype (G)			**			*			**			NS			***		
Radiation (R)			*			NS			**			NS			NS		
Source: sink (SS)			**			***			***			***			***		
G x R			NS			NS			NS			NS			NS		
G x SS			NS			NS			NS			NS			NS		
SS x R			NS			NS			NS			NS			NS		
G x R x SS			NS			NS			NS			NS			NS		

*, **, ***: significant at 0.05, 0.01 and 0.001 probability level, respectively, and NS: not significant.

The Rs treatment significantly reduced yield (15%) and MSW (12%), with respect to Rn, without significant effects on the rest of the sub yield components. There was not any G x R interaction, demonstrating that Rn similarly affected all the genotypes (Table 1).

Source: sink manipulations significantly affected all yield components. Thus, PR diminished yield by 29% as a consequence of reductions of ca. 40% in the seeds/pl and pods/pl (41%), respect to the C treatment. The yield component seeds/pod was not affected by PR. Conversely that was observed in the number of seeds, the PR treatment increased MSW ca 19%, counterbalancing, although partially, the reduction in seeds/pl and thereby, the yield reduction was less than that observed in the latter yield component (Table 1, Fig. 1). The D treatment reduced yield, although the magnitude was less than that observed in PR, as D reduced yield 19% with respect to C. The 17% reduction observed in seeds/pl in D was produced by a diminished seeds/pod and pod/pl of ca. 9% in each yield component (Table 1, Fig. 1).

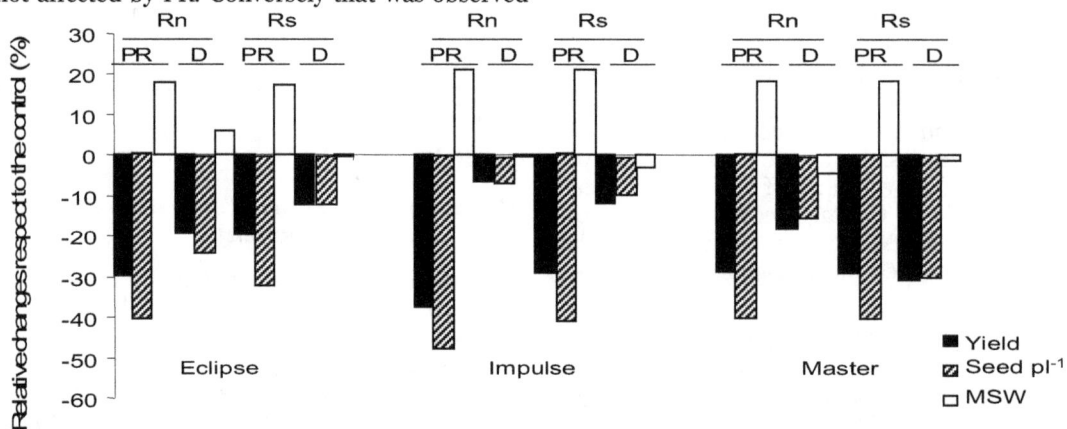

Fig. 1. Relative changes in yield per plant, seed number per plant and mean seed weight (MSW) with respect to the control treatment in both radiation (Rn: natural radiation; Rs: shading) and pod removal (PR) and defoliation (D) treatments (n = 3).

Yield components correlations

The variation in yield was mainly explained by changes in seed/pl (r^2=0.79, p<0.001). Variations in seed/pl were explained by variations in pods/pl (r^2=0.91, p<0.001) without significant association with seed/pod (r^2=0.08, p>0.1). MSW was negatively associated with seed/pl (r^2=-0.57, p<0.01). Although the trend was similar in Rn (r^2=-0.82, p<0.001) and Rs (r^2=-0.81, p<0.001), the points of MSW corresponding to the Rs were consistently lighter in respect to those of the Rn treatment. Thus, although the rate of reduction in grain weight per increase in seed per plant was similar in both treatments (Rn vs. Rs), MSW was reduced under shading than under control treatments for the same number of seeds/pl. The fact that MSW was lighter in Rs than in Rn explains the yield reductions for the same seed/pl.

Mean seed weight responses at different positions in the main raceme

Genotype (G), radiation (R), source: sink (SS) and position treatments significantly affected the weight of seeds positioned at different pod positions within the main raceme. The effect of G was similar to that observed in MSW as, with the exception of the lowermost position, Impulse showed the heaviest and Eclipse the lightest seeds. In relation to the position within the raceme, those seeds located in the lowermost and central pod sections were significantly, lighter (3.09 mg/seed) than those placed in the uppermost pods (3.29 mg/seed), while the seeds in the middle positions registered intermediate weights (3.14 mg/seed) (Table 2, Fig. 2). The SS treatments similarly affected seed weight independently of the position within the raceme. Thus, PR increased seed weight by 21% with respect to C in those seeds placed on lowermost and middle positions, while it increased the weight of the seeds located in the uppermost position onto the raceme by 15%.

Table 2. Effect of source: sink ratio and radiation levels on mean seed weight (mg/seed) from pods of different positions into the raceme. Different letters within the same genotype, main raceme section and radiation level indicate significant differences (p<0.05) among treatments.

Main Raceme Section	Radiation level	Treatments	Genotypes								
			Eclipse			Impulse			Master		
Lowermost	Rn	C	2.74	±	0.06 b	3.13	±	0.06 b	3.30	±	0.07 b
		PR	3.18	±	0.13 a	3.87	±	0.25 a	3.93	±	0.08 a
		D	2.98	±	0.09 a	3.24	±	0.08 b	3.21	±	0.10 b
	Rs	C	2.36	±	0.07 b	2.79	±	0.06 b	2.82	±	0.14 b
		PR	2.88	±	0.05 a	3.58	±	0.18 a	3.38	±	0.10 a
		D	2.47	±	0.10 b	2.82	±	0.04 b	2.92	±	0.08 b
Middle	Rn	C	2.72	±	0.04 b	3.28	±	0.10 b	3.23	±	0.14 b
		PR	3.20	±	0.07 a	4.26	±	0.36 a	3.94	±	0.04 a
		D	2.80	±	0.14 b	3.33	±	0.07 b	3.09	±	0.10 b
	Rs	C	2.47	±	0.06 b	3.00	±	0.08 b	2.93	±	0.10 b
		PR	2.87	±	0.12 a	3.77	±	0.19 a	3.33	±	0.11 a
		D	2.38	±	0.07 b	2.88	±	0.02 b	2.95	±	0.11 b
Uppermost	Rn	C	2.80	±	0.08 c	3.49	±	0.15 b	3.56	±	0,23 b
		PR	3.38	±	0.07 a	4.21	±	0.39 a	4.02	±	0,13 a
		D	2.97	±	0.02 b	3.51	±	0.10 b	3.35	±	0,07 b
	Rs	C	2.75	±	0.09 b	3.29	±	0.11 ab	3.12	±	0,16 b
		PR	3.21	±	0.10 a	3.45	±	0.44 a	3.62	±	0,16 a
		D	2.61	±	0.15 b	3.12	±	0.13 b	2.75	±	0,19 c
Genotype (G)			***								
Radiation (R)			***								
Treatment (T)			***								
Main Raceme Section (P)			***								
G x R			NS								
G x T			NS								
R x T			NS								
G x P			**								
T x P			NS								
R x P			NS								
G x T x P			NS								
G x T x R			NS								
G x R x P			*								
G x P x T			NS								
G x T x R x P			NS								
*, **, ***: significant at 0.05, 0.01.			and 0.001 probability level, respectively. NS: not significant								

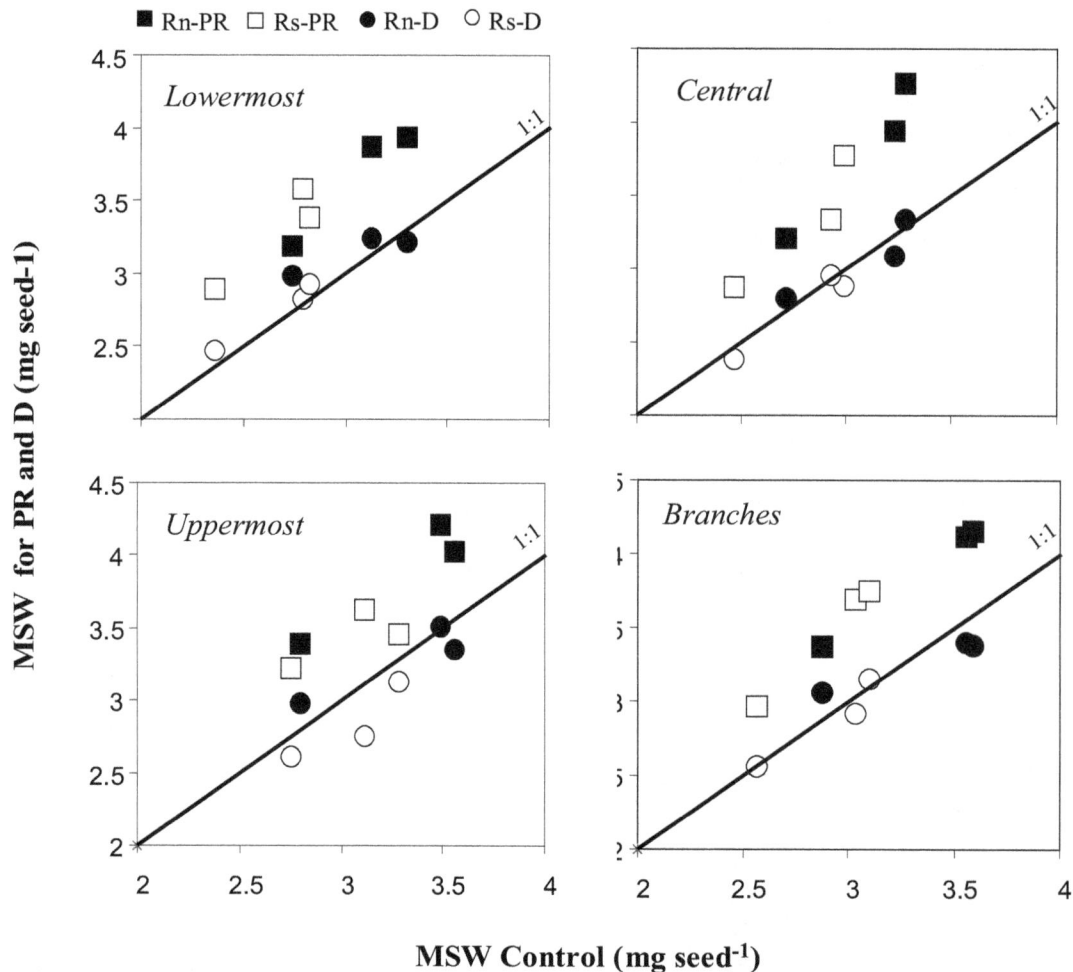

Fig. 2. Mean seed weight (MSW) for pod removal (PR) and defoliation (D) treatments against the control for different seed positions in the main raceme and branches for different radiation (R) and source:sink ratio treatment combinations.

The D treatment did not affect the seed weight when measured in seeds from different position of the raceme and for R (Table 2). The shadow treatment –Rs-, reduced seed weight ca. by 12% in respect to Rn, independently of the seed positions into the raceme (Table 2).

Similarly, to what was observed for MSW, the weight of seeds for different pods positions into the raceme showed a negative relationship when plotted against the number of seeds per plant corresponding to each position (Fig. 3). In general, no differences were detected in the slopes between the three sections and for all genotypes studied (data not shown). However, the slope for the uppermost seeds presented slight differences between Rn and Rs in Eclipse and Impulse, but not in Master. In contrast, middle and lowermost sections presented significant differences in the intercepts across Rn and Rs (data not shown), particularly for Eclipse and Impulse, although not so for Master (Fig. 3).

Seed growth dynamics

When dry weight dynamics of seeds for the applied treatments (G, R and SS) and pod positions were plotted against time, the points were fitted by a bi-linear model (see equation 1), showing coefficients of determination greater than 0.96 in all cases. The bi-linear model allowed the calculation of the rate of seed filling (SGR), total (TSFP) and effective (ESFP) duration of seed filling, and the maximum grain weight obtained in each seed position. Significant differences in the SGR were found for different treatments. When genotypes were compared, the SGR followed the same trend as MSW since SGR was higher in Impulse (0.00627 mg/°Cd), followed by Master (0.00615 mg/°Cd) and Eclipse (0.00602 mg/°Cd). PR, in line with that occurred in MSW, increased the SGR ca. by 16%, showing the uppermost seed positions the highest increases (i.e. 24%) when compared with C. Conversely, D did not affect SGR. Rs reduced SGR ca. by 14% in comparison to Rn, mostly affecting the seeds from the central (19%) and lowermost (15%) pod positions, while only slight reduction in Rs was observed in seeds from the uppermost positions (6%) (data not shown). TSFP and ESFP were slightly affected by the treatments. The largest difference in TSFP and ESFP among genotypes was ca. 40 °Cd, which represents approximately 4 days under the environmental conditions of the experiment. Radiation and source-sink treatments almost did not affect the duration of grain filling. Regardless of pod position along the raceme, variations in MSW were most closely related to variations in SGR ($P < 0.01$) rather than to ESFP for all genotypes. Eclipse and Impulse showed the best fit compared to Master (Fig. 4).

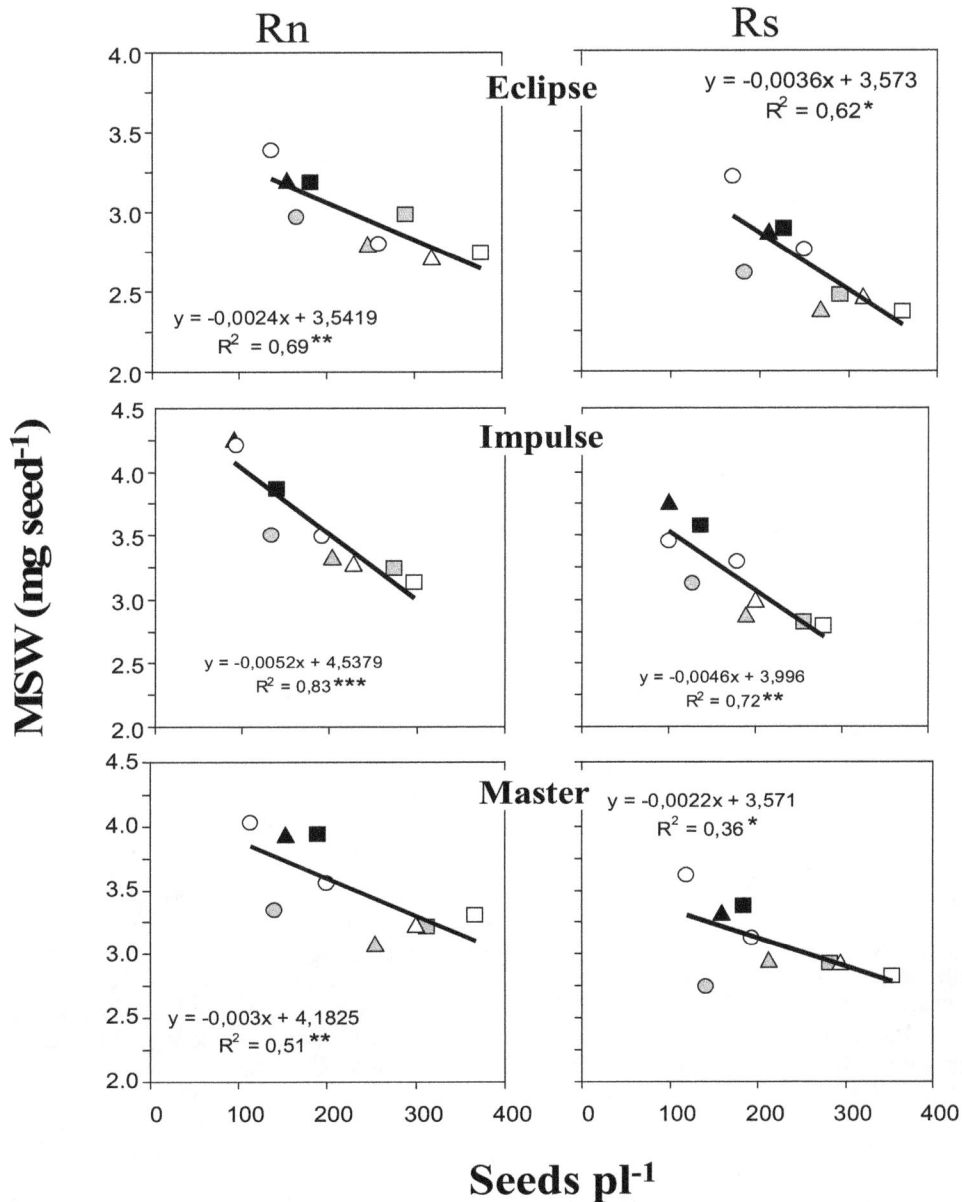

Fig. 3. Relationship between mean seed weight (MSW) and seed per plant for two radiation conditions across different positions in the main raceme and branches.

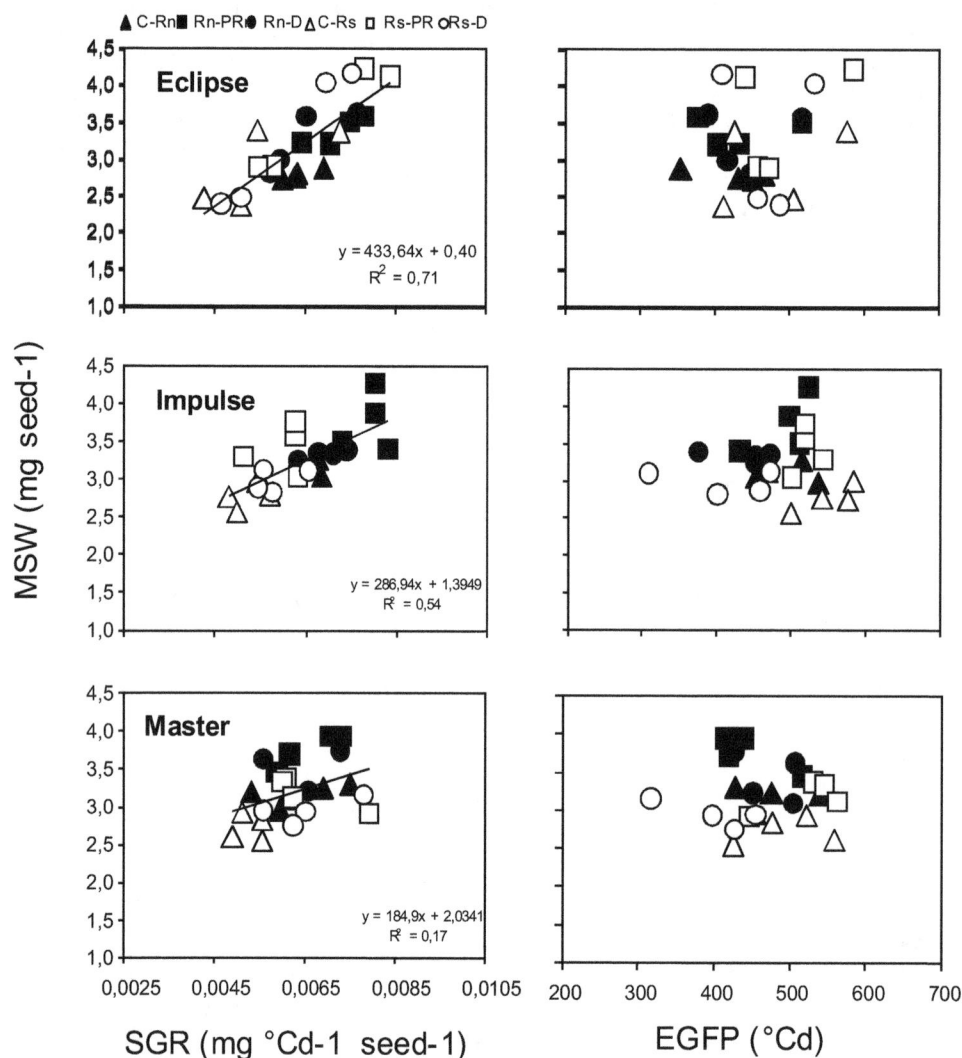

Fig. 4. Relationship between mean seed weight (MSW) and seed growth rate (SGR), and effective seed filling period (ESFP) across radiation conditions and source: sink treatments.

Discussion

As in many other crops, including cereals and oilseed, the variations in yield were associated with changes in the number of seeds set in the plant rather than with changes in mean seed weight (Egli, 1998; Habekotté, 1993). As reported in soybean (Gomez and Miralles, 2011), the number of seeds per plant was explained by variations in the number of pods per plant since the number of seeds per pod, although more variable in oilseed rape than in soybean, did not explain the changes in seeds per plant. Applying a simplistic and linear reasoning, the negative relationship between MSW and seeds/pl suggests a source limitation to fulfill the grains previously produced. The fact that in the present study MSW was (i) increased by 19% when sink was reduced ca. 40% and (ii) the lack of effects on seed weight when source was diminished by defoliation

suggests that even when there was response to seed weight when source was increased, there was not a complete limitation for assimilates. Borrás et al. (2004), taking several data from the published literature, showed that wheat was a crop mostly limited by sink as grain weight was not affected by changes in source: sink manipulation; while soybean and maize evidenced source limitation. The results of the present study suggest that oilseed rape presents an intermediate behavior between wheat and maize, as when source per seed was doubled (assuming the most simplistic approach), seeds rose to a half of those increases, suggesting a source:sink co-limitation, similar to that proposed in wheat by Miralles and Slafer (2007).

No differences in individual seed mass was found by defoliation treatments in both radiation level (Rn and Rs) when compared to C, particularly in pods developed later. This would be explained by two ways. First, in the most source: sink restrictive combination (Rs-D), 10% reductions in the number of seeds per pod was produced (Table 1), in turn; fewer seeds in the upper pods were exposed to more assimilate with respect to the C. An opposite situation was observed for seeds from branches (data not shown). The slight, but significant, increase in the number of seeds per pod, originated by the pod removal treatment produced in the three genotypes, can be related to assimilate and/or phytohormone stimulation. In oilseed rape, the flowering period is sequential and proceeds from basal to upper positions in main raceme. Therefore, by the time of treatment application upper pods were at an early stage of development and thereby were probably more susceptible to abortion than the peripheral ones (Keiller and Morgan, 1988; Leterme, 1988). The strong overlap of pod formation, seed set and seed filling periods allowed Rs or PR from the beginning of seed filling to have a strong impact on the number of seeds per pod.

Second, the lack of change in MSW when D was applied implicates that an important part of photoassimilates remobilized during seed filling comes from pods and stems. Therefore, in this situation, remobilization would sustain the growth of the seeds without any change produced in MSW. In a previous report, Habekotté (1993) showed that the level of carbohydrates at the end of seed fill represents ca. 20% of the carbohydrates level accumulated at the end of flowering. By contrast, Quilleré and Triboi (1987) described a contribution of remobilization to seed yield of 17.5%, at the most, in only one variety.

The degree of response observed in MSW when source per seed was increased, especially for those seeds set later on the flowering period (middle and uppermost pods), could also be attributed to changes in the potential sizes of the seeds. Thus, the earlier the manipulation, the higher the response with respect to control, as seed weight potential was expected to be more affected by source: sink manipulations when endospermatic cells and ovary size were growing (Lizana et al., 2010).

The analysis of the physiological components of seed growth indicated that the causes of the changes in seed weight by source: sink modifications were related to modifications in SGR rather than in ESFP, as suggested by Habekotté (1993), who reported reductions in SGR in oilseed rape as a result of shading treatments imposed around mid-flowering. Additional evidence for several other species showed that the final individual seed weight was largely correlated to the seed growth rate rather than to changes in seed filling duration (Miralles and Slafer, 1995; Alvarez-Prado et al., 2013).

Although the general pattern for explaining the variations in MSW consisted of changes in SGR, rather than in ESFP, a different response pattern observed when seeds from pods of different sections analyzed. The highest response to sink reduction in MSW, when compared to C in relative terms, observed in the lower and middle pod positions onto the raceme. However, the highest response in SGR corresponded to the grains places in pods of the uppermost position. An explanation for this may be related with the fact that at the time of treatment manipulation, ca. 350 °Cd (~20 days after onset of flowering), it might be possible that basal and middle pods were close to the beginning of the linear phase of growth, unlike the upper ones, which were initiating the seed growth i.e. lag phase. Munier-Jolain and Salon (2003) suggested that sucrose, together with endogenous phytohormones level during the embryo cell division phase, could determine its mitotic activity by modulating the expression of genes that participate in the cell cycle regulation. Therefore, if this is the case, when cell division starts in basal and middle pods of the raceme, competition for assimilates is low, as few pods are undergoing cell division. Hence, competition for assimilates supply increases in the reproductive tissues of basal and middle sections of the main raceme generating, in consequence, a restriction of assimilates and/or growth regulators in the upper section of the main raceme and of the branches, thereby limiting their cell division rate and growth.

Acknowledgments

We gratefully acknowledge financial support from FONCYT (Science and Technology Fund, Argentina)

References

Alvarez Prado, S., Gallardo, J.M., Serrago, R.A., Kruk, B.C. and Miralles, D.J. 2013. Comparative behavior of wheat and barley associated with field release and grain weight determination. *Field Crops Res.* 144: 28-33.

Berry, P.M. and Spink, J.H. 2006. A physiological analysis of oilseed rape yields: past and future. *J. Agric. Sci.* 144: 381–392.

Borrás, L., Slafer, G.A. and Otegui, M.E. 2004. Seed dry weight response to source–sink manipulations in wheat, maize and soybean: a quantitative reappraisal. *Field Crops Res.* 86: 131-146.

Diepenbrock, W. 2000. Yield analysis of winter oilseed rape (*Brassica napus* L.): a review. *Field Crops Res.* 67: 35-49.

Egli, D.B. 1998. Seed biology and the yield of grain crops. Cab International, Wallington, UK. p. 178.

Gabrielle, B., Denoroy, P., Gosse, G., Justes, E. and Andersen, M.N. 1998. A model of leaf area development and senescence of winter oilseed rape. *Field Crops Res.* 57: 209-222.

Gomez, N.V. and Miralles D.J. 2011. Factors that modify early and late reproductive phases in oilseed rape (*Brassica napus* L.): Its impact on seed yield and oil content. *Industrial Crops and Products* 34: 1277-1285

Habekotté, A. 1993. Quantitative analysis of pod formation, seed set and seed filling in winter oilseed rape (*Brassica napus* L.) under field conditions. *Field Crops Res.* 35: 21-33.

Jandel Scientific, 1992. Tbl curve fitting software. Version 3.0. Corte Madera USA: Jandel Scietific.

Jenner, C.F., Ugalde, T.D. and Aspinall, D. 1991. The physiology of starch and protein deposition in the endosperm of wheat. *Aust. J. Plant Physiol.* 18: 211–226.

Jullien, A., Mathieu, A., Allirand, J.M., Pinet, A., de Reffye, P., Cournède, P.H. and Ney, B. 2011. Characterization of the interactions between architecture and source: sink relationships in winter oilseed rape (*Brassica napus*) using the GreenLab model. *Ann. Bot.* 107: 765-779.

Keiller, D.R. and Morgan, D.G. 1988. Distribution of 14carbon-labelled assimilates in flowering plants of oilseed rape (*Brassica napus* L.). *J. Agric. Sci.* 111: 347–355.

Leach, J.E., Milford, G.F.J., Mullen, L.A., Scott, T. and Stevenson, H.J. 1989. Accumulation of dry matter in oilseed rape crops in relation to the reflection and absorption of solar radiation by different canopy structures. *Aspects Appl. Biol.* 23: 117-123.

Leterme, P. 1988. Croissance et développement du colza d'hiver: Les principales étapes. in: Cetiom (eds.), colza-physiologie et elaboration du rendement du colza d'hiver. Suppl. Inf. Techn. CETIOM, pp. 23-33.

Lizana, X. C., Riegel, R., Gomez, L.G., Herrera, J., Isla, A., McQueen-Mason, S.J. and Calderini, D.F. 2010. Expansins expression is associated with grain size dynamics in wheat (*Triticum aestivum* L). *J. Exp. Bot.* 61: 1147-1157.

Mendham, N.J., Shipway, P.A., and Scott R.K. 1981. The effects of delayed sowing and weather on growth, development and yield of winter oil-seed rape (*Brassica napus*). *J. Agril. Sci.* 96: 389-416.

Miralles, D.J. and Slafer, G.A. 1995. Individual grain weight responses to genetic reduction in culm length in wheat as affected by source-sink manipulations. *Field Crops Res.* 43: 55-66.

Miralles, D.J. and Slafer, G.A. 2007. Sink limitations to yield in wheat: how could it be reduced?. *J. Agril. Sci.* 145: 139-149.

Müller, J. and Diepenbrock, W. 2006. Measurement and modelling of gas exchange of leaves and pods of oilseed rape. *Agril. and Forest Meteor.* 139: 307-3221.

Munier-Jolain, N. and Salon, C. 2003 Can sucrose content in the phloem sap reaching field pea seeds (*Pisum sativum* L.) be an accurate indicator of seed growth potential? *J. Exp. Bot.* 54: 2457-2465.

Pechan, P.M. and Morgan, D.G. 1985. Defoliation and its effects on pod and seed development in oil seed rape (*Brassica napus* L.). *J. Exp. Bot.* 36: 458-468.

Peltonen-Sainio, P. and Jauhiaine or Jauhiainen, L. 2008. Association of growth dynamics, yield components and seed quality in long-term trials covering rapeseed cultivation history at high latitudes. *Field Crops Res.* 108: 101-108.

Quilleré, I. and Triboi, A.M. 1987. Dynamique des reserves carbonée chez le colza d'hiver: Impacts sur la crossance. 7, pp. 219-223.

Yates, D.J. and Steven, M.D. 1987. Reflexion and absorption of solar radiation by flowering canopies of oil-seed rape (*Brassica napus* L.). *J. Agric. Sci.* 109: 495-502.

8

EFFECT OF DIFFERENT SOWING DATES ON YIELD OF TOMATO GENOTYPES

M.F. Hossain[1*], N. Ara[2], M.R. Islam[3], J. Hossain[4] and B. Akhter[5]

Abstract

The experiment was conducted at Agricultural Research Station, Thakurgaon, Bangladesh during October 2009 to March 2010 to observe the effect of sowing dates on yield of tomato genotypes. Three sowing dates viz. October 1, October 15 and October 30 were considered as factor A and tomato variety viz., BARI Tomato-2, BARI Tomato-3, BARI Tomato-4, BARI Tomato-9 and BARI Hybrid Tomato-4 considered as factor B. The experiment was laid out in RCBD (Factorial) with three replications. Early flowering (52.40 days) as well as early fruit harvesting (119.13 days) was occurred in October 1 sowing, where as sowing on October 30 resulted in delayed flowering (71.73 days) and fruit harvesting (140.67 days), respectively. Number of fruits per plant was also the highest (27.40) in October 1 sowing and the lowest (13.73) was in October 30 sowing. Seed sowing of October 1 was found better in respect of yield (74.75 tha^{-1}) compared to October 15 (58.55 tha^{-1}) and October 30 (24.60 tha^{-1}) sowing. Among the variety, BARI Tomat-2 produced the highest (68.12 tha^{-1}) marketable yield followed by BARI Tomato-9 (56.16 tha^{-1}) and BARI Tomato-3 while BARI Tomato-4 gave the lowest (36.91 tha^{-1}) marketable yield.

Keywords: Tomato, Genotype, Sowing, Flowering, Fruit Setting and Yield

[1]Senior Scientific Officer, Regional Agricultural Research Station, BARI, Ishwardi, Pabna, Bangladesh
[2]Principal Scientific Officer, Regional Agricultural Research Station, BARI, Ishwardi, Pabna, Bangladesh
[3,4&5]Scientific Officer, Regional Agricultural Research Station, BARI, Ishwardi, Pabna, Bangladesh

*Corresponding author's email: farukgolap@gmail.com (M.F. Hossain)

Introduction

Tomato (*Lycopersicon esculentum* Mill.) is one of the most popular vegetables in Bangladesh, which is receiving increased of the growers and consumers and made its position within few of the highest cultivated vegetables. It is an essential component of human diet for the supply of vitamins, minerals and certain hormone precursors in addition to protein and energy (Boamah *et al.*, 2010; Kallo, 1993). In Bangladesh, congenial atmosphere remains for tomato production during October to March. It is mainly grown in winter season. High temperature decreases flower production and /or to bud and flower drop. However, differences between varieties in fruit set under high temperature have been reported (FAO, 1990). Went (1984) assured that fruit set was abundant only when night temperature was between 15°C and 20°C, which might over simplify the issue. The importance of temperature in fruit set was clearly evident. Curme (1992) reported that fruit set in certain varieties with temperature as low (7.2°C) and with temperature as high (26.6°C) (Schaible, 1990) had created more flexible situation in respect of the variety temperature interactions. Climate change is a major threat for crop production not only Bangladesh but also all over the world. The meteorological data for the last 10 years indicated that the crop suffer from cold injury during the month of January (Anonymous, 2007) which result shy yield of this crop. In some areas of the country particularly in the northwestern part, the night temperature falls even sometimes go below 5-6°C which results tremendous yield loss in tomato. By this time BARI released a good number of tomato variety but their characteristics against tolerance to cold temperature injury has yet not been studied. Therefore, the present study was undertaken to find out the effect of sowing date on flowering, fruit setting and yield of tomato genotype.

Materials and Methods

The experiment was carried out at the Agricultural Research Station, BARI, Thakurgaon during the cropping season of October 2009 to March 2010. Three different dates of sowing viz. October 1, October 15 and October 30 considered as factor A and tomato variety viz., BARI Tomato-2, BARI Tomato-3, BARI Tomato-4, BARI Tomato-9 and BARI Hybrid Tomato-4 considered as factor B. The experiment was laid out in RCBD (Factorial) with three replications. The unit plot size was 4 m x 1 m. The crop was fertilized with cow dung 10 tonha-1, urea 300 kgha-1, TSP 250 kgha-1, MOP 200 kgha-1, gypsum 100 kgha-1 and boric acid 12 kgha-1. Total amount of cow dung, TSP, and one third of MOP were applied during final land preparation. Urea and rest of MOP were applied in two equal installments at 21 and 35 days after transplanting. Thirty days old seedlings were transplanted in the main field according to treatments. Irrigation along with other intercultural operations and plant protection measures were taken as and when necessary. Data were collected on flowering, fruit setting and yield contributing characters and analyzed statistically.

Results and Discussion

Effect of sowing date

There was a significant difference among the different seed sowing dates in respect of yield and yield attributes (Table 1). The earliest 50% flowering was detected at October 1 (52.40 days) followed by October 15 and October 30. More or less similar findings were reported by Sam and Iglesias (1994) in tomato in Cuba who observed that early October proved to be suitable date to planting tomatoes in the field. Peyvast (2001) reported that the early sowing date significantly affected tomato inflorescence initiation. Similar trend was found in case of first harvesting. Cluster per plant of tomato were significantly influenced by sowing dates. The highest number of clusters per plant (25.60) was obtained from early sowing (October 1) followed by mid sowing (October 15). This result was almost similar to the findings of Haque et al. (1999). The maximum number of flowers per cluster (5.33) was recorded from October 1 sowing and October 30 produced the minimum (4.40) number of flowers per cluster. This result was agreed with the findings of Hossain et al. (1986) who reported that early sowing enhanced total number of flowers per plant. Almost similar trend was found in case of fruit setting per cluster (Table 1). Fruit size (4.76 cm × 4.88 cm) of early sowing was bigger than others sowing. The highest average fruit weight (66.33 g) was found in October 1 seed sowing followed by October 15 sowing (59.13). October 1 seed sowing scored the highest number of fruits per plant (27.70) and the late sowing scored the lowest number of fruits per plant (13.73). This result was agreed with findings of Taha et al., (1984). The crop grown from October 1 seed sowing produced the highest marketable fruit yield (74.75 t/ha) and the lowest (24.60 t/ha) was from October 30. This may be due to better translocation of photosynthesis from source to sink and higher accumulation of photosynthesis in the fruits. Plants of October 1 seed sowing get shorter cold condition at flowering and fruit development stage. On the contrary, other sowing dates get extreme cold condition at these stages resulted low yield. These results were agreed with the findings of Peyvast (2001) who reported that the earliest sowing date resulted in a significantly higher total fruit yield compared to the later sowing date. Singh and Tripanthy (1995) showed variation in yield of tomato when sown in different dates from June to August at Orissa of India.

Effect of tomato genotypes

Flowering, fruit setting and fruit yield of tomato were significantly influenced by different tomato genotypes (Table 2). The days required for 50% flowering was found the earliest (53.78 days) in BARI hybrid Tomato-4 while the flowering was detected delayed (63.67 days) in BARI Tomato-3. Similar trend was found in case of first harvesting. BARI Hybrid Tomato-4 produced the highest number of cluster per plant (23.00) and BARI Tomato-3 produced the lowest number of cluster per plant (18.44). The number of flower per cluster was also the highest (5.77) in BARI Hybrid Tomato-4 where as the lowest in BARI Tomato-3 (4.22). Number of fruit set per cluster was also detected maximum (4.56) in BARI Hybrid Tomato-4 and the minimum (3.22) was in BARI Tomato-3. The maximum fruit length (5.34 cm) was recorded from BARI Tomato-9, which was statistically identical to BARI Tomato-2 (4.94 cm) and the lowest fruit length (3.52 cm) was recorded from BARI Tomato-4. On the other hand, the maximum fruit diameter (4.79 cm) was recorded from BARI Tomato-2, which was statistically similar to BARI Toimato-3 and the minimum fruit diameter (3.74 cm) was counted from BARI Tomato-4. The highest average fruit weight (72.33 g) was obtained from BARI Tomato-2 followed by BARI Tomato-3 (68.89 g) and the lowest average fruit weight (39.67 g) was obtained from BARI Hybrid Tomato-4. The highest number of fruits (27.78) was recorded by the genotype BARI Hybrid Tomato-4 and the lowest number of fruits (18.33) was recorded in BARI Tomato-3. Similar results were reported by Mohammed (1995) and Mahmoud (2005). The highest marketable fruit yield was recorded (75.31 tha-1) from BARI Tomato-2 whereas the lowest fruit yield (36.91 tha-1) was recorded from BARI Tomato-4. This may be due to its maximum fruit size, average fruit weight and number of fruits per plant. These results were agreed with previous findings of Omara (1995).

Combined effect of sowing date and tomato genotypes

Combined effect of sowing date and tomato genotypes had influenced on different yield parameters (Table 3). The earliest (47 days) 50% flowering was found in BARI Hybrid Tomato-4 with October 1 seed sowing.

Table 1. Effect of sowing date on the yield and yield contributing characters of tomato

Time of sowing	Days to 50% flowering	No. of clusters/plant	No. of flowers/cluster	No. of fruits/cluster	Days to 1st harvest	Fruit length (cm)	Fruit diameter (cm)	Average fruit weight (g)	No. of fruits/plant	Yield (tha⁻¹)
S_1	52.40	25.60	5.33	4.20	119.13	4.76	4.88	66.33	27.40	74.75
S_2	56.67	19.20	4.67	3.67	132.67	4.43	4.48	59.13	24.20	58.55
S_3	71.73	15.33	4.40	3.26	140.67	4.21	3.76	46.13	13.73	24.60
LSD (0.05)	4.034	2.105	0.441	0.45	2.951	0.330	0.154	3.036	2.336	1.960
CV (%)	5.17	14.04	12.68	6.17	3.01	9.90	14.83	7.10	14.34	12.36

Table 2. Effect of genotypes on the yield and yield contributing characters of tomato

Genotypes	Days to 50% flowering	No. of cluster/plant	No. of flower/cluster	No. of fruits/cluster	Days to 1st harvest	Fruit length (cm)	Fruit diameter (cm)	Average fruit wt (g)	No. of fruits/plant	Yield (tha⁻¹)
V_1	60.78	20.33	4.67	3.67	128.56	4.94	4.79	72.33	22.33	68.12
V_2	63.67	18.44	4.22	3.22	134.33	4.64	4.78	68.89	18.33	55.12
V_3	60.67	18.77	4.33	3.33	131.44	3.52	3.74	43.56	19.88	36.91
V_4	62.44	19.67	5.00	3.78	140.33	5.34	4.54	61.56	20.55	56.16
V_5	53.78	23.00	5.77	4.56	119.44	3.88	4.00	39.67	27.78	48.81
LSD (0.05)	3.007	2.718	0.570	0.459	3.802	4.27	0.198	3.920	3.016	3.120
CV (%)	5.17	14.04	12.68	6.17	3.01	9.90	14.83	7.10	14.34	12.36

Table 3. Combined effect of sowing date and tomato genotypes on the yield and yield contributing characters of tomato

Treatment combination	Days to 50% flowering	No. of cluster/plant	No. of flower/cluster	No. of fruits/cluster	Days to 1st harvest	Fruit length (cm)	Fruit diameter (cm)	Average fruit wt (g)	No. of fruits/plant	Yield (tha⁻¹)
S_1V_1	54.00	24.33	5.67	4.33	115.33	5.53	5.33	82.67	30.00	97.21
S_1V_2	54.00	23.00	5.00	3.67	116.67	5.23	5.30	79.00	25.33	83.43
S_1V_3	53.00	26.33	5.00	4.00	119.67	3.53	4.27	46.67	22.67	44.16
S_1V_4	54.00	25.00	5.00	4.33	131.67	5.50	4.97	77.67	27.67	89.56
S_1V_5	47.00	29.33	6.00	4.67	112.33	4.00	4.53	45.67	31.33	59.34
S_2V_1	56.33	19.00	4.33	3.67	131.33	4.76	5.00	74.00	24.67	76.36
S_2V_2	58.00	18.33	4.00	3.00	135.67	4.67	4.76	71.67	20.33	60.56
S_2V_3	58.00	16.67	4.00	3.33	135.67	3.53	3.67	47.00	22.67	44.49
S_2V_4	57.33	18.33	5.00	3.67	140.67	5.40	4.76	60.67	23.67	59.03
S_2V_5	53.67	23.67	6.00	4.67	120.00	3.76	4.03	42.33	29.67	52.33
S_3V_1	72.00	17.67	4.00	3.00	139.00	4.53	4.03	60.33	12.33	30.79
S_3V_2	79.00	14.00	3.67	3.00	150.67	4.04	4.30	56.00	9.33	21.53
S_3V_3	71.00	13.33	4.00	2.67	139.00	3.50	3.30	37.00	14.33	22.07
S_3V_4	76.00	15.67	5.00	3.33	148.67	5.13	3.90	46.33	10.33	19.86
S_3V_5	60.67	16.00	5.33	4.33	126.00	3.86	3.27	31.00	22.33	28.75
LSD (0.05)	5.208	4.707	0.980	1.010	6.58	0.742	0.343	6.790	5.224	7.246
CV (%)	5.17	14.04	12.68	6.17	3.01	9.90	14.83	7.10	14.34	12.36

S_1= 1 October, S_2= 15 October, S_3= 30 October, V_1= BARI Tomato-2, V_2= BARI Tomato-3, V_3= BARI Tomato-4, V_4= BARI Tomato-9 and V_5= BARI Hybrid Tomato-4

Similarly, the number of cluster (29.33), flowers per cluster (6) and number of fruits per cluster (6) of BARI Hybrid Tomato-4 with October 1 sowing showed better performance than other combinations. However, fruit size was observed larger (5.53 cm × 5.33 cm) in BARI Tomato-2 with October 1 sowing compared to other combinations. The highest average fruit weight (82.67 g) was recorded from BARI Tomato-2 with October 1 seed sowing. On the other hand, the highest number of fruits per plant (31) was recorded from BARI Hybrid Tomato-4 with October 1 sowing followed by BARI Tomato-2 (30 g). The highest marketable fruit yield (97.21 tha^{-1}) was obtained from BARI Tomato-2 with October 1 sowing followed BARI Tomato-9 (89.56 tha^{-1}) whereas, the lowest yield (16.86 tha^{-1}) was recorded from BARI Tomato-9 with October 30 seed sowing.

Conclusion

The experiment revealed that the yield of tomato was significantly affected by different sowing dates and tomato genotypes. BARI Tomato-2 with October 1 seed sowing was suitable combination for maximum yield of tomato in northwestern part of Bangladesh.

References

Anonymous. 2007. The year book of Agricultural Statistics of Bangladesh. Bangladesh Bur. Stat., Ministry of Planning, Govt. People's Republic of Bangladesh. p. 10.

Boamah, P.O., Sam-Amoah, L.K. and Owusu-Sekyere, J.D. 2010. Effect of irrigation interval on growth and development of tomato under sprinkler. *Asian J. Agric. Res.* 4: 196-203.

Curme, J.H. 1992. Effect of low night temperatures on tomato fruit set. Proc. Plant Sci. Symp., Campbell Soup. Co. pp. 99-108.

FAO. 1990. FAO Production Yearbook. Basic Data Unit, Statistics Division, Food and Agriculture Organization United Nation, Rome, Italy. p. 90.

Haque, M.A., Hossain, A.K.M.A. and Ahmed, K.U. 1999. Varietal response of different seasons and temperature in respect of yield and yield components. *Bangladesh Hort.* 26: 39-45.

Hossain, M.M., Karim, M.M., Haque, M.M. and Hossain, A.M.A. 1986. Performance of some tomato lines planted at different dates. *Bangladesh Hort.* 14 (1): 25-28.

Kallo, G. 1993. Tomato. *In:* Genetic improvement of vegetable crops. Oxford, England: Pergamon Press. p. 6.

Mahmoud, S.M. 2005. The effect of cultivars, seedbed preparation and plant density on the growth and yield of toamto (*Lycopersicon esculentum* Mill.) in Forest Location. *J. Agril. Sci.* 5 (2): 152-158.

Mohammed, B.M. 1995. Vegetable production in central Sudan. Integrated Pest Management Medani, Sudan. p.106.

Omara, S. 1995. Tomato experiments, International Institute for Promotion Horticultural Exports. Khartoun, Sudan. p. 17.

Peyvast, G.H. 2001. Study of some quality and quantity factors of tomato. *J. Veg. Crop Prod.* 10: 15-22.

Sam, O. and Iglesias, L. 1994. Characterization of the flowering-fruiting process in tomato cultivars in two sowing seasons. *Cultivos Tropicals* 15 (2): 34-43.

Schaible, L.W. 1990. Fruit setting response of tomatoes to high night temperatures. Campbell Soup Co., camden, N.J., pp. 89-98.

Singh, D.N. and Tripanthy, P. 1995. Growth and yield of tomato genotypes and technology. *Indian J. Agril. Sci.* 65 (12): 863-865.

Taha, A.A., Abdelfattah, Hassan, M.S. and Ali, A.W. 1984. Effect of sowing date and stage of maturity at harvest on yield and quality of tomato for export. Acta. Hort. 143. Tropical Horticulture VIII. 6 (5): 665-669.

Went, F.W. 1984. Plant growth under controlled conditions. II. Thermoperiodicity in growth and fruiting of the tomato. *American J. Bot.* 31: 135-150.

9

EFFECT OF SEASONAL AND ENVIRONMENTAL VARIATION ON YIELD AND YIELD COMPONENTS OF HYBRID MAIZE

M.K. Islam[1], M.S. Mahfuz[2*], M.A.I. Sarker[1], S. Ghosh[3] and A.S.M.Y. Ali[3]

Abstract

The experiment was conducted during kharif II in 2006 & 07, rabi in 2006-07 & 07-08 and kharif I in 2007 and 2008 at ARS, Burirhat, Rangpur in RCB design to understand the influence of season and location specific environment effect on growth and yield of hybrid maize and selected suitable variety(s). Four hybrid maize varieties V_1=BARI hybrid Maize (BHM)-2, V_2=BHM-3, V_3= BHM-5 and V_4= Pacific-984 (as check) were tested during kharif II in 2006 while seven varieties V_1= BHM-2, V_2= BHM-3, V_3= BHM-5 and V_4= Pacific-984, V_5= Pacific-60, V_6= Pacific-11 and V_7= Prolin were during the other seasons except rabi, 07-08, where variety Pacific-555 was used in V_7 instead of Prolin. Seeds were sown on August 30 and August 11 for kharif II of 2006 and 2007, respectively, November 28 and 17 for rabi 2006-2007 and 2007-08 and March 08 and 12 for for kharif I 2007 and 2008. The crops were harvested on January 25, 2007 and January 12, 2008 in the two consecutive kharif II seasons; May 10 and April 29 in rabi 2006-07 and 2007-08, and July 01 and 05 in kharif I 2007 and 2008, respectively. Yield parameters were mostly varied significantly. The highest yield was obtained from BHM-5 (9.03 t ha⁻¹), which was followed by Pacific-984 (8.89 t ha⁻¹), BHM-3 (8.81 t ha⁻¹) and BHM-2 (8.58 t ha⁻¹) in kharif II, 2006 while in kharif II, 2007 the highest significant yield was noted in Pacific-984 (9.22 t ha⁻¹). In rabi, 2006-07, significant highest grain yield was obtained from Pacific-60 (11.03 t ha⁻¹), which was statistically identical with Prolin (10.20 t ha⁻¹). The yield of Prolin was also statistically identical with Pacific-11 (10.01 t ha⁻¹), BHM-5 (10.00 t ha⁻¹), BHM-3 (9.92 t ha⁻¹) and BHM-2 (9.51 t ha⁻¹). Comparatively lower temperature during ear initiation (mean 29.6°C in rabi and 31°C in kharif II) and silking (mean 18.2°C in rabi and 20.2°C in kharif II) contributed much for higher trend of yield in rabi over kharif. In kharif I, 2007, the highest yield (9.55 t ha⁻¹) was recorded from Pacific-60, which was identical to Pacific-984 (9.25 t ha⁻¹), BHM-5 (9.11 t ha⁻¹) and BHM-3 (8.89 t ha⁻¹). All the BARI hybrid maize varieties were suitable to grown in kharif I, kharif II and rabi season although BHM-3, BHM-5, Pacific-60 and Pacific-984 were better in Kharif I and Pacific-60, BHM-5, Prolin and Pacific-555 were found better in Rabi season.

Keywords: Season, Environment, Yield, Maize

[1]Senior Scientific Officer, Bangladesh Agricultural Research Institute, Burirhat, Rangpur, Bangladesh
[2]Regional Farm Broadcasting Officer, Agriculture Information Service, Rangpur, Bangladesh
[3]Scientific Officer, Bangladesh Agricultural Research Institute, Burirhat, Rangpur, Bangladesh

*Corresponding author's email: shohag_agri@yahoo.com (M.S. Mahfuz)

Introduction

Among the cereals maize ranks 3rd position after rice and wheat. Maize can be grown throughout the year. High yield per hectare, high food value and multipurpose use as well as its high demand expanding maize cultivation day by day. There is a demand of 1.2 million tons of maize only to feed the poultry industries of Bangladesh (BARI, 2001). Again, for starch industries, other industries and for human consumption maize has a great demand. The yield of hybrid maize is about 20-30% higher than that of open pollination composite ones, which encouraged the farmers to prefer hybrid maize. Now, 87% of the total maize area is under hybrid maize. Recently BARI has developed 5 hybrid maize varieties. High quality protein maize (QPM) rich in tryptophen and lycin associated with recently evolved BARI hybrid maize-5 and other predominant BARI hybrid maize varieties along with other commercial hybrid maize varieties, the performance of the same across the seasons as well as across the locations is yet to be studied.

The environmental scientist working with maize have concluded from long term weather data that higher mean seasonal temperature was negatively correlated with grain yield and final grain yield was dependent on the rate and duration of grain growth and dry matter accumulation (BARI, 2001; Law and Brander, 1999). Increase in temperature decreased net photosynthesis and optimum temperature ranged from 28°C to 37.5°C for higher photosynthesis (Brandner and Salvucci, 2002). Physiological parameter like NAR depends on temperature due to enzyme activity rubisco and at 28°C rubisco oxygenase activity was 100% (Edwards et al., 2001). However, the relationship among physiological parameters with the yield and yield components in three contrasting season has not been worked out in detail. The present investigation was carried out to understand the influence of season and location specific environment effect on growth and yield of hybrid maize and selected suitable variety (s) adaptive for specific areas.

Materials and Methods

The experiment was conducted during kharif II season of 2006 & 07, rabi in 2006-07 & 07-08, kharif I in 2007 at sandy loam soils of ARS, Burirhat, Rangpur following RCB design with 3 replications. The treatments included 4 hybrid maize varieties namely: V_1=BARI hybrid maize (BHM)-2, V_2=BHM-3, V_3= BHM-5 and V_4 = Pacific-984 (as check) in kharif II, 2006 and seven varieties viz. V_1= BHM-2, V_2= BHM-3, V_3= BHM-5 and V_4= Pacific-984, V_5= Pacific-60, V_6= Pacific-11 and V_7= Prolin in other seasons except rabi, 07-08 where variety Pacific-555 was used in V_7 instead of Prolin. Seeds were treated with Vitavex 200 @ 5 g kg⁻¹ seed for 10 minutes in airtight bag then kept open. Blanket application of cow dung @ 10 t ha⁻¹ was made 7 days before sowing. The field was fertilized with 200-60-110-45-5-2 kg NPKSZn and B/ha, respectively. One-third amount of N as Urea and rest full amount of all other fertilizer were applied in experimental unit plot one day before sowing. Seeds of different hybrids maize varieties were accommodated in 4.5 m × 6.0 m sized plot. Two seeds/hill were sown on August 30 and August 11 for kharif II, 2006 and 2007, respectively, November 28 and 17 for rabi 2006-07 and 07-08, and March 08 and 12 for kharif I, 2007 and 2008, at spacing 75 cm × 20 cm. Necessary gap filling were made by re-sowing within 8 days of sowing (DAS). Rest 2/3rd Urea in equal splits was top dressed (TD) at 30 and 60 DAS. Weeding and mulching were done properly; final earthling up was made after 2nd TD, which was followed by irrigation. Futher irrigation was made at 75 and 90 DAS. Despite that plant received about 405

mm and 554 mm of precipitation in kharif II 2006 and 2007 respectively. In rabi, the crop was irrigated 4 times at 60, 90, 110 and 120 DAS. The crop also received rainfall at 2nd decades of December and February, slight in the month of March and 3rd decade of April in rabi, 2006-07. The crop in kharif I was irrigated 4 times at 10 days interval starting form 25 DAE. The crop also received a total of around 880 mm rainfall in Kharif I, 2007. Diazinon 60 EC @ 2 ml L⁻¹ of water was sprayed at 72 DAS. Guard was placed to avert bird (crow, parrot etc) damage. Minor variation about the maturity among the hybrid maize varieties was observed. The crops were harvested on January 25, 2007 and January 12, 2008 respectively for the two consecutive kharif II seasons of 2006 and 2007, May 10 and April 29 for rabi 2006-07 and 07-08, and July 01 and 05 for kharif I, 2007 and 2008.

Data on yield and yield components were taken properly, analyzed statistically and presented in tables 1 and 2.

Results and Discussion

Traditionally hybrid maize is sown either rabi or kharif I season. Here the hybrid maize varieties were grown during kharif II season. In lieu of sowing in August 30 if the crop could be sown on mid July. The crop could be received about 600 mm rainfall and also assumed to be harvested within November 2006. Dowswell et al. (1996) also opined that, in most tropical environments, maize requires 600-700 mm of moisture well distributed over the growing season. Forty five per cent of total maize area in the developing tropical countries of the world covers 36.7 million hectares. The environment is characterized by a high mean temperature (around 28°C) during the growing season (FAO, 1988). They have a high mean maximum temperature (around 32°C) and a high mean minimum temperature (around 22°C). At Rangpur region of Bangladesh there prevailed mean high temperature (around 33°C) and high mean minimum temperature (around 26°C) during growing season of maize. Sultan (2006) opined that the temperature 24° to 30°C is suitable for maize production. He also opined that fluctuation of the temperature results in fluctuation in field duration of maize.

Table 1.a . Performance of hybrid maize varieties during kharif II season of 2006 season at ARS, Burirhat, Rangpur

Treatments	Plant height (cm)	No. of leaves/ plant	Days to 50% tasseling	Days to 50% silking	Ear height (cm)	Cob length (cm)	No. of cobs/ plant	No. of seeds/ cob	1000 grain wt. (g)	Yield (t ha⁻¹)
V_1=BHM-2	211a	14	62a	65a	101a	20.3a	1.20a	442b	325a	8.58
V_2=BHM-3	216a	13	61a	64a	96a	19.7a	1.10b	487ab	318a	8.81b
V_3=BHM-5	196ab	12	55b	61ab	80b	16.6b	1.29a	474b	298ab	9.03a
V_4=Pacific 984	186b	13	53b	57b	78b	19.7a	1.02b	540a	285b	8.89b
CV (%)	5.15	4.60	1.98	2.18	4.27	2.81	2.70	5.76	3.05	6.32

Means in a column having similar or no letter did not differ significant at 5% level of significance

Table 1.b. Performance of different hybrid maize during rabi season of 20006-07 at ARS, Burirhat, Rangpur

Treatment	Plant height (cm)	No. of leaves/ plant	Days to 50% tasseling	Days to 50% silking	Ear height (cm)	No. of cob/ plant	Cob length (cm) Pri.	Cob length (cm) Sec.	No. of grain/cob Pri.	No. of grain/cob Sec.	1000 grain wt. (g) Pri.	1000 grain wt. (g) Sec.	Yield (t ha⁻¹)
T₁=BHM-2	252a	16.7a	109a	112a	145a	1.29ab	16.7ab	10.5	384d	209c	381ab	331b	9.51b
T₂=BHM-3	255a	15.3b	112a	114a	148/a	1.09cd	16.6ab	9.2	430c	218c	397a	322bc	9.92b
T₃=BHM-5	223b	14.9b	110a	112a	130b	1.32a	16.2b	10.5	441bc	258ab	348b	285c	10.00b
T₄=Pacific-984	211b	13.5cd	111a	113a	104d	1.03d	16.8ab	10.0	474ab	253b	360ab	290bc	9.67b
T₅=Pacific-60	250a	14.7bc	104b	107b	136b	1.13cd	17.9a	10.7	491a	266ab	375ab	372a	11.03a
T₆=Pacific-11	224b	14.7b	105b	107b	134b	1.21bc	16.8ab	10.3	417cd	280a	390a	303bc	10.01b
T₇=Prolin	225b	13.3d	104b	106b	113c	1.03d	15.5b	8.0	501a	198c	365ab	293bc	10.20ab
CV (%)	3.55	3.27	1.3	.88	3.13	3.98	3.15	11.09	2.93	5.78	3.66	7.16	4.92

Means in a column having similar or no letter did not differ significant at 5% level of significance

Table 1.c. Performance of different hybrid maize during kharif I season of 2007 at ARS, Burirhat, Rangpur

Treatment	Plant height (cm)	No. of leaves/ plant	Days to 50% tasseling	Days to 50% silking	Ear height (cm)	No. of cobs/ Plant	Cob length (cm)	No. of seeds/ cob	1000 seeds wt. (g)	Yield (t ha⁻¹)
V₁=BHM-2	245.07a	15.80a	56.67a	58.67a	130.40a	1.08ab	17.66	396d	328.00a	8.47b
V₂= BHM-3	237.20a	13.87b	55.67ab	58.00ab	107.80b	1.08ab	18.84	430bcd	308.70ab	8.89ab
V₃= BHM-5	200.87b	12.47bc	54.33bcd	56.67abc	84.00bc	1.09a	17.36	460bc	294.00b	9.11ab
V₄=Pacific-984	213.93b	12.73bc	53.33cd	55.00c	85.73bc	1.05ab	16.93	531a	295.00b	9.25ab
V₅=Pacific-60	212.27b	12.87bc	54.00bcd	56.00bc	90.60bc	1.08ab	16.88	476b	322.20a	9.55a
V₆=Pacific-11	193.00b	12.80bc	53.00d	55.33c	92.33b	1.07ab	16.25	419cd	315.90ab	8.54b
V₇=Prolin	167.07c	11.80c	55.33abc	58.00ab	66.67c	1.02b	16.97	425bcd	306.90ab	8.49b
CV (%)	4.06	5.05	2.19	2.39	9.62	2.16	8.07	4.49	3.95	4.64

Means in a column having similar or no letter did not differ significant at 5% level of significance

All the parameters varied significantly in kharif II season except number of leaves/plant and yield in 2006-07 and days to 50% silking in 2007-08. The highest grain yield (9.03 t ha⁻¹) was associated with BARI hybrid maize-5 in 2006-07 while in 2007-08, the highest significant yield was noted in Pacific-984 (9.22 t ha⁻¹), which was statistically similar with other varieties except BHM-2 and Prolin (Table 1.a & 2.a). Yields by BHM-3 (8.30 t ha⁻¹), Pacific-11 (8.04 t ha⁻¹), Prolin (7.57 t ha⁻¹) and BHM-2 (7.47 t ha⁻¹) were also identical. Higher number of cob/plant, seeds/cob and 1000-grain wt. contributed to the higher yields.

Table 2.a. Performance of different hybrid maize during kharif II season of 2007 at ARS, Burirhat, Rangpur

Treatment	Plant height (cm)	No. of leaves/ plant	Days to 50% tasseling	Days to 50% silking	Ear height (cm)	No. of cobs/ Plant	Cob length (cm)	No. of seeds/ cob	1000 seeds wt. (g)	Yield (t ha⁻¹)
V₁=BHM-2	248.80a	15.87a	60.33abc	62.33	126.93a	1.08ab	18.60a	421.00c	321.67a	7.47b
V₂= BHM-3	249.60a	14.33bc	61.67ab	63.67	134.00a	1.11a	18.44a	450.67bc	323.00a	8.30ab
V₃= BHM-5	238.47ab	14.80b	62.67a	63.67	129.73c	1.10ab	17.95ab	440.67c	310.00ab	9.01a
V₄=Pacific-984	215.20b	13.20de	60.00abc	62.33	101.73c	1.02c	18.63a	531.33a	311.00ab	9.22a
V₅=Pacific-60	234.33ab	14.00bcd	58.00c	61.00	123.07ab	1.04bc	16.29bc	476.00b	312.00ab	9.08a
V₆=Pacific-11	217.80b	13.53cde	57.33c	60.00	124.40ab	1.05bc	16.22bc	419.33c	300.00b	8.04ab
V₇=Prolin	216.77b	12.80e	59.00bc	61.67	107.67bc	1.01c	14.60c	427.00c	298.00b	7.57b
CV (%)	3.85	2.98	3.03	3.60	5.66	3.50	4.27	2.70	3.09	6.02

Means in a column having similar or no letter did not differ significant at 5% level of significance

Table 2.b. Performance of different hybrid maize during rabi season of 2007-08 at ARS, Burirhat, Rangpur

Treatment	Plant height (cm)	No. of leaves/ plant	Days to 50% tasseling	Days to 50% silking	Ear height (cm)	No. of cobs/ Plant	Cob length (cm)	No. of seeds/ cob	1000 seeds wt. (g)	Yield (t ha⁻¹)
T₁ = BHM-2	203.33	14.57	106.00	108.00	104.53	1.13	16.78ab	423.00a	370.00ab	9.53b
T₂ = BHM-3	218.00	14.87	104.00	106.00	113.33	1.17	16.20bc	415.00a	343.33c	9.70b
T₃ = BHM-5	186.80	14.30	104.00	106.00	99.07	1.10	17.72a	414.33a	371.67ab	11.08a
T₄ = Pacific-984	205.53	14.57	103.33	105.33	102.93	1.20	16.15bc	363.33b	351.67bc	9.72b
T₅ = Pacific-60	212.80	14.60	103.33	106.00	107.80	1.30	16.02bc	386.00ab	380.00a	11.10a
T₆ = Pacific-11	204.73	15.20	104.33	106.33	107.67	1.13	15.54c	356.67b	370.00ab	9.88b
T₇ = Prolin	219.07	15.03	103.67	106.00	115.60	1.13	16.11bc	391.67ab	360.00abc	10.00ab
CV (%)	5.18	3.84	2.12	3.14	9.50	8.57	3.45	4.52	3.60	6.03

Means in a column having similar or no letter did not differ significant at 5% level of significance

Table 2.c. Performance of different hybrid maize during kharif I season of 2008 at ARS, Burirhat, Rangpur

Treatment	Plant height (cm)	No. of leaves/ plant	Days to 50% tasseling	Days to 50% silking	Ear height (cm)	No. of cobs/ Plant	Cob length (cm)	No. of seeds/ cob	1000 seeds wt. (g)	Yield (t ha⁻¹)
T_1 = BHM-2	230 a	14.93 a	56.16a	58.30ab	120a	1.12a	16.5	370b	330a	7.8b
T_2 = BHM-3	229 a	14.75 a	55.9a	59.03a	110ab	1.04b	17.0	375a	310b	8.6ab
T_3 = BHM-5	220 ab	14.30 bc	55.2b	57.10b	101b	1.08ab	16.8	415a	305b	9.0a
T_4 = Pacific-984	225 a	14.70 ab	55.3ab	57.06bc	95b	1.10a	16.7	400ab	320ab	8.7ab
T_5 = Pacific-60	210 b	14.50 b	53.05c	55.48c	92b	1.09ab	16.7	398ab	340a	9.4a
T_6 = Pacific-11	217 b	13.86 c	56.52a	58.75a	125a	1.03b	16.4	372b	308b	8.3ab
T_7 = Pacific-555	215 b	14.60 b	53.80c	55.67bc	98b	1.10a	17.1	380ab	320ab	9.1a
LSD (0.05)	10.88	18.41	2.23	0.56	12.85	-	-	-	-	-
CV (%)	2.77	3.8	2.36	3.05	6.4	3.18	4.62	5.03	3.70	4.8

Means in a column having similar or no letter did not differ significant at 5% level of significance

Table 1.b revealed that, in rabi 2006-07, the yield and all the yield-contributing characters except secondary cob length were significantly varied. Significant highest cob length (17.9 cm) along with significant highest number of grains/primary cob (491), higher number of grains/secondary cob (266), significant higher 1000 grain wt. of primary cob (375 g) and highest weight (372 g) of secondary cob contributed much to pacific-60 for giving significantly highest yield (11.03 t ha⁻¹) which was statically identical to Prolin (10.20 t ha⁻¹). The yield of Prolin was also statistically identical with other hybrid maize. In rabi, 2007-08, cob length, number of seeds/cob, 1000 seed weight and yield were varied significantly (Table 2.b). Pacific-60 was also highest yielder (11.10 t ha⁻¹), which was identically followed by BHM-5 (11.08 t ha⁻¹) and Pacific-555 (10.00 t ha⁻¹) but significantly higher than the other varieties. Cumulative favourable effects of higher number of seeds/cob 386, 414.33 & 391.67 and higher 1000 seed weight 380, 371.67 & 360, respectively contributed much for higher yields in Pacific-60, BHM-5 and Pacific-555. The yield trend with hybrid maize during rabi season was comparatively higher than that of kharif II maize. This might be due to (about 40 cm) taller plant stand associated with rabi maize. Again, average 2 more number of leaves/plant was observed in rabi maize than kharif II. Longer period taken for days to 50% tasseling (around 108 days) and days to 50 % silking (around 111 days) in rabi, 2006-07 over kharif II maize (around 58 and 62 days, respectively). Comparatively lower temperature during ear initiation and silking of rabi maize (mean maximum around 29.6°C and mean minimum around 18.2°C) over kharif II maize (mean maximum 31°C and mean minimum 20.2°C) contributed much for higher trend of yield in rabi, 2006-07 over kharif II, 2007. Similar trend was also observed in the kharif II, 2008. BARI (2001) opined that increase in temperature decreased spikelet fertility and grain number was found to be dependent to temperature and radiation regimes during the period from ear initiation to silking.

In kharif I, all the parameters varied significantly except cob length in 2008. The highest yield in 2007 (9.55 t ha⁻¹) was recorded from Pacific-60, which was identical to Pacific-984 (9.25 t ha⁻¹), BHM-5 (9.11 t ha⁻¹) and BHM-3 (8.89 t ha⁻¹) but significantly higher than the rest varieties (Table-1.c). Except Pacific-60, all other varieties produced identical yield. Significantly, higher number of cobs/plant (1.08) and 1000 grain weight (322.20g) attributed to highest yield in Pacific-60. Similar trend was also observed in Pacific-984, BHM-5 and BHM-3 where either number of cobs per plant or number of seeds per plant alone or along with 1000-grain weight favoured the higher yield. Similar trend was observed in kharif I, 2008 where the highest yield (9.4 t ha⁻¹) was recorded from Pacific 60 identically followed by Pacific 555 (9.1 t ha⁻¹), BHM 5 (9 t ha⁻¹) BHM 3 (8.6 t ha⁻¹) and Pacific 11 (8.3 t ha⁻¹). BHM 2 produced significantly the lowest yield (7.8 t ha⁻¹).

Conclusion

All the BARI hybrid maize varieties including check (Pacific-984) can be grown in kharif I and kharif II season although BHM-3, BHM-5, Pacific-60 and Pacific-984 are better in Kharif I. Again, considering the performance during rabi season, all the varieties can be grown as their yields were either higher or identical to the check variety Pacific-984 although Pacific-60, BHM-5, Prolin and Pacific-555 were found better.

References

BARI. 2001. CIMMYT sub-topical intermediate yellow-QPM hybird trail. Annual report of Bangladesh Agricultural Research Institute, Joydebpur, Gazzipur. pp. 44-67.

Brandner, S.J.C. and Salvucci, M.E. 2002. Sensivity of photosynthesis in a C₄ plant, maizc, to heat Stress. *Plant Physiol.* 129 (4): 1773-1780.

Dowswell, C.R., Paliwal, R.L. and Cantrell, R.P 1996. Maize in third world. Winrock development oriented literature series. Steven A. Breth, series editor published in co operation with Winrock International Institute for Agricultural Development. p. 42.

Edwards, G.E, Furbank, R.T, Hatch, M.D. and Osmond, C.B. 2001. What does it takes to be C_4? Lessons from the evolution of C_4 photosynthesis. *Plant Physiol.* 125: 46-49.

FAO. 1988. FAO Production Year books, CIMMYT maize program. Food and Agricltural Organization, Rome, Italy.

Law, R.D. and Brandner, S.J.C. 1999. Inhibition and acclimation of photosynthesis to heat stress. *Plant Physiol.* 120: 173-181.

Sultan. 2006. Seed production technique of hybrid maize, Training Manual of "Hybrid Vhuttar Bis Utpadaner Kalakawshal". Bangladesh Agricultural Research Institute, Joydebpur, Gazipur. p. 21.

10

THE ROLE OF COW DUNG AND KITCHEN MANURE COMPOSTS AND THEIR NON-AERATED COMPOST TEAS IN REDUCING THE INCIDENCE OF FOLIAR DISEASES OF *Lycopersicon esculentum* (Mill)

A. Ngakou[1]*, H. Koehler[2] and H.C. Ngueliaha[1]

Abstract

Compost teas are fermented watery extracts of composted materials used for their beneficial effect on plants. A study was conducted in the field to compare the efficacy of cow dung and kitchen manure composts and their derived non-aerated compost teas on disease symptoms expression and severity of *Lycopersicon esculentum*. The experimental layout was a complete randomised block design comprising six treatments, each of which was repeated three times: the negative control plot (Tm-); the positive control or fungicide plot (Tm+); the cow dung compost plot (Cpi); the kitchen manure compost plot (Cpii); the compost tea derived cow dung plot (Tci); and the compost tea derived kitchen manure plot (Tcii). Compost tea derived cow dung was revealed to be richer in elemental nutrients (N, P, K) than compost tea from kitchen manure, and significantly ($p < 0.0001$) enhanced fruit yield per plant. Similarly, the two composts and their derived compost teas significantly ($p < 0.0001$) reduced the incidence and severity of disease symptoms compared to the controls, with the highest efficacy accounting for cow dung compost and compost tea. Although the non-aerated compost teas were not amended with micro-organisms, these results suggest that the two compost teas in use were rich enough in microbial pathogen antagonists, and therefore, are perceived as potential alternatives to synthetic chemical fungicides. Future work will attempt to identify these microbial antagonists with highly suppressive activity in the non-aerated compost teas.

Keywords: Compost, Compost Tea, *Lycopersicon esculentum*, Fungicide, Disease Symptoms

[1]Department of Biological Sciences, University of Ngaoundere, P.O.Box 454 Ngaoundéré, Cameroon
[2]Center for Environmental Research and Sustainable Technology (UFT), University of Bremen, Germany

*Corresponding author's email: alngakou@yahoo.fr (A. Ngakou)

Introduction

In Cameroon, agriculture represents up to 20% of the national internal raw products (Anonymous, 2008). The total area under tomato production in tropical Africa is about 300,000 ha with an estimated annual production of 2.3 million tons (Van der Vossen *et al.*, 2004). In Cameroon, Adamawa is the third national producing region, with 51 700 tons after west and centre (AGRI-STAT, 2010). The fruit is rich in vitamins A, C, thiamine, riboflavin, and niacin as well as some minerals like potassium and sodium (Janes, 1994). Moreover, lycopene content in tomato is higher than in other fruits or vegetables, and at least 85 % of its input into the human body comes from tomato or tomato derived products (Bramley, 2000; Chalabi *et al.*, 2004). Tomato culture thus constitutes an income generation activity for many of the growers in rural and urban areas. However, its production is affected by numerous diseases which are the main causes of decrease or total yield losses (Welke, 2005). The present day agriculture is not more sustainable in most parts of the country, because of the dependence of growers on chemical fertilizers and pesticides from the North. The powerful message that derives from all thoughts and dialogues is to move towards organic farming systems, and/or biological agriculture. Farming systems attempt to provide a balanced environment, in which the maintenance of soil fertility and the control of pests and diseases are achieved by the enhancement of natural processes and cycles, with moderate inputs of energy resources, while maintaining an optimum productivity. Uncontrolled use of chemical agricultural techniques have resulted in a great increase in productivity, but at the cost of negative impacts including soil degradation, resistance to pesticides (Barbier and Catin, 1994), detention of soil health and environmental pollution (Lachance and Rouleau, 2004). Some pesticides or contaminants may accumulate in fruits and tubers (Sonchieu, 2002). As an answer to these problems, recent research has favoured the organic farming with uncontaminated quality compost (Znaïdi, 2002; Ngakou *et al.*, 2008), from which compost tea can be extracted. Compost tea represents not only a mean of completing the role of compost in the fertilization process, but also a strategy of cultivation of useful microflora to prevent stem and floral diseases (Deschênes, 2007). Hence, there is growing interest in the potential for using composts tea to stimulate root and vegetative growth, prevent and control diseases in field crops, increase crop yields and quality. Information concerning its use and effectiveness is slowly increasing (Litterick *et al.*, 2004; Haggag and Saber, 2007). Previous work conducted in the region of Adamawa focused on the relative effects of compost and non-aerated compost tea on *Solanum tuberosum* production (Ngakou *et al.*, 2012).The chemical components were not assessed. Considering the role of compost tea that depends on the source of composting material and its chemical composition, this research is attempted to assess and evaluate its potential as biological control agent against tomato disease symptoms in the field. The outcomes of this study will enable making up our mind on whether compost tea could be used as a potential substitute of chemical fertilizers and pesticides.

Materials and Methods

Experiments were carried out at Dang-Ngaoundéré in the Adamawa region (Cameroon) located at 7°25'119"N, 13°33'415"E at 1106 m altitude above the sea level. During the field work period from June to October 2011, the average daily rainfall, temperature and relative humidity were 228.14mm, 28.24°C and 80.44%, respectively, with a mean sunlight of 139.22 h per day (CMSN, 2011).

Biological and chemical materials

Seeds of tomato (*Lycopersicon esculentum*) of the « Rio grande » variety were purchased from a phytosanitary stores in the local market. Compost made separately from cow dung and kitchen manure was produced by the research team of the corresponding author as described by Ngakou *et al.* (2008). Compost tea was extracted (Fig. 1) as described by Ngakou *et al.* (2012) from a mixture of compost/tap water in a 1/15 (Kg/L) ratio. After extraction, compost teas were stored in 1.5 L bottles at 4°C in the refrigerator (Fig. 2), and applied to the field at a rate of 5L/plot.

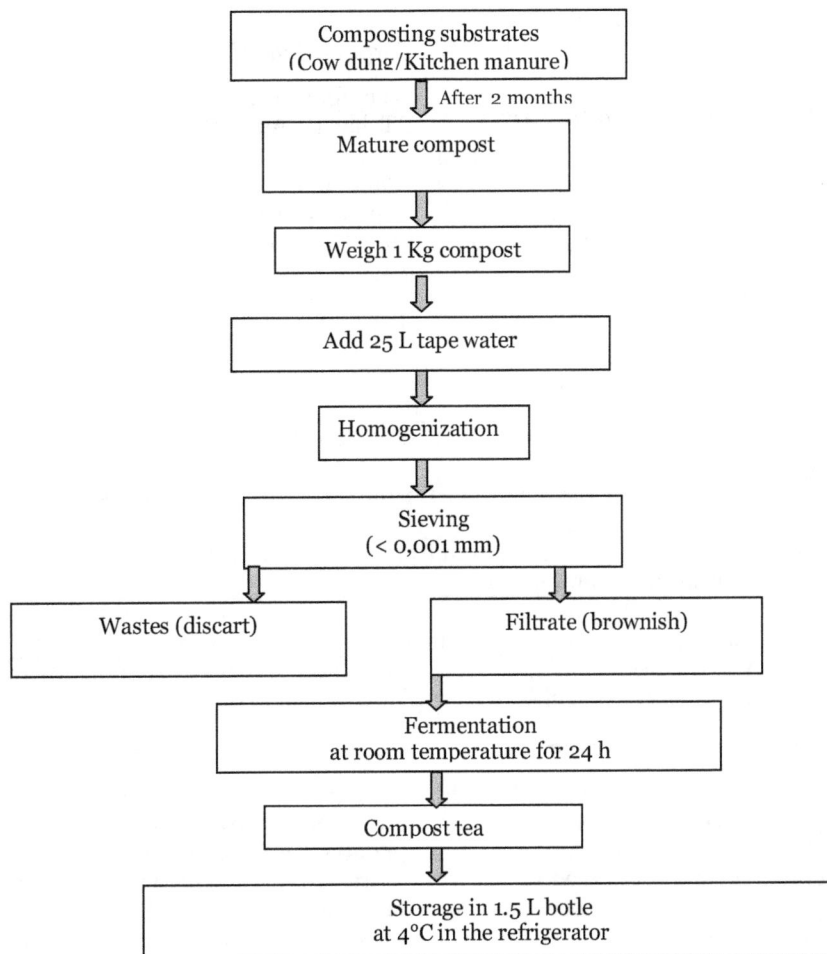

Fig. 1. Extraction process of non-aerated compost tea from a compost

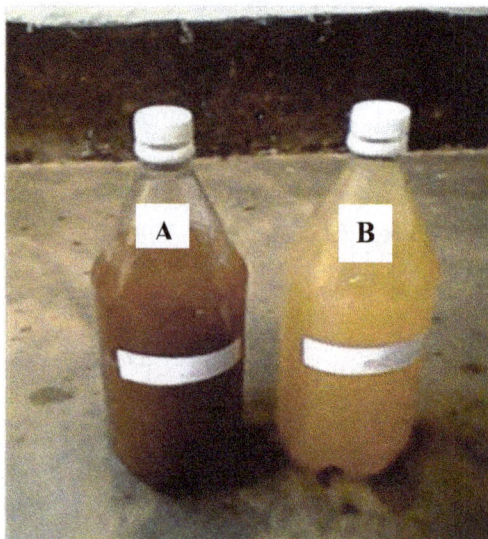

Fig. 2. Sample of compost tea from cow dung (A) and kitchen manure (B)

The fungicide used as positive control was the so called «Fongistar 72% WP», commonly used by tomatoes growers and available in local phytosanitary shops. The fungicide Fongistar 72% WP (25 g) was dissolved in 16 L tap water and applied at a rate of 1.5 L per elemental unit.

Pretreatments and rearing of plantlets

Direct sowing of tomato seeds is not often recommended, since the germination of seeds and health of plantlets is not always guaranteed. Hence, plantlets were reared through a nursery. In the nursery, three plots of 2 m² each were prepared as pretreatments:

Cd: plot covered with 2cm layer of cow dung compost,

Km: plot covered with 2 cm layer of kitchen manure (Km) compost

Tm-: negative control plot covered with 2cm layer of top soil.

Experimental design and treatments

The following 6 treatments were investigated (Table 1). An experimental unit repeated thrice for a treatment was 4.60 x 1.40 m² plot. Each plot contains 20 plants, organized in two rows, and seperated 0.40 m apart. The distance from plant to plant in each row was 1 m. Thus, the experimental field was made up of 18 plots, arranged in randomized complete block design (RCBD).

Table 1. Treatments and their significances

Tm+	Positive control derived from Tm- nursery plantlets, to which fungicide was applied at a rate of (25 ml/plantlet)
Cpi	Plots on which each plantlet from Cd nursery pre-treated plants received 50g of Cd compost at transplanting
Cpii	Plots on which each plantlet derived from Km nursery pre-treated plants received 50g of km compost at transplanting
Tci	Plots on which each plantlet derived from Cd nursery pre-treated plants received thrice Cd compost tea at a rate of 50 ml/plantlet after every two weeks as from transplanting
Tcii	Plots on which each plantlet derived from Km nursery pre-treated plants received thrice Km compost tea at a rate of 50 ml/plantlet after every two weeks as from transplanting

Chemical analysis of compost tea

Total Nitrogen: mineralization with Kjeldahl method (AFNOR, 1984), nitrogen determination according to Hantzsch reaction (Devani et al., 1989). Total Phosphorus: as described by Rodier (1978). Total potassium: atomic absorption flame photometer (Electrothermal AAS, Dortmund, Germany) after 1/100 dilution of sieved extracts and spectrophotometric lecture (Elmer, 1994).

Evaluation of growth parameters

Tomato survival rate after transplantation was evaluated per treatment at 7 and 21 days after planting (number survived/number transplanted expressed in %). The height of 10 tomato plants per plot was assessed from the soil level to the apex at flowering using a graduated ruler. The dates of 50% flowering and fructification were also determined.

Determination of disease symptoms

The disease symptoms expressed by tomatoes were assessed at 7, 21 and 35 days after transplantation based on counts of the number diseased plants per treatment and symptoms observed on plant leaves and stems. At 65 days after transplantation, the number of diseased fruits was also assessed by counts on 10 plants per plot. Main symptoms per treatment were assessed, and rated according to their frequencies on plant parts.

Yield assessment

Tomato yield was determined by counting and weighing fruits collected from 10 plants randomly selected plants per elementary unit and per treatment.

Statistical analysis

Data were statistically analyzed by Analysis of Variance (ANOVA) using the Statistical Package for the Social Science (SPSS) program, version 10.0. Means were compared between treatments using the Least Significant Difference (LSD) procedure. Correlations between parameters were assessed using the same SPSS program.

Results and Discussion

Chemical composition of tea

The compost tea was subjected to chemical analysis for the elements nitrogen (N), phosphorus (P) and potassium (K), which are the most requested for plant growth. Results analyzed by ANOVA indicate that the concentrations of N, P and K in cow dung (Cd) compost tea (Tci) were greater than those of Kitchen manure (Km) compost tea (Tcii) treatments, although no significant difference was found for potassium (Table 2). Such elevated concentrations were recently reported from anaerobic compost tea used to control gray mold of tomato with 10.9 mg/L for N, 72 mg/L for P and 421 mg/L for K (Koné et al., 2010). The differences between elemental concentrations of compost tea were previously justified by the variation of the efficacy of compost extracts with both the extraction procedure, the nature of the starting substrate, the duration of the composting process, and the maturity of compost (Brinton, 1995).

Table 2. Elemental composition (N, P, K) of cow dungs and kitchen manure derived compost teas

Compost teas	Concentration of elements (mg/L)		
	N	P	K
Tci	14.49a	26.3a	116.0a
Tcii	12.46b	23.53b	101.2a
LSD	5.03	0.11	ns

For each elemental component value of the same column affected by the same letter are not significantly different at the considered level of probability; Tci: Cow dungs compost tea; Tcii: Kitchen manures compost tea.

Effect of treatments on survival of tomato plants

Nursery plants transplanted to the experimental units were assessed for their ability to survive under different treatments (Fig. 3). Data from this figure show that at 7 days after planting (DAP), the number of plants surviving the chemical fertilizer treatment (Tm+) was slightly but significantly (p = 0.004) lower (18) than that of the other treatments (all 20 survived). Twenty-one DAP, the numbers of plants surviving the negative (Tm-) and positive (Tm+) control treatments were significantly (p = 0.034) lower compared to those of compost treatments (Cpi and Cpii). Although the numbers of plants suviving the compost tea treatments (Tci and Tcii) were higher than those of the control, it was not enough to display significant difference. The decrease in the number of control plants from the 7th to the 21st DAP would signify that plants surviving until day 7 DAP died in the following two weeks until 21 DAP, meaning that compost and compost tea contributed to increase the survival of the plants. In a related trial conducted on Solanum tuberosum, compost was reported to positively influence by 80% the development of potato plants (Ngakou et al., 2012).

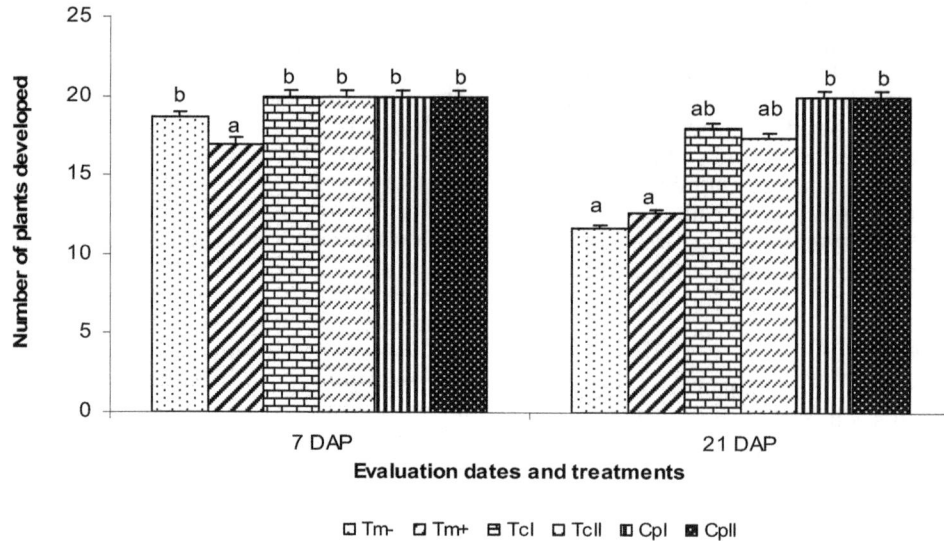

Fig. 3. Effect of compost and compost tea on the development of tomato plants

For each planting date, bars affected by the same letter are not significantly different at the p-value considered
Tm-: Negative control plot on which no compost or fungicide was provided to plantlets; Tm+: Positive control
plot on which only a fungicide was provided to plantlets at a rate of 25 ml/plantlet; Cpi: Plot on which each
plantlets received 50 g of Cd compost at transplanting; Cpii: Plot on which each plantlet received 50g of Km
compost at transplanting; Tci: Plot on which each plantlet received thrice Cd compost tea at a rate of 50
ml/plantlet after every two weeks as from transplanting; Tcii: Plot on which each plantlet received thrice Km
compost tea at a rate of 50 ml/plantlet after every two weeks.

Similarly, among the different compost types tested on tomato in the field, compost from horse dung was revealed to be able to stimulate the germination of seeds and development of plant roots (Levy and Taylor, 2003). Compost has been reported to be a best amendment for tomato plant growth (Meunchang *et al.*, 2006), whereas for other authors, compost tea from animal could be better than compost from plants as growth stimulator (Weltzien, 1991; Al-Dahmani *et al.*, 2003; Haggag and Saber, 2007).

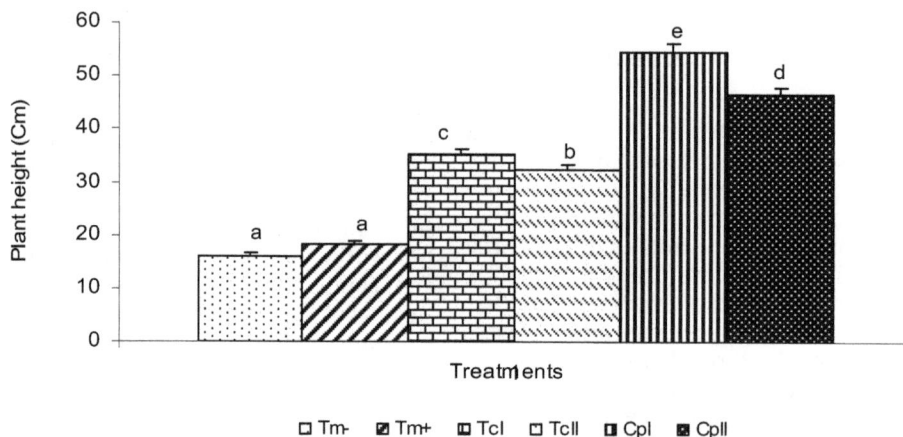

Fig. 4. Height of tomato plants at flowering as influenced by treatments

For each planting date, bars affected by the same letter are not significantly different at the p-value considered
Tm-: Negative control plot on which no compost or fungicide was provided to plantlets; Tm+: Positive control
plot on which only a fungicide was provided to plantlets at a rate of 25 ml/plantlet; Cpi: Plot on which each
plantlets received 50 g of Cd compost at transplanting; Cpii: Plot on which each plantlet received 50 g of Km
compost at transplanting; Tci: Plot on which each plantlet received thrice Cd compost tea at a rate of 50
ml/plantlet after every two weeks as from transplanting; Tcii: Plot on which each plantlet received thrice Km
compost tea at a rate of 50 ml/plantlet after every two weeks.

Changes in plant heights at flowering under different compost and compost tea receipts

As far as the height of plants is concerned (Fig. 4), the positive and negative control treatments did not have an effect on this parameter. However, composts as well as Cd and Km compost teas significantly increased the height of tomato plants, more than that of the controls, but with a more pronounced effect of compost over the compost tea treatments. Ngakou *et al.* (2012) revealed 80.5% increased of *Solanum tuberosum* height at 65 DAP as the response of spray of plants with compost tea. Compost tea applied on *Vigncola* was also reported to reduce by 60% the stem diameter lesions caused by *Plasmopara viticola*, thus allowing increase in vertical growth of plants (Larbi, 2006). This height improvement of tomato plants by compost could be explained by provision of assimilable nutrients to plants and reduction of the soil acidity through buffer effect (Santerre, 1999). Compost tea activity may reflect the quality of the substrate used, as well as the production process (Brinton, 1995). Its active components (bacteria, fungi) ensure the protection of plants against pathogens, thus favouring its development (Deschênes, 2007). These observations over the controls could be attributed to the beneficial effects of compost on soil fertility, in addition to accumulation of organic carbon (Zinati *et al.*, 2001), that is able to positively influence the plant growth (Chen *et al.*, 1994).

Variation of fruit yield as affected by different treatments

Values estimating the number of fruits revealed a significant difference between treatments (Table 3). Compost and compost tea significantly ($p < 0.0001$) contributed to increase the healthy fruits number compared to the control treatments, the best contribution accounting for Cd compost. Conversely, the negative control significantly ($p < 0.0001$) increased the number of diseased fruits/plant (56%) more than any other treatment, with the best-reduced number of diseased fruit/plant accounting for Cd derived compost tea (95.9%).

Cow dung's compost ranked first in the increment of fruits yield/plant with 26 tomato fruits/ plant, among which 23 were healthy and 3 diseased, whereas the lowest fruit yield was attributed to the negative control treatment, with a total amount of 2 fruits/plant thus 1 healthy and 1 diseased. The two compost tea increased the number of healthy fruits by 84.26% compared to the diseased fruits (15.74%). Kitchen manures compost was revealed to enhance by 95.7% the yield of *Solanum tuberosum* tuber (Ngakou *et al.*, 2012). Compost tea may act within the nearest environment of the plant roots as a protection layer against pathogens, but also as the interface where minerals are uptaken by the host plant roots (Deschênes, 2007).

Table 3. Status of fruits per plant as influenced by compost and compost tea at 65 DAP

Treatments	Healthly fruit/plant	Diseased fruit/plant	Total fruits/plant
Tm-	1.30a	0.73a	2.3a
Tm+	3.40ab	0.36a	5.76ab
Tci	6.50b	0.40a	8.90b
Tcii	5.73b	0.90a	7.63b
Cpi	23.26d	2.93b	26.20d
Cpii	10.96c	3.50b	14.46c
p-value	p < 0.0001	p < 0.0001	p < 0.0001
LSD	4.43	2.03	4.60

Values of the same colomn affected by the same letter are not significantly different at the p-value considered.

Tm-: Negative control plot on which no compost or fungicide was provided to plantlets; Tm+: Positive control plot on which only a fungicide was provided to plantlets at a rate of 25 ml/plantlet; Cpi: Plot on which each plantlets received 50 g of Cd compost at transplanting; Cpii: Plot on which each plantlet 50 g of Km compost at transplanting; Tci: Plot on which each plantlet received thrice Cd compost tea at a rate of 50 ml/plantlet after every two weeks as from transplanting; Tcii: Plot on which each plantlet received thrice Km compost tea at a rate of 50 ml/plantlet after every two weeks.

The negative and positive control plants were the most affected by fruit looses, with respectively 86.95% and 40.97% damaged fruits. In fact, plants treated with fungicide are protected against a wide range of pathogens, while the negative control plants are unable to ensure this protection. It is then obvious to count more diseased fruits on negative control plants. The application rate of compost tea is very important since it can contribute, not only to reduce pathogen attacks, but also increase crop yield (Hoikink and Grebus, 1994). A weekly application was shown to decrease the incidence of *Botrytis* on latus, leading to a substantial reduction disease severity and increase of healthy and commercial latus (McQuilken *et al.*, 1994). As far as compost treatment is concerned very few diseased fruits were observed.

During thecomposting process, a population of antagonistic microorganisms is also developed, and confers to compost the potential to protect plant against diseases or pathogen attacks (Scheuerell and Mahaffee, 2002).

Plant symptoms analysis

Symptoms as appeared on plant parts were recorded at 7, 21 and 35 DAP for each of the treatments (figure 5). Plants showing symptoms at 7 DAP from treatment Cpi, Cpii, Tci, Tcii and Tm+ were significantly (p = 0.030) lower in number than those of the negative control (Tm-), although compost was more efficient in reducing disease symptoms on plants than any other treatment.

Whereas the number of symptomatic plants amended with compost was maintained stable at 21 and 35 DAP, plants sprayed with Cd and Km compost teas were significantly (p = 0.006) reduced with time, while the number of diseased plants from control treatments were instead enhanced. These results are in agreement with findings of Znaïdi (2002), Deschênes (2007), who reported inhibition or suppression of disease symptoms by compost and compost tea based on competition between the active micro-organisms of compost tea and their plant pathogen antagonists. Compost teas have been revealed to

be natural inhibitors of plant diseases, due to beneficial micro-organisms or active chemical substances they contain (Siddiqui et al., 2008), with animal derived compost tea being more efficient than kitchen manures compost tea (Haggag and Saber, 2007). Conversely, other authors believe that compost types have no direct influence on compost tea, and thus, compost tea from both sources may act on plant pathogen with the same efficiency (Scheuerell and Mahaffee, 2006). However, the results from this study have clearly demonstrated that cow dung's derived compost tea was more effective in reducing tomato disease symptoms than kitchen manures compost tea (confer Fig. 5 at 35DAP). The mechanisms by which compost tea control plant pathogens have been reported to be through beneficial microorganisms, either by their ability to compete with plant pathogens for space and nutrients (Al-Mughrabi et al., 2008), for available food supply, the so called "general suppression or microbiostasis" (Chen et al., 1987; Boehm and Hoikink, 1992), through destruction of the pathogen by parasitism (El-Masry et al., 2002), influence on the phyllosphere (Larbi, 2006), and/or by induction of systemic resistance to plant pathogens (Zhang et al., 1998).

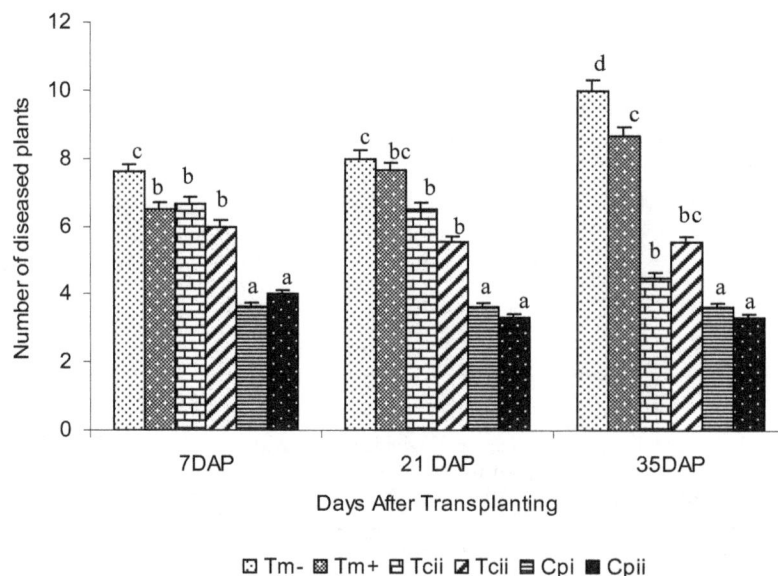

Fig. 5. Number of symptomatic plants per treatment

For each planting date, bars affected by the same letter are not significantly different at the p-value considered Tm-: Negative control plot on which no compost or fungicide was provided to plantlets; Tm+: Positive control plot on which only a fungicide was provided to plantlets at a rate of 25 ml/plantlet; Cpi: Plot on which each plantlets received 50 g of Cd compost at trans-planting; Cpii: Plot on which each plantlet received 50 g of Km compost at transplanting; Tci: Plot on which each plantlet received thrice Cd compost tea at a rate of 50 ml/plantlet after every two weeks as from transplanting; Tcii: Plot on which each plantlet received thrice Km compost tea at a rate of 50 ml/plantlet after every two weeks.

Links between observed symptoms and diseases

This was done by observations and counting of diseased plants by determining the individual symptoms appearing on plant leaves, stem, fruits and collar for each treatment at 7 and 21 DAP.

Symptoms identified were foliar and stem spots, apical and total shedding of leaves, stunted of plants, spoilage at the bottom of fruits, black spots on fruits, fissure on fruits, apical shedding, yellowish of leaves, and stem spots/necrosis (Fig. 6).

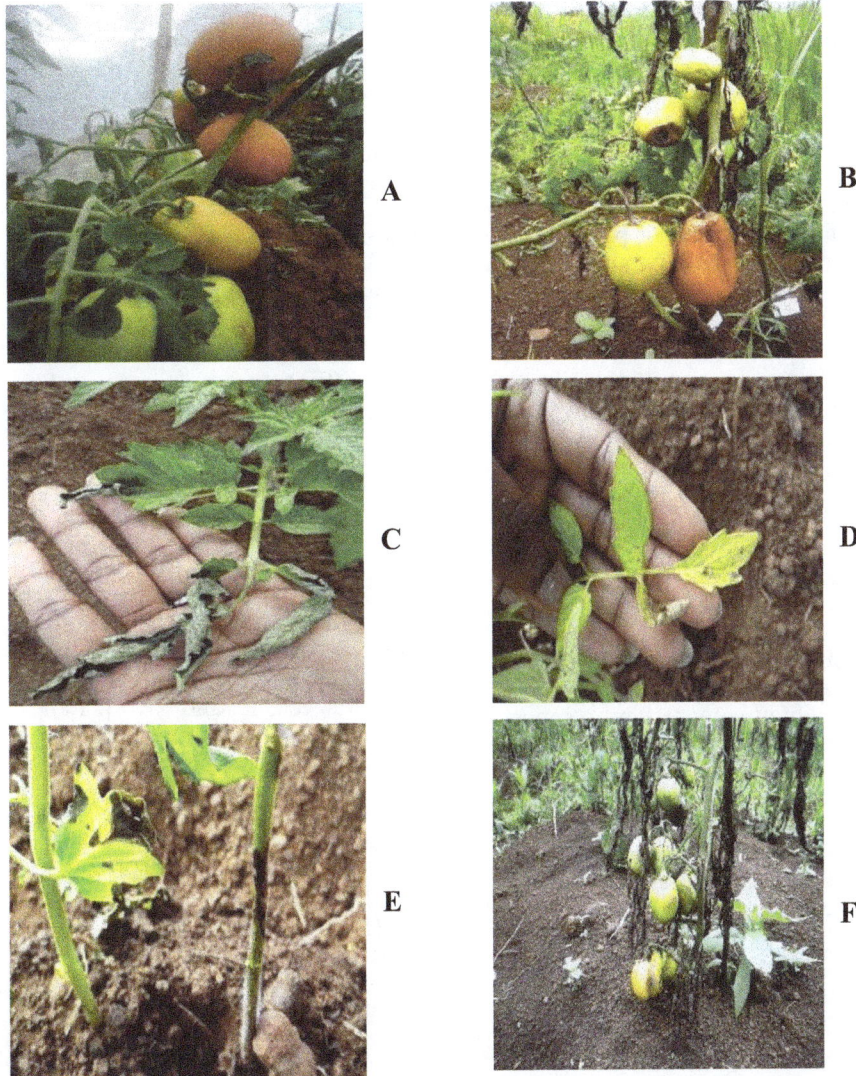

Fig. 6. Status of some disease symptoms compared to a healthy plant

Healthly plant and fruits (A); Diseased plant and fruits (B): Apical leaves shadding (C); leaves yellowish (D); Stem spot or necrosis (E); Total shadding of plant (F).

Table 4. Variation of main symptoms as influenced by treatments

Observed symptoms	Treatments					
	Tm-	Tm+	Tci	Tcii	Cpi	Cpii
Stem and foliar spots	+ + + +	+ + +	+ +	+ +	+	+
shadding	+ + + + +	+ + +	+ +	+ +	+	+
stunted	+ + + + +	+ + +		+		
Spoliage at the bottom of fruits	+ + + + +	+ +	+	+	+	+
Spots on fruits	+ + + + +	+ +	+ +	+ +	+	+
Yellowish of leaves	+ + + +	+ + +	+ +	+ +	+	+
Total spoilage of fruits	+ + + +	+ +	+	+	+	+

[0-2] diseased plants or fruits/plant: +;   [2-4] diseased plants or fruits/plant: ++; [4-6] diseased plants or fruits/plant: +++; [4-8] diseased plants or fruits/plant: ++++ ; [8-10] diseased plants or fruits/plant: +++++

All treatments presented all the symptoms, but symptoms such as shedding, black spots on fruits, yellowish of leaves, spoilage of fruits were less observed on Tci, Tcii, Cpi et Cpii treated plants than on Tm- and Tm+ plants. Stunted of plants was completely inhibited by Cpi and Cpii treatments (Table 4), confirming the findings of Elad and Shtienberg (1994) who postulated that the type of the compost does not directly affect the efficacy of the tea, and hence, both manure-based composts and other compost types can provide an efficient tea to control plant pathogens. A positive and significant correlation was found between the number of plant leaves and height ($r = 0.79$; $p < 0.0001$) on one hand, and the number of flowers ($r = 0.61$; $p < 0.0001$) on the other, implying that these parameters are closely related.

Conclusion

From this study, it is established that cow dung compost and compost tea are more effective and efficient in reducing diseases incidence on tomato plants than kitchen manure compost and compost tea. Hence, this is accomplished through stimulation of growth, reduction of disease symptoms, and suppression of foliar pathogens. They act by both their chemical composition and microorganisms and could therefore be used at appropriate proportions as biological fertilizers/pathogen antagonists. This result complies with those reported in a previous study on suppressive activity of compost and compost tea on *Solanum tuberosum* in the same agro-ecological zone. Since a biological control system requires a better knowledge of the inhibitive agent action mechanism, further researches will be undertaken to identify all the microorganisms involved thereof, and determine their suppressiveness mechanisms.

Acknowledgement

The authors wish to thank the Master students (2011 promotion) who have contributed to the production of cow dung and kitchen manure compost.

References

AFNOR. 1984. Animal nutrients. Determination of calcium by atomic absorption spectrophotometry method. French Norm, NF, Afnor, Paris, France, V18-108.

AGRI-STAT. 2010. *Annuaire des statistiques du secteur agricole campagne 2007 et 2008.* MINADER/DESA AGRI-STAT N°15, 111p.

Al-Dahmani, J.H., Abassi, P.A., Miller, S.A. and Hoitink, H.A.J. 2003. Suppression of bacterial spot of tomato with foliar sprays of compost extracts under greenhouse and field conditions. *Plant Disease* 87: 913–919.

Al-Mughrabi, K.I., Bertheleme, C., Livingston, T., Burgoyne, A., Poirier, R. and Vikram, A. 2008. Aerobic compost tea, compost and a combination of both reduce the severity of common scab (*Streptomyces scabiei*) on potato tubers. *J. Plant Sci.* 3: 168–175.

Anonymous, 2008. Tomato: when tomato is served. Ministry of Agriculture, Yaounde, Cameroon. 28p.

Barbier, B. and Catin, M. 1994. *Promotion of sustainable agricultural systems in pays in Sudano-Sahelian African countries,* FAOCTA CIRAD, Dakar, Senegal, 31p.

Boehm, M.J. and Hoikink, H.A.J. 1992. Sustenance of microbial activity in potting mixes and its impact on severity of *Pythium* root rot on pointsettia. *Phytopathol.* 82: 259-263.

Bramley, P.M. 2000. Is lycopene beneficial to human health? *Phytochem.* 54: 233–236.

Brinton, W. 1995. The control of plant pathogenic fungi by use of compost teas. *Biodynamics* 197: 12-15.

Chalabi, N., Le Corre, L., Mauizis, J.C., Bignon, Y.J. and Bernard-Gallon, D. 2004. The effects of lycopene on the proliferation of human breast cells and BRCA1 and BRCA2 gene expression. *European J. Cancer* 40: 1768–1775.

Chen, W., Hoikink, H.A.J. and Schmitthenner, A.F. 1987. Factors affecting suppression of *Pythium* damping-off in container media amended with composts. *Phytopathol.* 77: 775-760.

Chen, Y., Maggen, H. and Riv. J. 1994. Humics substances originating from rapidly decomposing organic matter properties and effects on plants growth. In: Senesi, N., Miano, T.M. (eds.). Humic substances in the global environment and implications on human health, Elsevier Science, Amsterdam, The Netherlands. pp. 427-445.

CMSN, (Meteo Secondary Centre Ngaoundere), 2011. Climatologic abstract. N°05/2011/MET/ FKKNYMYX, Meteo Station Ngaoundere.

Deschênes, A. 2007. Compost tea for vigorous and health cultures. The pelerin garden, St Andre Kamouraska. 5p.

Devani, M.B., Shioshoo, C.J., Shal, S.A. and Suhagia, B.N. 1989. Microchemical methods. Spectrophotometrical method for microdetermination of nitrogen in Kjeldahl digest. *J. Ass. Off. Anal. Chem.* 72 (6): 953-956.

Elad, Y. and Shtienberg, D. 1994. Effect of compost water extracts on grey mould (*Botrytis cinerea*). *Crop Prot.* 13: 109–114.

El-Masry, M.H., Khalil A.I., Hassouna, M.S. and Ibrahim, H.A.H. 2002. *In situ* and *in vitro* suppressive effect of agricultural composts and their water extracts on some phytopathogenic fungi. *World J. Microbiol. Biotechnol.* 18: 551–558.

Elmer, F.J. 1994. Nonlinear Dynamics of Atomic Force Microscopes. *Helv Phys. Acta* 67: 213.

Haggag, W.M. and Saber, M.S.M. 2007. Suppression of early blight on tomato and purple blight on onion by foliar sprays of aerated and non-aerated compost teas. *J. Food Agric. Env.* 5: 302–309.

Hoikink, H.A.J. and Grebus, M.E. 1994. Status of biological control of plant diseases with compost. *Compost Sci. Util.* 2 (2): 6-12.

Janes, H.W. 1994. Tomato production under protected cultivation. *Encyclopedia Agric. Sci.* 4: 337–349.

Koné, S.B., Dionne, A., Tweddell, R.J., Antoun, H. and Avis, T.J. 2010. Suppressive effect of non-aerated compost teas on foliar fungal pathogens of tomato. *Biol. Control* 52: 167-173.

Lachance, P. and Rouleau, D. 2004. Growth without herbicide: factors of success. Lavoisier. (ed.). Paris, France. 125p.

Larbi, M. 2006. Influence of compost quality and their extracts on plant protection against fungal diseases. PhD Thesis. Faculty of Science, University of Neuchâtel Institute Botany, Switzerland. 137p.

Levy, J.S. and Taylor, B.R. 2003. Effects of pulp mill solids and three composts on early growth of tomatoes. *Bioresour. Tech.* 89 (3): 297-305.

Litterick, A.M., Harrier, L., Wallace, P., Watson, C.A. and Wood, M. 2004. The role of uncomposted materials, composts, manures and compost extracts in reducing pest and disease incidence and severity in sustainable temperate agricultural and horticultural crop production – a review. *Critical Rev. Plant Sci.* 23: 453–479.

McQuilken, M.P., Whipps, J.M. and Lynch, J.M. 1994. Effects of water extracts of a composted manure-straw mixture on the plant pathogen *Botrytis cinerea*. *World J. Microbiol. Biotechnol.* 10: 20-26.

Meunchang, S., Panichsakpatana, S. and Weaver, R.W. 2006. Inoculation of sugar mill byproducts compost with N-2-fixing bacteria. *Plant and soil* 280 (1-2): 171-176.

Ngakou, A., Koundou, N. and Koehler, H. 2012. The relative effects of compost and non-aerated compost tea in reducing disease symptoms and improving tuberization of *Solanum tuberosum* in the field. *Int. J. Agric. Res. Rev.* 2 (4): 504-512.

Ngakou, A., Megueni, C., Noubissié, E. and Tchuenteu, T.L. 2008. Evaluation of the physico-chimical properties of cattle kitchen manures-derived compost and their effect on field grown *Phaseolus vulgaris*. *Int. J. Suitain. Crop Prod.* 3 (5): 13-32.

Rodier, J. 1978. Analysis of water: chemistry, physico-chemistry, bacteriology, biology. 6th edition. Dunod Technic, Paris, France. 1136 p.

Santerre, C. 1999. *Advantages of composting.* Ministry of environment of the new Brunswick. 60p.

Scheuerell, S.J. and Mahaffee, W.F. 2002. Compost tea: principles and prospects for plant disease control. *Compost Sci. Util.* 10: 313–338.

Scheuerell, S.J., Mahaffee, W.F. 2006. Variability associated with suppression of gray mold (*Botrytis cinerea*) on geranium by foliar applications of non-aerated compost teas. *Plant Disease* 90: 1201–1208.

Siddiqui, Y., Meon, S., Ismail, M.R. and Ali, A. 2008. *Trichoderma*-fortified compost extracts for the control of *choanephora* wet rot in okra production. *Crop Prot.* 27: 385–390.

Sonchieu, J. 2002. Study of the Manebe, carotene, and vitamin C contents in tomato at Dang-Ngaoundere. Master thesis, Department of food and Nutrition, ENSAI. University of Ngaoundere, Cameroon. 52 p.

Van der Vossen, H.A.M., Non-Wondim, R. and Messian, C.M. 2004. *Lycopersicon esculentum* Mill. pp. 373–379. In: *Plant Resources of Tropical Africa 2.* Vegetables (Grubben, G.J.H. and Denton, O.A. (eds). PROTA Foundation Backhuys Publishers, Netherlands. 667p.

Welke, S.E. 2005. The effect of compost extract on the yield of strawberries and the severity of *Botrytis cinerea*. *J. Sustain. Agric.* 25: 57–68.

Weltzien, H.C. 1991. Biocontrol of foliar fungal diseases with compost extracts. pp. 430-450. In: Andrews, J.H. and Hirano, S.S. (eds.), *Microbial Ecology of Leaves.* Springer-Verlag, New York.

Zhang, W.D., Han, Y., Dick, W.A., Davis, K.R. and Hoitink, H.A.J. 1998. Compost and compost water extract-induced systemic acquired resistance in cucumber and *Arabidopsis*. *Phytopathol.* 88: 450–455.

Zinati, G.M., Li, Y.C. and Bryan, H.H. 2001. Accumulation and distribution of copper, iron, manganese and zinc in calcareous soil amended with compost. *J. Env. Sci.Health* 36 (2): 229-243.

Znaïdi, A. 2002. Study on the evaluation of composting differents types of organic matters and the effect of biological compost tea on plant diseases. *Master thesis*, Degree N°286. Mediterranean Organic Agriculture, CIHEAM Mediterranean Agronomic Institute, Tunisia. 94p.

11

GENETIC DIVERSITY OF SOME CHILI (*Capsicum annuum* L.) GENOTYPES

M.J. Hasan[1], M.U. Kulsum[2], M.Z. Ullah[3], M. Manzur Hossain[4] and M. Eleyash Mahmud[5]

Abstract

A study on genetic diversity was conducted with 54 Chili (*Capsicum annuum* L.) genotypes through Mohalanobis's D^2 and principal component analysis for twelve quantitative characters viz. plant height, number of secondary branch/plant, canopy breadth , days to first flowering, days to 50% flowering, fruits/plant, 5 fruits weight, fruit length, fruit diameter, seeds/fruit, 1000 seed weight and yield/plant were taken into consideration. Cluster analysis was used for grouping of 54 chili genotypes and the genotypes were fallen into seven clusters. Cluster II had maximum (13) and cluster III had the minimum number (1) of genotypes. The highest inter-cluster distance was observed between cluster I and III and the lowest between cluster II and VII. The characters yield/plant, canopy breadth, secondary branches/plant, plant height and seeds/fruit contributed most for divergence in the studied genotypes. Considering group distance, mean performance and variability the inter genotypic crosses between cluster I and cluster III, cluster III and cluster VI, cluster II and cluster III and cluster III and cluster VII may be suggested to use for future hybridization program.

Keywords: Chili, *Capsicum annuum*, Cluster and Genetic diversity

[1]Principal Scientific Officer, Hybrid Rice Division, Bangladesh Rice Research Institute, Joydebpur, Gazipur, Bangladesh
[2]Scientific Officer, Hybrid Rice Division, Bangladesh Rice Research Institute, Joydebpur, Gazipur, Bangladesh
[3]Field Coordinator (Vegetable), AVRDC, Bangladesh
[4]Assistant Information Officer (Crop Production), Agriculture Information Services, Khamarbari, Farmgate, Dhaka-1215
[5]PhD fellow BSMRAU and Plant Breeder, Energypac Agro Ltd. Monipur, Gazipur, Bangladesh

*Corresponding author's email: jamilbrri@yahoo.com (M.J. Hasan)

Introduction

Chili (*Capsicum annuum* L.) is grown worldwide both as a spice and as a vegetable crop and world's second most important solanaceous vegetable after tomato. Chili is an important commercial crop of Bangladesh and grown for its green fruits as vegetable and in ripe dried fruits as spice and throughout the world because of its pungency and pleasant flavors. However, consumption in small amount enriches our diet as and considered good sources of minerals, vitamins and other food components. Almost all the varieties of low and medium pungency cultivated on a field scale in Bangladesh are belonged to *Capsicum annuum*. A number of cultivars are grown in Bangladesh differing in habit, yield and consumer's preference and in size, shape, color and pungency of the fruit. Chili is grown practically all over Bangladesh in Rabi and Kharif season and area under cultivation is 259383 acres. Most of the varieties cultivated in Rabi season and production is about 176134 metric ton but the yield is only 1.68 metric ton per hectare (BBS, 2011). The lack of improved genotypes is the main constraint to low yield. Assessment of different desirable traits spread over diverse genotypes is important to rapid

advance in yield improvement of any crop. The importance of genetic diversity in the improvement of a crop has been studied in both self and cross-pollinated crop (Gaur *et al.,* 1978). The plant breeders are always interested to know the genetic divergence among the varieties available due to reasons that crosses between genetically diverse parents are likely to produce high heterotic effect (Ramanujam *et al.,* 1974) and crosses involving distantly related parents within the same species produce wide spectrum of variability. A logical way to start any breeding program is to collect precise information on the nature and degree of genetic divergence that would help the plant breeder in choosing the right type of parents for purposeful hybridization in heterosis breeding (Patel *et al.,* 1989). Moreover, evaluation of genetic diversity is important to know the source of genes for a particular trait within the available germplasm (Tomooka, 1991). Genetic divergence is a basic requirement for effective selection within the existing population or population arising out of hybridization. More diverse the parents within a reasonable range, better are the chances of improving economic characters under consideration in the offspring. Mohalanobis's D^2 statistic of multivariate analysis is recognized as a powerful tool in quantifying the degree of genetic divergence among the populations. Therefore, the present study was undertaken to assess the genetic diversity in 54 genotypes of chili and to identify suitable donors for a successful breeding program in this crop. Farhad et al. (2010) and Dutonde et al. (2008) have also been conducted similar studies in chili.

Materials and Methods

The experiment was conducted at Research and Development Farm of Energypac Agro Ltd., during Rabi season of 2011-12. The experimental farm is located at Monipur, about 20 km away from Gazipur chowrasta having 24.00° N latitude and 90.25° E longitude with an elevation of 8.4 meter from the sea level. A total 54 genotypes of chili (*Capsicum annuum* L.), collected from Bangabandhu Sheikh Mujibur Rahman Agricultural University (BSMRAU), local and exotic sources were included in this study. The experiment was laid out in Randomized Block Design with three replications having plot size of 4.0 sq m providing a spacing of 60 × 40 cm in 1 m wide bed. Manures and fertilizers were applied as per recommended dose. Seeds were sown on November, 2011 in separate plots. The seedling emerged 9-12 days after sowing. Thirty five days old seedlings were transplanted in the experimental plot. Weeding was done as and when necessary to keep the crop free from weeds. Irrigation was given to the plants when necessary. Data on plant height (cm), number of secondary branch/plant, canopy breadth (cm), days to first flowering, days to 50% flowering, fruits/plant, 5 fruits weight (g), fruit length (cm), fruit diameter (cm), seeds/fruit, 1000 seed weight (g) and yield/plant (g) were recorded on individual plant basis from the 10 plants selected at random per plot. Mean data of each character was subjected to multivariate analysis viz. Principal Coordinate Analysis (PCO) Principal Component Analysis (PCA), Cluster Analysis and Canonical Variate Analysis using GENSTAT5 (Digby et al., 1989; Jager et al., 1983) program in computer.

Results and Discussion

The computations from distance matrix gave non-hierarchical clustering among 54 chili genotypes and grouped them into seven clusters (Table 1). Cluster II contained the highest number of chili genotypes (thirteen), followed by cluster I constituted by ten chili genotypes. Cluster III was composed of single genotype BSMRAU Sel-7 indicated that this genotype is totally different from other genotypes used in this study. Cluster IV, VI and VII constituted of nine genotypes each. Cluster V comprises of three chili genotypes.

Karad et al. (2002) reported eight clusters with 40 genotypes. Manju and Sreelathakumary (2004) reported six clusters with 32 accessions. Senapati et al. (2003) reported six clusters in 20 diverse genotypes of chili and Amarul Junior et al. (2005) reported eight distinct grouped in 50 accessions of chili.

Table1. Distribution of 54 Chili genotypes in different clusters grown in Rabi season of 2011-12

Cluster no.	No. of Genotypes	Number of population	Chili genotypes
I	3, 10, 21, 33, 39, 41, 42, 50, 51, 54	10	Premium LT, BSMRAU Sel-1, Bindu, BSMRAU Sel-20, BSMRAU Sel-26, BSMRAU Sel-28, BSMRAU Sel-29, Fire Cracker, PP-7, PP-10
II	5, 6, 7, 8, 15, 17, 20, 25, 35, 43, 47, 48, 52	13	Premium short, Sonic, RK-219, Hot Pepper 86235, BSMRAU Sel-6, BSMRAU Sel-8, Sonic Super, BSMRAU Sel-13, BSMRAU Sel-22, BSMRAU Sel-30, Fire Volcano, , HPCT-1946, PP-8
III	16	1	BSMRAU Sel-7
IV	19, 22, 26, 29, 31,32, 37, 40, 46	9	OP-collection, BSMRAU Sel-10, BSMRAU Sel-14, Bindu-2, BSMRAU Sel-18, BSMRAU Sel-19, BSMRAU Sel-24, BSMRAU Sel-27, Bindu-3
V	2, 27, 45	3	HPCT-1947, BSMRAU Sel-15, BSMRAU Sel-32
VI	1, 4, 9, 23, 34, 38, 44, 49, 53	9	PP 4651, Premium long, HP86235, BSMRAU Sel-11, BSMRAU Sel-21, BSMRAU Sel-25, BSMRAU Sel-31, RK-221, PP-9
VII	11, 12, 13, 14, 18, 24, 28, 30, 36	9	BSMRAU Sel-2, BSMRAU Sel-3, BSMRAU Sel-4, BSMRAU Sel-5, BSMRAU Sel-9, BSMRAU Sel-12, BSMRAU Sel-16, BSMRAU Sel-17, BSMRAU Sel-23

Intra (bold) and inter cluster distances shown in Table 2. The inter-cluster distances were larger than the intra cluster distances. The inter-cluster distance was maximum between clusters I and III (28.820) indicating wide genetic diversity between these two clusters followed by the distance between cluster III and VI (24.738), cluster II and cluster III (21.415) and cluster III and cluster VII (18.351). Genotypes from these four clusters if involve in hybridization may occur a wide spectrum of segregating population as genetic diversity is very distinct among the groups. The selection of diverge genotype from cluster would produce a broad spectrum of variability for morphological and quality traits studied which may enable further selection and improvement.

Table 2. Intra (bold) and inter cluster distances (D²) for 54 chili genotypes

Clusters	I	II	III	IV	V	VI	VII
I	**1.056**	7.917	28.820	16.541	13.828	4.492	10.975
II		**0.837**	21.415	8.875	7.282	3.830	3.760
III			**0.000**	13.257	17.936	24.738	18.351
IV				**0.485**	6.287	12.364	6.011
V					**0.827**	9.758	5.800
VI						**0.929**	6.779
VII							**0.622**

The minimum inter-cluster distance was observed between cluster II and cluster VII (3.760) followed by cluster II and cluster VI (3.830) and cluster I and cluster VI (4.492) indicating that the genotypes of these clusters were genetically close. Cluster mean value of 12 different characters shown in Table 3. Difference in cluster means existed for almost all the characters studied. Highest mean value for plant height (cm), number of secondary branch/plant, canopy breadth (cm), fruits/plant, 5 fruits weight (g), fruit length (cm), seeds/fruit and yield/plant (g)was observed in cluster III that means the genotype fallen in cluster III having the genetic potentiality to contribute better for yield maximization of chili genotypes. Cluster V possessed genotypes with maximum number of fruits coupled with dwarf plant stature and earliness indicating selection of genotypes from these cluster for future chili breeding program have positive impact for short plant type, earliness and number of fruits. Cluster I had the genotypes that showed lowest mean value for almost all the characters studied indicating selection of genotypes from these cluster for future chili breeding program have no positive impact except for dwarfness.

Table 3. Cluster mean values of 12 different characters of 54 Chili genotypes

Characters	I	II	III	IV	V	VI	VII
Plant height (cm)	77.8	87	100	87.8	85	85.8	85.1
Secondary branches/plant	7	9.8	12	9.6	8.7	8.1	7
Canopy breadth (cm)	51.2	65.3	100	63.1	80	62.8	65.7
Days to first flowering	83.1	85.3	75	81.6	74.7	86.1	80.7
Days to 50% flowering	90.2	93.3	83	89.9	82.7	93.7	88.4
Fruits/plant	105.7	217.1	404	382	538	196.4	264.4
5 fruits weight (g)	16.4	20.8	25	19.8	10.7	16.6	20
Fruit length (cm)	5.6	6.6	7.6	6.4	6.4	7.1	7.1
Fruit diameter (cm)	1.1	1.4	1.3	1.3	0.8	1.1	1.3
Seeds/fruit	49.8	53.8	82	55.2	49.3	50.9	53.4
1000 seed weight (g)	3.8	3.8	3.7	3.8	3.8	3.8	3.8
Yield/plant (g)	334.1	829	2210	1413.6	1125.7	595.6	1023.6

Relative contribution towards divergence presented in Table 4. Vector-1 and Vector-2 values were obtained from principal component analysis. In first axis Vector-1, all the studied characters had positive impact towards divergence except fruits/plant. In Vcetor-1 the important characters responsible for divergence in the major axis of differentiation were 5 fruits weight (0.4587), plant height (0.3737), fruit diameter (0.3476) and seeds/fruit (0.3437). In vector-2, fruits/plant (0.4653), yield/plant (0.4255) canopy breadth (0.3122), secondary branches/plant (0.2277) and plant height (0.1315) had better positive impact towards divergence. The character that showed positive value in both Vectors contributed most towards divergence. That means yield/plant, canopy breadth, secondary branches/plant, plant height and seeds/fruit contributed most for divergence in the studied genotypes. Present findings are accordance with the findings of Mubarak Begum (2002) and Prabhudeva (2003).

Table 4. Relative contributions of the twelve characters of 54 Chili genotypes to the total divergence

Characters	Vector-1	Vector-2
Plant height (cm)	0.3737	0.1315
Secondary branches/plant	0.3007	0.2277
Canopy breadth (cm)	0.2645	0.3122
Days to first flowering	0.2233	-0.4275
Days to 50% flowering	0.2341	-0.4085
Fruits/plant	-0.0632	0.4653
5 fruits weight (g)	0.4587	-0.1369
Fruit length (cm)	0.2472	0.0545
Fruit diameter (cm)	0.3476	-0.0788
Seeds/fruit	0.3437	0.0858
1000 seed weight (g)	0.1528	-0.2258
Yield/plant (g)	0.2398	0.4255

Genotypically distant parents are able to give high heterosis. Therefore, considering group distance, mean performance and variability the inter genotypic crosses between cluster I and cluster III, cluster III and cluster VI, cluster II and cluster III and cluster III and cluster VII may be suggested to use for future hybridization program. Superior genotype BSMRAU Sel-7 fallen in cluster III had shown best results on canopy breadth, secondary branches/plant, fruit weight, fruit length, seeds/fruit and yield/plant. Hence, these characters should be given prime importance for further improvement in yield in further breeding program

References

Amarul Junior, A.T., Rodrigues, R., Sudre, C.P., Riva, E.M. and Karasawa, M. 2005. Genetic divergence between 'chilli' and sweet pepper accessions using multivariate techniques. *Horticultura Brasileira* 23 (1): 22-27.

BBS. 2011. Year book of Agricultural Statistics of Bangladesh. Planning Division, Ministry of Planning, Govt. of the People's Republic of Bangladesh, Dhaka. pp. 23-27.

Digby, P.N., Galway and Lane, P. 1989. Genstat 5: A second course. Oxford Sci. Publication, Oxford. pp. 103-108.

Dutonde, S.N., Bhalekar, M.N., Patil, B.T., Kshirsagar, D.B. and. Dhumal, S.S. 2008. Genetic diversity in chili (*Capsicum annuum* L.). *Agric. Sci. Digest* 28 (1): 45-47.

Farhad, M.I., Hasanuzzaman, M., Biswas, B.K., Arifuzzaman, M. and Islam, M.M. 2010. Genetic divergence in chilli (*Capsicum annuum* L.). *Bangladesh Res. Publ. J.* 3 (3): 1045-1051.

Gaur, P.C., Gupta, P.K. and Kishore, H. 1978. Studies on genetic divergence in potato. *Euphytica* 27: 361-368.

Jager, M.I.D., Garethojones, I.D. and Griffith, E. 1983. Component of partial resistance of wheat seedlings to *Septoria nodorum*. *Euphytica* 32: 575-584.

Karad, S.R., Raikar, G.R. and Navale, P.A. 2002. Genetic divergence in chilli. *J. Maharashtra Agril. Univ.* 27 (2): 143-145.

Manju, P.R. and Sreelathakumary, I. 2004. Genetic divergence in hot chilli (*Capsicum chinense* Jacq.). *Capsicum and Eggplant Newsl.* pp. 69-72.

Mubarak Begum, S. 2002. Evaluation of chilli germplasm for productivity, its component traits and resistance to some biotic stresses. M.Sc. (Agri.) Thesis, University of Agricultural Sciences, Dharwad. p. 29.

Patel, M.Z., Reddi, M.V., Rana, B.S. and Reddy, B.J. 1989. Genetic divergence in safflower (*Carthamus tinctorius* L.). *Indian J. Genet.* 49 (1): 113-118.

Prabhudeva, S.A. 2003. Variability genetic diversity and heterosis study in chilli (*Capsicum annuum* L.). M.Sc. Thesis, University of Agricultural Sciences, Dharwad. pp. 31-32.

Ramanujam, S., Tiwary, A.S. and Mehra, R.B. 1974. Genetic divergence and hybrid performance in mungbean. *Theor. Appl. Genet.* 44 (5): 211-214.

Senapati, B.K., Sahu, F.K. and Sarkar, G. 2003. Genetic divergence in chilli. *Crop Res. Hisar.* 26 (2): 314-317.

Tomooka, N. 1991. Genetic diversity and landrace differentiation of mungbean, (*Vigna radiata* L.) Wilczek, and evaluation of its wild relatives (The subgenus *Ceratotropics*) as breeding materials. Tech. Bull. Trop. Res. Centre, Japan No. 28. Ministry of Agriculture, Forestry and Fisheries. Japan. p. 1.

12

KNOWLEDGE AND PERCEPTION OF EXTENSION WORKERS TOWARDS ICT UTILIZATION IN AGRICULTURAL EXTENSION SERVICE DELIVERY IN GAZIPUR DISTRICT OF BANGLADESH

F.A. Prodhan[1*] and M.S.I. Afrad[2]

Abstract

The primary purpose of the study was to assess the extent of knowledge and perception of extension workers towards ICT utilization and to determine the relationship between the selected characteristics of the respondents and knowledge and perception of extension workers towards ICT utilization in extension service delivery. The study was conducted in Gazipur district and comprised proportionate random sample of 90 extension workers from five upazila of Gazipur district. A pre-tested interview schedule was used to collect data from the respondents. To measure the knowledge on ICT utilization 35 statements were selected regarding 7 ICT with five possible answer of each tools and a score of one was given to the right answer and zero to the wrong answer alternatively to measure the perception of the respondents rated each of 10 statements ICT utilization in agriculture on a 5-point Likert type scale and the total of these ratings formed perception index. The result of the study showed that out of seven ICT tools the knowledge of extension workers was highest in case of MS Word this was followed by internet/ web service and the lowest knowledge was found in case of Geographical Information System. It is observed that an overwhelming majority (88.9%) of agricultural extension workers in the study area had low to medium knowledge towards ICT utilization. Findings reveal that the respondents had top most perception on the ICT utilization in respect of 'Extension work can be greatly enhanced by ICT' followed by on 'The benefits of ICT use outweigh the financial burden involved'. The result also indicated that more than fourth-fifth (84.4%) of the respondents had medium to high perception towards ICT utilization. There were significant relationship between service experience and use of the information sources of the respondents with their knowledge towards ICT utilization conversely innovativeness, cosmopoliteness and job satisfaction of the respondents showed positive significant relationship with their perception towards ICT utilization.

Keywords: Knowledge, Perception, ICT and Extension Workers

[1]Lecturer, Department of Agricultural Extension and Rural Development, Bangabandhu Sheikh Mujibur Rahman agricultural University, Gazipur-1706, Bangladesh
[2]Professor, Department of Agricultural Extension and Rural Development, Bangabandhu Sheikh Mujibur Rahman agricultural University, Gazipur-1706, Bangladesh

*Corresponding author's email: foyez_bsmrau @yahoo.com (F.A. Prodhan)

Introduction

Agriculture is a major economic activity in Bangladesh. It currently employs about 50 per cent of the country's labor force and contributes about 20 per cent of country's gross domestic product (GDP) (World Bank, 2008). In Bangladesh, about 70 per cent of poor people live in the rural areas, and these poor people are concentrated in the agricultural sector (Raihan, 2012). The role of agricultural extension service has traditionally been to provide the important link between agricultural research and farming communities, especially for technology transfer in support of agricultural and rural development (FAO, 2004). Agricultural extension service over the years has been working via different approaches, methodologies and program to ensure farmers adoption of improved technologies with little success (Aker, 2010). Agricultural extension program can provide much-needed help in the form of practical field-advice, innovations from scientists and practitioners, and sound commodity-marketing principles. Yet the enormous demand for smallholder training can never be met through conventional extension methods. Arokoyo (2005) reported that a strong agricultural extension linkage complimented by flawless information flow enhanced by the effective use of information and communication technologies (ICT) would significantly boost agricultural production and improve rural livelihoods in developing countries. Agriculture is one of the vital sectors in which ICT can be used reasonably in transferring the modern agricultural technologies to the farmers. ICT has many potential applications in agricultural extension most especially in accessing required information and knowledge (McNamara, 2009). Information and communication technology in agriculture includes internet, radio/community radio, television, wireless communication tools, cell phone, audio visuals, digital camera, Geographic Information System (GIS), Global Positioning System (GPS) and other technologies which direct the agricultural activities towards precision agriculture (Mirzaei, 2003). ICT can bring new information services to rural areas where farmers, users, will have much greater control. It follows from the foregoing that ICT are information transmission technology built on the potential of electronic communication devices such as computers and telecommunication equipment, for connecting and accessing various ends in the information pathway (Aboh, 2008). ICT as an extension tool could enhance the flow of information in the application of agricultural extension services. Agricultural extension, which depends largely on information exchange between and among farmers, has been identified as one area in which ICT can have a particularly significant impact (Ballantyne and Bokre, 2003). The introduction of various relevant ICTs in agricultural information dissemination could help farmers' access market information, land resources and services, management of pest and diseases, rural development program and help in broadening the orientation of farmers in production activities thereby causing a major turnaround in the agricultural sector as it is doing in many other sectors (Meera *et al.*, 2004). If extension agents are crucial to agricultural and rural development and use of various ICTs promises boost for agricultural and rural development, then a study on the knowledge and perception of extension agents on ICT utilization in extension service delivery is important. Considering this in view, the present study was undertaken with the following objectives:

1) To identify the socio-demographic characteristics of the extension workers;
2) To assess the extent of knowledge and perception of extension workers towards ICT utilization in extension service delivery; and
3) To determine the relationship between the selected characteristics of the respondents and knowledge and perception of extension workers towards ICT utilization in extension service delivery.

Methodology

The study was conducted in Gazipur district, which consist of five upazilas viz. Kapasia, Sreepur, Kaliakoir, Kaligonj and Gazipur Sadar upazila. The Sub Assistant Agriculture Officers (SAAOs) of the Gazipur district were the target population of this study. The total target population in Kapasia, Sreepur, Kaliakoir, Kaligonj and Gazipur Sadar upazilas, of Gazipur district were 214, out of this population, a number of 90 (42%) SAAOs were selected as the sample of the study following the proportionate random sampling. A pre-tested interview schedule was used to collect data from the respondents. Selected characteristics of the respondents' viz. age, service experience, training exposure, job satisfaction, innovativeness, cosmopoliteness, use of information sources and aspiration were considered as independent variables of the study. To measure the knowledge on ICT utilization 35 statements were selected regarding 7 ICT tools. These were presented to the respondents with 5 possible answers for each ICT tool. A score of one was given to the right answer and zero to the wrong answer. The possible obtainable scores ranged between 35 and 0 respectively. Based on the total score obtained by the respondents of each ICT tool Knowledge Index was worked out by using the following formula.

$$\text{Knowledge Index (KI)} = \frac{\text{Scores obtained}}{\text{Obtainable scores}} \times 100$$

To measure perception of the respondents towards ICT utilization a 5-point Likert type scale ranging from 1 as 'strongly disagree' to 5 as 'strongly agree' used for the measurement. The respondents were asked to rate 10 statements based upon a five-point Likert type scale towards ICT utilization in agriculture. A perception score of a respondent was obtained by summing up the weights for his 10 statements regarding ICT utilization in agricultural extension. The perception score of a respondent could range from 10 to 50, while 10 indicating low perception and 50 indicating highest level of perception. In order to measure the extent of perception on the statement towards ICT utilization Perception Index (PI) was used as developed by Biswas (2004).

$$\text{Perception Index (PI)} = \frac{f_1X_1 + f_2 X_2 + \ldots + f_nXn}{N} \times 100$$

$$= \times 100 \; \frac{\sum_{i=0}^{n} f_nX_n}{N}$$

Where,

X_i = scale value at the ith priority of the statement
f_i = Frequency of responses on that statement
n = number of statements in the parameter
N = number of respondents
i = 1, 2, 3,......n

Results and Discussion

Personal characteristics of the respondents

Characteristics profile of the extension workers were determined and presented in Table 1. The findings indicate that most of the respondents were middle aged (56.7%) followed by short-term service experience (62.2%). It is also found that majority of them had low training experience (47.8%) while two-fifth (40%) had no training, high innovativeness (55.6%), medium cosmopoliteness (70%) and medium job satisfaction (56.7%). It was also indicate that most of them used medium information sources (64.4%) and had medium aspiration (45.6%).

Table 1. Distribution of the respondent according to their socio-economic characteristics

Character	Measuring unit	Categories	No. of respondents	Per cent	Mean	SD
Age	Actual year	Young aged (up to 35)	28	31.1		
		Middle aged (36-45)	51	56.7	37.04	6.68
		Old (>45)	11	12.2		
Service experience	Year of service	Short term (up to 10 years)	15	62.2		
		Medium term (11- 20 years)	56	27.8	10.24	6.17
		Long term (>20 years)	25	10		
Training exposure	Number	No training	36	40		
		Low training(1 to 3)	43	47.8	1.41	-
		Medium training (4-5)	11	12.2		
Innovativeness	Scores	Low (up to18)	19	21.10		
		Medium (19-30)	21	23.30	24.49	6.97
		High (>30)	50	55.6		
Cosmopoliteness	Scores	Low (up to 15)	16	17.8		
		Medium (16 to 25)	63	70	19.67	4.37
		High (>25)	11	12.2		
Job satisfaction	Scores	Low (up to 22)	23	25.5		
		Medium (23to 34)	51	56.7	28.44	6.66
		High (>34)	16	17.8		
Use of information sources	Score	Low (up to 18)	18	20		
		Medium (19-28)	58	64.4	23.03	5.12
		High (>18)	14	15.6		
Aspiration	Score	Low (up to 12)	28	31.1		
		Medium (13-16)	41	45.6	14.12	2.80
		High (>16)	21	23.3		

Knowledge on ICT utilization

The Knowledge of the agricultural extension workers towards ICT utilization were conceptualized as consisting of seven ICT tools namely i)Internet/ Web services, ii) MS Word, iii) MS Excel, iv) MS Power point, v) Analytical packages SPSS, vi) Geographical Information System (GIS) and vii) Web based searched engine. For each ICT tools of knowledge the respondents' actions were arbitrarily judged from 0 to 5 continuums. The salient features of different ICT tools have been presented in Table 2 and Table 3.

Table 2. Rank order of the ICT tools regarding knowledge of the respondents based on theirKnowledge Index (KI) and mean score

Sl. No.	ICT tools	Knowledge Index(KI)	Mean	Rank order
1.	Internet/Web service	78	3.89	2nd
2.	MS Word	80	3.99	1st
3.	MS Excel	56	2.78	5th
4.	Microsoft Power Point	59	2.94	4th
5.	Analytical package SPSS	53	2.63	6th
6.	Geographical Information System (GIS)	51	2.54	7th
7.	Web based search engine	62	3.10	3rd

The knowledge of extension workers was highest in case of MS Word tool (KI=80) with mean score 3.89 (Table 2) because more than fourth-fifth (84.4%) of the respondents know that bold, italic, underline, alignment, spelling and grammar can be checked through it and 83.3 per cent of the respondents be familiar with that it is word processor (Table 3). This was followed by Internet/Web service (KI=78) with mean score 3.89 (Table 2) as 88.9 percent respondents be aware of that it helps to get information, data and images almost instantaneously (Table 3) this might be the fact that of above findings. In case of web based search engine the knowledge Index

(KI) of the respondents was 62 with mean score 3.10 ranked third. The reason behind this 70 percent of the respondents know about that it includes 'Google', 'Yahoo', or 'Alta Vista' to find the information on the internet and for this browser software like 'Internet Explorer' or Mozilla Firefox is required on computer (Table 3).

Table 3. Distribution of the respondents based on their knowledge on seven ICT tools

ICT tools	Statements		Yes	No
Internet/ Web service	i)	Internet is a worldwide network of networks.	76 (84.4)	14 (15.6)
	ii)	It is a global network connecting millions of computers.	77 (85.6)	13 (14.4)
	iii)	It makes contact with people by chat facility.	58 (64.4)	32 (35.6)
	iv)	It helps to get information, data and images almost instantaneously	80 (88.9)	10 (11.1)
	v)	Unlike online services, which are centrally controlled, the internet is decentralized by design.	50 (55.6)	40 (44.4)
MS Word	i)	It is word processor	75 (83.3)	15 (16.7)
	ii)	Microsoft Word is a powerful tool to create professional looking	72(80)	18 (20)
	iii)	Bold, italic, underline, alignment, spelling and grammar can be checked through it.	76 (84.4)	14 (15.6)
	iv)	Word formatting is done through this software.	69 (76.7)	21 (23.3)
	v)	It creates a new, blank file bases on the default template.	67 (74.4)	23 (25.6)
MS Excel	i)	Excel is a data processor	48 (53.3)	42 (46.7)
	ii)	Microsoft Excel also serves as database management.	56 (62.2)	34 (37.8)
	iii)	It stores data in the forms of columns and rows	60 (66.7)	30 (33.3)
	iv)	In excel there are 256 columns and 65536 rows.	39 (43.3)	51 (56.7)
	v)	Excel helps in entering values, text, formulas etc.	47 (52.2)	43 (47.8)
Microsoft Power Point	i)	It is presentation program developed by Microsoft.	68 (75.6)	22 (24.4)
	ii)	It consists of a number of individual pages or slides.	61(67.8)	29 (37.2)
	iii)	Professional quality slide show can be made through this	46 (51.11)	44 (48.9)
	iv)	It helps in interesting pictures, diagrams etc.	59 (65.5)	31 (34.4)
	v)	Animation can be done through this.	33 (36.7)	57 (63.3)
Analytical package SPSS	i)	It is among the widely used programs for statistical analysis.	53 (58.9)	37 (41.1)
	ii)	It is used for market, health government, education and other research purpose.	41 (45.6)	49 (54.4)
	iii)	It manages millions of rows effortlessly, so you will never again have to break up your data and analyze it piecemeal	47 (52.2)	43 (47.8)
	iv)	This eliminates duplicate records and jumbled cases	33 (36.7)	57 (63.3)
	v)	SPSS has measures of significance that allow confident, actionable conclusions.	45 (50)	45 (50)
Geographical Information System (GIS)	i)	GIS is for analysis and management of spatial data and mapping.	49 (54.4)	4 1(45.6)
	ii)	It helps to get information like populations, literacy level, density of roads, land etc.	47 (52.2)	43 (47.8)
	iii)	It helps in storing and retrieving.	40 (44.4)	50 (55.6)
	iv)	It also helps in comparing spatial data to support some analytical process.	47 (52.2)	43 (47.8)
	v)	Use of GIS in socioeconomic research, characterization of production system according to adoption of improved technologies	46 (51.1)	44 (48.9)
Web based search engines	i)	A Web search engine is a tool designed to search for information on the World Wide Web	49 (54.4)	41 (45.6)
	ii)	It helps to search items in the search statement	62 (68.9)	28 (31.1)
	iii)	It includes 'Google', 'Yahoo', or 'Alta Vista' to find the information on the internet	63 (70)	27 (30)
	iv)	For this browser software like 'Internet Explorer' or Mozilla Firefox is required on computer	63 (70)	27 (30)
	v)	Information consists of web pages, images, and other types of files	48 (53.3)	42 (46.7)

Note: 1) Figures in the parenthesis indicate percentage 2) Multiple responses are possible

The extension workers had moderate knowledge in case of Microsoft Power Point (KI=59) and MS Excel (KI=56) with mean score 2.94 and 2.78, respectively (Table 2). In case of Microsoft Power point three-fourth (75.6%) of the respondents be acquainted with that Microsoft Power Point is presentation program developed by Microsoft and 67.8 per cent of respondents have knowledge of that Power Point presentations consist of a number of individual pages or slides while in case of MS Excel 66.7 per cent respondents are slightly friendly with that MS Excel stores data in the forms of columns and rows (Table 3). The extension workers had very poor knowledge in case of Analytical package SPSS (KI=51) and Geographical Information System (GIS) (KI=51) with mean score 2.63 and 2.54, respectively (Table 2). This is because only 58.9 per cent respondents know that SPSS is among the widely used programs for statistical analysis and 54.4 per cent respondents be aware of that GIS is for analysis and management of spatial data and mapping.

Best on total score of the respondents regarding seven ICT tools the overall knowledge score of respondents were classified into three categories namely low (less than mean - S.D.), medium (between mean ± S.D.) and high (more than mean + S.D.) has been presented in Fig. 1. Observed knowledge scores of the respondents ranged from 15 to 32 against possible range from 0 to 35 with mean and standard deviation were 21.88 and 3.04, respectively.

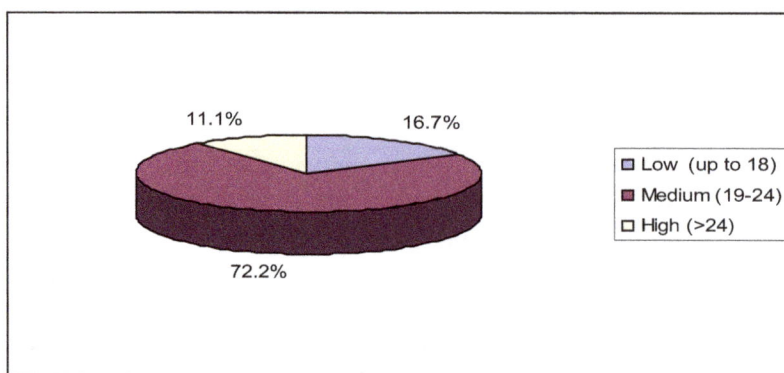

Fig. 1. Distribution of the respondents based on their overall knowledge score towards ICT utilization

Data presented in Figure 1 demonstrate that 16.7 per cent of the respondents had low knowledge while about three-fourth (72.2%) of the respondents had medium knowledge on ICT and only 11.1 had high knowledge towards ICT utilization. However, it is observed that an overwhelming majority (88.9%) of agricultural extension workers in the study area had low to medium knowledge. Therefore, one may apprehend that due to moderate and poor knowledge on most of the ICT tools except MS Word and Internet/Web service most of the respondents of the study area might have low to medium knowledge.

Perception towards ICT utilization

A rank order of 10 statements regarding ICT utilization according to their Perception Index (PI) has been presented in Table 4 for a clear understanding of the comparative perception of the respondents on the statements.

Table 4. Rank order of the statement regarding ICT utilization according to their perception index

Sl#	Perceptional Statements	SA	A	UD	DA	SDA	PI	Ranking
1.	Knowledge of ICT use has a great significance in agricultural development	31	25	23	8	3	381	3rd
2.	It's a waste of time bringing ICT into agriculture	28	28	18	9	7	368	5th
3.	ICT use influences rate of adoption	29	29	12	18	2	372	4th
4.	Extension work can be greatly enhanced by ICT	36	24	19	9	2	392	1st
5.	The benefits of ICT use outweigh the financial burden involved	28	32	22	6	2	387	2nd
6.	ICT removes a lot of cost, barriers and saves time in transferring technology	18	27	20	19	6	336	7th
7.	Infrastructural facilities to support ICT use is not available	13	19	40	17	1	329	9th
8.	ICT use is good but not necessary in extension service delivery	27	23	16	23	1	358	6th
9.	The financial burden of using ICT is unbearable	6	25	31	27	1	309	10th
10.	ICT are too complicated and hard to use	11	32	23	22	2	331	8th

Data contained in Table 4 indicate that the respondents had top most perception on the ICT utilization in respect of 'extension work can be greatly enhanced by ICT' was the highest (PI=392) followed by on 'the benefits of ICT use outweigh the financial burden involved' was the second highest (PI=387) and 'knowledge of ICT use has a great significance in agricultural development' (PI= 381) was third. Perception was lowest (PI=309) in respect of 'The financial burden of using ICT is unbearable'. The respondents showed 'infrastructural facilities to support ICT use are not available' was the second lowest (226). Perception Indices (PI) towards ICT utilization of 10 statements were between 392 to 309 maximum possible perception index (PI) is 450 and that of six statements between 350 and 400. However, none of the statements has between 401 and 450; these facts indicate that perception of the respondents was considerably moderate in comparison to the maximum possible level on all the statements.

Extension workers' perception score was further analyzed and the summary presented in Table 5. Data contained in Table 5 indicate that the highest proportion (64.4%) of the respondents fell in medium perception category compared to 20 percent in high and 15.6 per cent in low perception category.

Table 5. Distribution of respondents according to their extent of perception

Category	Respondents		Mean	SD
	Frequency	Per cent		
Low (up to 31)	14	15.6		
Medium (32-39)	58	64.4	35.62	4.26
High (>39)	18	20.0		
Total	90	100.0		

The findings reveal that only 15.6 per cent of the extension workers in the study area had low perception towards ICT utilization. More than fourth-fifth (84.4%) of the respondents had medium to high perception towards ICT utilization because most of the respondents be aware of that extension work can be greatly enhanced by ICT and Knowledge of ICT use has a great significance in agricultural development. This may have positive impact in transferring the modern agricultural technologies to the farmers for sustainable agricultural production.

Relationship between the selected characteristics of the respondents and extent of knowledge and perception towards ICT utilization

The variables studied were subjected to correlation analysis using Pearson product moment correlation to find out the relation between the independent variables and dependent variables (knowledge and perception on various ICTs). Results in Table 5 reveals that there was no significant relationship between the age, aspiration, innovativeness, cosmopoliteness, training exposure and job satisfaction towards ICT utilization but service experience and use of the information sources of the respondents showed positive and significant relationship with their knowledge towards ICT utilization. This implies that more service experience and more use of information sources of the respondents leads to a tendency towards more knowledge towards ICT utilization in agricultural extension service delivery i.e., extension workers with medium and high service experience and use of information sources had more knowledge towards ICT utilization. This might be due to high service experience and large number use of source of information help one able to become more knowledgeable and more innovative and also keep up to date with the latest technology.

Table 6. Relationship between the selected characteristics of the respondents and their knowledge and perception towards ICT utilization

Characteristics of the extension workers	Knowledge (r value)	Perception (r value)
Age	0.173(NS)	0.143(NS)
Service experience	0.261*	0.193(NS)
Training exposure	-0.134(NS)	0.123(NS)
Innovativeness	0.151(NS)	0.578**
Cosmopoliteness	-0.028(NS)	0.236*
Job satisfaction	0.060(NS)	0.236*
Use of information sources	0.273*	.0058(NS)
Aspiration	0.189(NS)	0.197(NS)

*Significant at 0.05 level of probability, ** Significant at 0.01 level of probability, NS = Non significant*

On the other hand, out of eight selected characteristics of the respondents' innovativeness, cosmopoliteness and job satisfaction showed positive significant relationship with their perception towards ICT utilization indicating that respondents having higher innovativeness, cosmopoliteness and job satisfaction more likely to have more perception towards ICT utilization in agricultural extension service delivery. Because through cosmopoliteness quality, an individual becomes aware of the recent information and innovativeness of an individual helps to adopt new ideas and technology. High job satisfaction creates positive response toward various facets of one's job, which make an individual more competitive to perform his duty.

Conclusion

Agricultural development can be accelerated and livelihoods can be supported by ICT in so many ways. The dramatic changes that have taken place in the last decade in ICT have touched almost every field of human activity, and agriculture is not an exception. ICT has tremendous potential to revolutionize the way information, knowledge and new technology is managed, developed and delivered to farmers through ICT utilization in agriculture. Findings of the study reveals that knowledge of extension workers among the seven ICT tools was highest in case of MS Word this was followed by Internet/Web service and web based search engine. The extension workers had moderate knowledge in case of Microsoft Power Point and of MS Excel while the extension workers showed very poor knowledge in case of Analytical package SPSS and Geographical Information System (GIS). It was also found that an overwhelming majority (88.9%) of agricultural extension workers in the study area had over all low to medium knowledge towards ICT utilization. Among the ten statements regarding extension workers perception towards ICT utilization, 'extension work can be greatly enhanced by ICT' was the highest followed by on 'the benefits of ICT use outweigh the financial burden involved' was the second highest score. Furthermore it was observed that more than fourth-fifth (84.4%) of the respondents had medium to high perception towards ICT utilization because most of the respondents be aware of that extension work can be greatly enhanced by ICT and knowledge of ICT use has a great significance in agricultural development. Service experience and use of the information sources of the respondents showed positive and significant relationship with their knowledge towards ICT utilization conversely innovativeness, cosmopoliteness and job satisfaction showed positive significant relationship with their perception towards ICT utilization. The study recommends that adequate information on various relevant ICT suitable for extension service delivery should be given to extension agents so that they can improve their knowledge and on it help them to develop more positive perception that will enhance the future use of these ICT in extension service delivery.

References

Aboh, L.C. 2008. Assessment of the frequency of ict tools usage by agricultural extension agents in Imo State, Nigeria. *J. Agric. Social Res.* 8: 21-30.

Aker, J.C. 2010. Dial 'A' for Agriculture: using information and communication technologies for agricultural extension in developing countries. Tuft University, Economics Department and Fletcher School, Medford MA02155. p. 37.

Arokoyo, T. 2005. ICTs application in agricultural extension service delivery. *In*: S.F. Adedoyin (Ed.), Agricultural Extension in Nigeria. llorin, Nigeria: Agricultural Extension Society of Nigeria. pp. 245-251.

Ballantyne, P. and Bokre, D. 2003. ICTs: transforming agricultural extension? Report of an iNARSe-discussion, Retrieved October 15, 2007 from: http://www.livelihoods.org/info/docs/inars _Supersummary.pdf

Biswas, S. 2004. Women's empowerment and demographic changes. Bangladesh Academy for Rural Development, Kotbari, Comilla. pp. 147-152.

FAO. 2004. Institute building to strengthen agricultural extension. 27th FAO Regional Conference for Asia and the pacific Beijing, china, May, 17th – 21st, 2004. pp. 22-28.

McNamara, K. 2009. Improving agricultural productivity and markets: The role of information and communication technologies. agriculture and rural development notes, Issue 47, April, The World Bank, Washington DC. p. 4.

Meera, S.N., Jhamtani, A. and Rao, D.U.M. 2004. Information and communication technologies in agricultural development: A comparative analysis of three projects from india. Agricultural Research and Extension Network. New Delhi, India. p. 135.

Mirzaei, A. 2003. Country Report of Islamic Republic of Iran. The study meeting on application of information technology for effective agricultural extension. New Delhi, India p. 13.

Raihan, S. 2012. Economic reforms and agriculture in Bangladesh: Assessment of impacts using economy-wide simulation models / Selimraihan; International Labour Organization; ILO Country Office for Bangladesh. Dhaka: ILO, 2012. Available at: http://www.ilo.org/wcmsp5/groups/public/ ---asia/---ro-bangkok/---ilo- dhaka/documents/publication/wcms_2040 89.pdf

World Bank. 2008. World Development Report 2008: Africulture for development (Washington, DC). Available at: http://siteresources.worldbank.org/INTWD R2008/Resources/WDR_00_book.pdf

13

IMPROVEMENT OF POTATO BASED CROPPING PATTERNS BY INCLUSION OF SHORT DURATION MUNGBEAN AND T. *AMAN* RICE IN MONGA PRONE AREAS OF RANGPUR

M.K. Islam[1], M.S. Mahfuz[2], S. Ghosh[3], A.S.M.Y. Ali[3] and M.Z. Hasnat[4]

Abstract

The experiment was carried out to compare the improved cropping patterns against the farmers existing potato based cropping patterns having no mungbean/brown manure crop for higher yield, economic return and income generation in agricultural field in the off period following RCBD design with three replications at farmer's field at Paikan Gangachara, Rangpur district during September-October. The treatments (cropping patterns) were T_1 = T. *aman* rice (BINA 7) - Potato - Mungbean (BARI mungbean 6) *(Improved pattern)*, T_2 = T. *aman* rice (BR11) - Potato - Fallow *(Farmers pattern)*, T_3 = T. *aman* rice (BINA 7) - Potato + Maize intercrop - Mungbean *(Improved pattern)* and T_4 = T. *aman* rice (BR11) - Potato / Maize relay *(Farmers pattern)*. The highest yield (4.16 t ha⁻¹) was recorded in T_2 (BR11) which is statistically at par with T_4 (4.15 t ha⁻¹) but higher than the other treatments. Early planting sole potato (T_1) gave highest yield (26.10 t ha⁻¹) which was significantly higher than all other treatments. Late planting sole potato (T_2), intercropped early potato (T_3) and relay potato (T_4) showed similar yield (23.61 – 24.79 t ha⁻¹). Intercropped (T_3) and Relay (T_4) maize did not vary significantly in the studied parameters and yields were 8.21 and 7.92 t ha⁻¹, respectively. Mungbean after sole potato (T_1) gave higher number of pods/plant (17.25), and yield (1.47 t ha⁻¹) which is significantly higher than those of T_3 (14.89 and 1.28 t ha⁻¹, respectively). Highest gross return (GR) (Tk. 417720) and gross margin (GM) (Tk. 220220) were calculated in improved pattern T_3 and the lowest of those (Tk. 289670, Tk. 146020) in farmers pattern T_2. The other improved pattern T_3 was the second highest performer considering GR and GM. But BCR (2.21) was highest in T_1 and second highest in T_3. The results indicated that the improved patterns (T_1, T_3) were better than farmers pattern (T_2, T_4). The improved pattern (T_1) gave GR Tk. 67890 and GM Tk. 51785 higher than farmers pattern (T_2). Similarly, the other improved pattern (T_3) showed Tk. 51870 and Tk. 37395 higher than farmers pattern (T_4). The improved pattern T_1 and T_3 created 45 working day job for the labour for harvesting early matured rice in the Monga/ jobless period (October) while farmers' pattern gives only 3 days work. The mungbean included improved cropping patterns can be suggested for increased production, economic return and Monga mitigation (work opportunity in off period) in Rangpur.

Keywords: Improvement, Pattern, Potato, Maize, Mungbean, Monga, Labour

[1]Senior Scientific Officer, Bangladesh Agricultural Research Institute, Burirhat, Rangpur, Bangladesh
[2]Regional Farm Broadcasting Officer, Agriculture Information Service, Rangpur, Bangladesh
[3]Scientific Officer, Bangladesh Agricultural Research Institute, Burirhat, Rangpur, Bangladesh
[4]Information Officer, Agriculture Information Service, Dhaka, Bangladesh

*Corresponding author's email: shohag_agri@yahoo.com (M.S. Mahfuz)

Introduction

Cultivable land of Bangladesh is reducing day by day because of new industries and human settlement. On the contrary, demand for food is increasing due to population pressure. Crop production has to be increased to feed the ever-increasing mouths. Gangachara is one of the Monga (seasonal unemployment) prone areas of Rangpur. A huge area is under potato cultivation in this area and farmers are practicing different potato based cropping patterns like T. *aman* rice - Potato - Fallow, T. *aman* rice - Potato/ Maize, T. *aman* rice – Potato - Jute etc. They use long duration T. *aman* variety BR 11 causing delay in

potato planting in December, which ultimately reduces yield and favours crop failure due to late blight disease. November is best for potato planting which can avert the disease by maturing crop before the severe infestation period. In these patterns, there is no work for the poor in October - November in agricultural field in Rangpur region. Poor people starve at that time which is called Monga in local dialect and government has to take extra measure to feed them. Short duration T. *Aman* rice (BINA Dhan 7) can facilitate timely/early potato planting (November) which is important for good yield. It also can create job opportunity for the poor in Monga period (September-October) for harvesting rice. On the other hand, a good soil should have organic matter content more than 3.5 per cent, while more than 60% of our cultivated soil contains organic matter at low level (<1.7%). Recycling of organic matter is essential for maintaining soil fertility (BARC, 2001). Declining soil fertility is a major reason for lower crop yield in Bangladesh (Chowdhury et al., 2003). Inclusion of green manure crop in the cropping pattern is a solution to soil degradation. Adjustments in cropping patterns are necessary on a large scale incorporating green manuring and grain legume crops to improve soil health (BARC, 2001). If maize is intercropped with potato, it can facilitate two months time to grow mungbean before T. Aman rice. Intercropping system maximizes complementary use of growth resources (Krishna and Raikhelkar, 1997) and enhances the total productively (Umrani et al., 1984). Potato single stand requires high levels of inputs, which are in most instances beyond the economic capacity of farmers (Luis and Manrique, 1996). Moreover, it improves soil health by reducing pathogenic microorganisms survive in soil by creating disruption of life cycle (Dey, 2001). Under some conditions, intercropping can usefully contribute to the control of pest or disease populations and the reduction of yield loss (Trenbath, 1993). As the field duration of the variety BARI Mungbean 6 is short (55 days) and pods mature almost at a time,

after harvesting of pod biomass incorporation into soil can help enrich soil by adding organic matter. A huge foreign currency can also be saved for importing pulse from abroad. Keeping the views in mind the experiment was undertaken to accommodate mungbean into farmers cropping pattern for increased yield and economic return, to generate work in agricultural field in the off period during September-October and to increase pulse production and save foreign currency for import.

Methodology

The experiment was conducted at farmer's field at village Paikan under Gangachara Upozilla of Rangpur district following RCBD design with three replications. Twenty-five decimal lands for each pattern were selected. The improved cropping patterns were compared with the farmers existing potato based patterns having no mungbean/ brown manure crop. The treatments (cropping patterns) were - T_1 = T. aman rice (BINA 7) - Potato - Mungbean (BARI mungbean 6) (Improved pattern), T_2 = T. aman rice (BR11) - Potato - Fallow (Farmers pattern), T_3 = T. aman rice (BINA 7) - Potato + Maize intercrop - Mungbean (Improved pattern) and T_4 = T. aman rice (BR11) - Potato / Maize relay (Farmers pattern). The patterns initiated in July 2011 and planting and harvesting time of crops are given below in the Table 1.

Table 1. Planting and harvesting time of crops

T_1	T. aman rice (BINA 7)	Potato	Mungbean	-
	21 July - 21 Oct	05 Nov - 06 Feb	21 Feb - 21 Apr	
T_2	T. aman rice (BR11)	Potato	Fallow	-
	05 August - 05 Dec	10 Dec - 03 Mar		
T_3	T. aman rice (BINA 7)	Potato	Maize	Mungbean
	21 July - 21 Oct	5 Nov - 20 Jan	13 Nov - 12 Apr	6 Apr - 02 Jun
T_4	T. aman rice (BR11)	Potato	Maize	
	05 August – 05 Dec	10 Dec - 01 Mar	01 Feb - 04 Jun	

After harvesting, mungbean biomass (brown manure) was incorporated into soil to improve and maintain soil fertility. Recommended doses of fertilizer and management practice were maintained. Intercropping maize was sown in between potato rows at 7 days after sowing of potato. Relay maize was sown on the side of potato ridge at 50 days after sowing of potato. Plant spacing was 60 cm × 25 cm for potato and intercrops and relay maize, 75 cm × 20 cm for sole maize, 30 cm × 10 cm for mungbean and 20 × 15 cm for T. aman rice. Data on yield and yield components was taken and data were analyzed statistically following Gomez and Gomez (1984) and mean separation was done by LSD. Economic analysis was also done. The total benefits of the patterns were calculated.

Results and Discussion

T. Aman rice

Plant height, number of effective tiller/hill, 1000-grain weight and yield increased significantly due to treatments (Table 2). The BR11 rice produced significantly higher plant height (106.2 cm in T_4 and 106.0 cm in T_2) than BINA dhan 7 (99.07 cm in T_3 and 98.90 cm in T_1). Similar trend was observed in other parameters, which differed significantly due to treatment effect. The highest yield (4.16 t ha⁻¹) was recorded in T_2 which is statistically at par with T_4 (4.15 t ha⁻¹) but significantly higher than the other treatments T_1 (3.98 t ha⁻¹) and T_3 (3.96 t ha⁻¹). Treatment T_1 and T_3 is similar in producing yield. Higher number of effective tiller and 1000-grain weight favours higher yield in BR11.

Table 2. Yield and yield contributing characters of T. *aman* rice in farmer's field, Rangpur

Treatments (Cropping patterns)	Plant height (cm)	Effective tiller/hill (no)	Panicle length (cm)	Grains/ panicle (no)	1000 grain weight (g)	Yield (t ha⁻¹)
T_1 = BINA7-Pot-Mung	98.90b	8.40b	23.60	69.00	22.17b	3.98b
T_2 = BR11-Pot-Fallow	106.0a	8.73a	23.30	69.40	24.63a	4.16a
T_3 = BINA7-P+M-Mung	99.07 b	8.46b	23.77	67.00	21.97b	3.96b
T_4 = BR11-Pot/Maize	106.2a	8.50ab	23.97	69.67	25.13a	4.15a
CV (%)	3.2	1.5	1.77	2.33	4.17	2.12
Level of Significance	*	*	NS	NS	*	*
LSD (0.05)	6.56	0.26			1.96	0.17

Significant at 0.05 level of significance
Letters in a column having similar or no letter did not differ significantly

Potato

All the parameters except number of stem/ hill varied significantly (Table 3). Early planting sole potato (T_1) gave highest yield (26.10 t ha⁻¹) which was significantly higher than all other treatments. Late planting sole potato (T_2), intercropped early potato (T_3) and relay potato (T_4) showed similar yield (23.61 – 24.79 t ha⁻¹). Highest plant height (68.97 cm), and number (8.76) and weight (466.7 g) of tuber/ hill contributed to the highest yield in T_1. Treatment T_2, T_3 and T_4 performed identically in producing plant height (65.21-65.27 cm), and number (7.16-7.26) and weight (413.3-430.0 g) of tuber/ hill which were significantly lower than T_1.

Table 3. Yield and yield contributing characters of potato in farmer's field, Rangpur

Treatments (Cropping Patterns)	Plant height (cm)	Stem/hill (no)	Tuber/hill (no)	Weight Tuber/hill (g)	Yield (t ha⁻¹)
T_1 = BINA7-Pot-Mung	68.97a	3.30	8.76a	466.7 a	26.10a
T_2 = BR11-Pot-Fallow	65.23b	3.30	7.26b	430.0 b	24.79b
T_3 = BINA7-P+M-Mung	65.21b	3.33	7.20b	413.3 b	23.65b
T_4 = BR11-Pot/Maize	65.27b	3.30	7.16b	423.3 b	23.61b
CV (%)	2.71	4.7	8.67	4.2	3.21
Level of Significance	*	NS	*	*	*
LSD (0.05)	3.58	0.31	1.32	36.36	1.57

Significant at 0.05 level of significance
Letters in a column having similar or no letter did not differ significantly

Maize

Intercropped (T_3) and Relay (T_4) maize did not vary significantly in the parameters studied (Table 4).

Table 4. Yield and yield contributing characters of maize in farmer's field, Rangpur

Treatments (Cropping patterns)	Plant population /m² (no)	Plant height (cm)	Cobs/ plant (no)	Cob length (cm)	Grains/ cob (no)	1000 grain weight (g)	Yield (t ha⁻¹)
T_1 = BINA7-Pot-Mung	-	-	-	-	-	-	-
T_2 = BR11-Pot-Fallow	-	-	-	-	-	-	-
T_3 = BINA7-P+M-Mung	5.33	226.33	1.03	16.97	513.33	28.5	8.21
T_4 = BR11-Pot/Maize	5.27	230.67	1.13	17.3	520	28.57	7.92
CV (%)	4.32	5.59	6.5	6.05	4.32	3.39	3.02
Level of Significance	NS	NS	NS	NS	NS	NS	NS
LSD	0.80	44.88	0.24	3.64	78.42	3.99	0.85

Significant at 0.05 level of significance
Letters in a column having similar or no letter did not differ significantly

Mungbean

Number of pods/plant and yield showed significant variation due to treatment (Table 5). Mungbean after sole potato (T_1) gave higher number of pods/plant (17.25), and yield (1.47 t ha^{-1}) which is significantly higher than those of T_3 (14.89 and 1.28 t ha^{-1}, respectively). Higher number of pods/plant contributed to higher yield in T_1.

Table 5. Yield and yield contributing characters of mungbean in farmer's field, Rangpur

Treatments (Cropping patterns)	Plant height (cm)	Pods/ plant (no)	Pod length (cm)	Seeds/pod (no)	1000 seed wt (g)	Yield (t ha^{-1})
T_1 = BINA7-Pot-Mung	53.3	17.25 a	7.03	7.32	41.90	1.47 a
T_2 = BR11-Pot-Fallow	-	-	-	-	-	-
T_3 = BINA7-P+M-Mung	55.07	14.89 b	7.01	6.65	40.67	1.28 b
T_4 = BR11-Pot/Maize	-	-	-	-	-	-
CV (%)	3.63	3.61	5.8	3.55	2.08	3.19
Level of Significance	NS	*	NS	NS	NS	*
LSD	6.91	2.03	1.43	0.87	3.01	0.15

*Significant at 0.05 level of significance
Letters in a column having similar or no letter did not differ significantly

Economic Analysis

Economic analysis of T. aman rice

Closer or similar gross return (GR), gross margin (GM) and benefit cost return (BCR) was noticed among the treatments indicating that BR11 and BINA dhan 7 offers more or less equal monetary benefit (Table 6). GR ranged from Tk. 63360 to 66400, GM from Tk. 30125 to 32000 and BCR from 1.91 to 1.93.

Table 6. Economic analysis of T. aman rice in farmer's field, Rangpur

Treatments (Cropping patterns)	Yield (t ha^{-1})	Gross return (Tk ha^{-1})	Production cost (Tk ha^{-1})	Gross margin (Tk ha^{-1})	BCR
T_1 = BINA7-Pot-Mung	3.98	63680	33235	30445	1.92
T_2 = BR11-Pot-Fallow	4.16	66560	34560	32000	1.93
T_3 = BINA7-P+M-Mung	3.96	63360	33235	30125	1.91
T_4 = BR11-Pot/Maize	4.15	66400	34560	31840	1.92

Selling Price = Tk. 16 per kg

Economic analysis of potato

Highest GR (Tk. 235080), GM (Tk. 129760) and BCR (2.23) were calculated in improved pattern T_1 and the lowest of those (Tk. 212490, Tk. 106620 and 2.01) was found in the farmers pattern T_4 (Table 7). Potato in improved pattern (T_1, T_3) showed better performance than farmers pattern (T_1, T_4).

Table 7. Economic analysis of potato in farmer's field, Rangpur

Treatments (Cropping patterns)	Yield (t ha^{-1})	Gross return (Tk ha^{-1})	Production cost (Tk ha^{-1})	Gross margin (Tk ha^{-1})	BCR
T_1 = BINA7-Pot-Mung	26.12	235080	105320	129760	2.23
T_2 = BR11-Pot-Fallow	24.79	223110	111460	111650	2.00
T_3 = BINA7-P+M-Mung	23.65	212850	100320	112530	2.12
T_4 = BR11-Pot/Maize	23.61	212490	105870	106620	2.01

Selling Price = Tk. 09 per kg

Economic analysis of maize

Higher GR (Tk. 90310), GM (Tk. 49935) and BCR (2.24) were calculated in improved pattern T_3 than those (Tk. 87120, Tk. 44525 and 2.05) of the farmers pattern T_4 (Table 8).

Table 8. Economic analysis of maize in farmer's field, Rangpur

Treatments (Cropping patterns)	Yield (t ha^{-1})	Gross return (Tk ha^{-1})	Production cost (Tk ha^{-1})	Gross margin (Tk ha^{-1})	BCR
T_1 = BINA7-Pot-Mung	-	-	-	-	-
T_2 = BR11-Pot-Fallow	-	-	-	-	-
T_3 = BINA7-P+M-Mung	8.21	90310	40375	49935	2.24
T_4 = BR11-Pot/Maize	7.92	87120	42595	44525	2.05

Selling Price = Tk. 11 per kg

Economic analysis of mungbean

Mungbean after sole potato (T_1) showed higher GR (Tk. 58800), GM (Tk. 35230) and BCR (2.49) than those (Tk. 51200, Tk. 27630 and 2.17) of T_3 (Table 9).

Table 9. Economic analysis of Mungbean in farmer's field, Rangpur

Treatments (Cropping Patterns)	Yield (t ha^{-1})	Gross return (Tk ha^{-1})	Production cost (Tk ha^{-1})	Gross margin (Tk ha^{-1})	BCR
T_1 = BINA7-Pot-Mung	1.47	58800	23570	35230	2.49
T_2 = BR11-Pot-Fallow	-	-	-	-	-
T_3 = BINA7-P+M-Mung	1.28	51200	23570	27630	2.17
T_4 = BR11-Pot/Maize	-	-	-	-	-

Selling Price = Tk. 40 per kg

Economic analysis of whole pattern

Highest GR (Tk. 417720) and GM (Tk. 220220) were calculated in improved pattern T_3 and the lowest of those (Tk. 289670, Tk. 146020) in farmers pattern T_2 (Table 10) The other improved pattern T_3 was the second highest performer considering GR and GM. But BCR (2.21) was highest in T_1 and second highest in T_3. The result indicated that the improved patterns (T_1, T_3) were better than farmers pattern (T_2, T_4).

Table 10. Economic analysis of the whole patterns in farmer's field Rangpur

Treatments (Cropping patterns)	Gross return (Tk ha^{-1})	Production cost (Tk ha^{-1})	Gross margin (Tk ha^{-1})	BCR
T_1 = BINA7-Pot-Mung	357560	162125	195435	2.21
T_2 = BR11-Pot-Fallow	289670	146020	143650	1.98
T_3 = BINA7-P+M-Mung	417720	197500	220220	2.12
T_4 = BR11-Pot/Maize	365850	183025	182825	2.00

Return of improved patterns over the farmers patterns

The improved pattern (T_1) gave GR Tk. 67890 and GM Tk. 51785 higher than farmers pattern (T_2) (Table 11). Similarly, the other improved pattern (T_3) showed Tk. 51870 and Tk. 37395 higher than farmers pattern (T_4).

Table 11. Return of improved patterns over the farmers patterns in farmer's field, Rangpur

Treatments (Cropping patterns)	Gross return (Tk ha^{-1})	Production cost (Tk ha^{-1})	Gross margin (Tk ha^{-1})	BCR
T_1 over T_2 (BINA7-Pot-Mung over BR11-Pot-Fallow)	67890	16105	51785	4.22
T_3 over T_4 (BINA7-P+M-Mung over BR11-Pot/Maize)	51870	14475	37395	3.58

Job Creation

The improved pattern T_1 and T_3 created 45 working day (1 day=8 hours work) job for a labour for harvesting early matured rice in the Monga/ jobless period (October) while farmers' pattern gives only 3 days work (Table 12). Again, T_3 gave total 440 days work throughout the year which was the highest, followed by T_4 (392 days), T_1 (372 days) and the lowest in T_2 (305 days). It indicates that the improved patterns created higher job opportunity than the farmers' pattern.

Table 12. Labour distribution in monga period (day ha⁻¹)

Treatments	Labour use (day ha⁻¹) in monga period (October)	Total labour
T_1 = BINA7-Pot-Mung	45	108 + 189 + 75 = 372
T_2 = BR11-Pot-Fallow	3	108 + 197 + 0 = 305
T_3 = BINA7-P+M-Mung	45	108 + 266 + 75 = 440
T_4 = BR11-Pot/Maize	3	108 + 197 + 80 = 392

Conclusion

The result showed that mungbean could successfully be accommodated in potato based cropping pattern. Intercropping potato + maize, rather than relay cropping potato with maize not only increase total productivity, but also facilitate two and half month time to grow mungbean before T. a*man* cultivation. The Improved (mungbean included) patterns increased total production and monetary return. Short duration T. *Aman* rice created job opportunity for the poor labour in Monga/ off period (October). Pulse production increased. The improved pattern T_1 = BINA dhan 7-Potato-Mungbean can be suggested against farmers pattern T_2 = BR11 – Potato - Fallow and the improved pattern T_3 = BINA dhan 7 - Potato + Maize intercropping – Mungbean can be suggested against farmers pattern T_4 = BR11-Pot/Maize. The mungbean included improved cropping patterns can be suggested for increased production, economic return and Monga mitigation (work opportunity in off period) in Rangpur.

References

BARC. 2001. Impact of land degradation of Bangladesh: A changing scenario in agriciltural land use. BARC, Farmgate, Dhaka. pp. 39-102.

Chowdhury, D.A., Shafiqul Islam, M. and Muktadir Alam, M. 2003. Integrated nutrient management for potato- *T. aus-T. Aman* cropping system. OFRD, BARI, Gazipur and SFFP (Integrated Soil Fertility and Fertilizer Management Project), DAE, Farmgate, Dhaka. pp. 1-5.

Dey, T.K. 2001. Occurrence and management of bacterial wilt of potato and survivability of *Talstonia soloni*. Ph.D. Thesis. Dept. Pl. Path. Bangabandhu Sheikh Mujibur Rahman Agricultural University, Salna, Gazipur. P.185.

Gomez, K.A. and Gomez, A.A. 1984. Statistical Procedures for Agricultural Research. John Wiley and sons, New York. p. 28.

Krishna, A. and Raikhelkar, S.V. 1997. Crop complementary in maize (*Zea mays*) when intercropped with different legumes. *Indian J. Agric. Sci.* 67: 291-294.

Luis, A. and Manrique. 1996. The potato in multiple cropping systems. *J. Plant Nutr.* 19 (10): 215-243.

Trenbath, B.R. 1993. Intercropping for the management of pests and diseases. *Field Crops Res.* 34 (3-4): 381-405.

Umrani, N.K., Sinadi and Dhonde, P.M. 1984. Studies on intercropping of pulses in sorghum. *Indian J. Agric. Sci.* 29 (1): 27-30.

14

INSTITUTIONAL PROVISIONS FOR ADMINISTRATION OF RURAL DEVELOPMENT PROGRAMMES: EXPERIENCE FROM FADAMA 111 DEVELOPMENT PROGRAMME IN TARABA STATE, NIGERIA

M.U. Dimelu[1,2*], F.H. Bonjoru[1], A.I. Emodi[1] and M.C. Madukwe[2]

Abstract

The study examined institutional provisions in the implementation of Fadama 111 Development Project in Taraba State, Nigeria during 2008-2013. All the staff of the project (57) from eight out of 16 local government areas participated in the programme was used in the study. Data were collected with questionnaire and analysed using descriptive statistics. The results showed strong linkages of the state Fadama coordinating office with government parastaltals and organizations at different levels of the project implementation. There were strong adherence to rules and regulations guiding staff recruitment, financial management, preparation of local development plan, environmental compliance and friendliness, and group formation. The project was constrained by several institutional factors namely delay in the payment of counterpart fund by the government (M=3.39), lack of transport and other logistic supports (M=3.06), lack of payment of counterpart fund by the government (M=3.04) and others. The study recommends that policy makers and development planner should ensure functional mechanisms that could foster and enhance linkages, and support adherence to rules and regulations prescribed for implementation of development programmes.

Keywords: Fadama Project , Fadama User Group, Fadama Community Association, Linkage, Institution

[1]Department of Agricultural Extension, University of Nigeria Nsukka, Nigeria
[2]Department of Agricultural Extension and Economics, River State University, Port Harcout, Nigeria

*Corresponding author's email: mabeldimelu@yahoo.com (M.U. Dimelu)

Introduction

The National Fadama Development Programme (NFDP) was incepted following success stories from small-scale irrigation projects carried out by the Agricultural Development Programmes (ADPs) in Fadama areas. Fadama is a Hausa word, meaning flood plains. According to Akinola (2003), the first phase of Fadama Development Programme (Fadama I) was implemented between 1992 and 1998 and it promoted production of arable crops in Fadama areas. The project adopted the small-scale irrigated farming system (SSIFS), as the preferred option because of its cost saving features when compared with large-scale irrigation projects in Nigeria. It was designed with the major aim of harnessing the substantial surface and underground water resources for small-scale irrigated agriculture through private sector participation. Following the successes recorded in Fadama I Project, the second phase of NFDP (Fadama II) was declared loan disbursement effective on the 27th May, 2004 with the actual disbursement to beneficiaries in September 2005. Unlike Fadama I, which covered the cultivation of only few arable crops, Fadama II Project emphasized both farm and non-farm activities linked to Fadama

resources, as well as conflict resolution among fadama resource users. Its development objective was to sustainably increase the incomes of all Fadama resource users (those who depend directly or indirectly on Fadama resources) (FMA and WR, 2008). Fadama II concentrated on a number of agricultural areas termed components. These components include rural infrastructure investments, pilot asset acquisition, demand driven advisory services and improved mechanisms to avoid and manage conflicts among Fadama resource users.

The successes recorded in both phases of the project informed the implementation of the third phase (Fadama-III) with addition of more components and sub components. For instance, rural finance/livelihood, public ADP and adaptive research support component were introduced with specialists in rural development and other supportive staff to handle credit issues for the Fadama III beneficiaries. Furthermore, sub components such as sustainable land management (SLM), the Fadama user equity fund (FUEF) were established from which beneficiaries borrow money at a reduced interest rate to finance the purchase of productive assets. Thus, effective implementation of the project demands adequate institutional provisions and arrangements.

The institutional mechanisms/design provides essential links, collaborations and roles for all the components and organizations involve in the implementation processes. Under Fadama III, the institutional arrangement from the federal level to the local government level is expected to be well defined. At each level, components and organizations are to explore the linkages for effective execution of roles and complementarities in functions in the system. In addition, rules and regulations guiding every operation should be clearly stated. These rules may include those on sub-project design and preparation; environmental compliance and friendliness; financial management; group formation; preparation of local development plan, staff recruitment and so on.

Kanshahu (2000) asserts that for a project to be successfully implemented and have a sustained impact on the lives of the beneficiaries, it must be well managed. He stresses that good management is facilitated by adequate institutions, that is, by supportive organizations, laws, policies, procedures and rules. In other words, inadequate and disregard for institutional provisions can undermine the effectiveness of any rural development project. Therefore, the study aimed to examine the institutional provisions in the implementation of Fadama 111 Development Project. It ascertains the extent of adherence to the rules and regulations guiding the implementation of the project and identifies perceived institutional issues that constrained effective implementation of Fadama III project in the state.

Methodology

The study was carried out in Taraba State of Nigeria. Taraba State lies approximately between latitude $6^{0}30^{11}$ and $9^{0}36^{11}$ north and longitude $9^{0}10^{11}$ and $11^{0}50^{11}$ east and covers a land mass of 60,291.82 km^2 (Taraba State Official Diary, 2012). As a result of its agrarian nature, a good proportion of the population is engaged in farming as occupation. About three quarters (75%) of the people are crop farmers, livestock farmers and fish farmers while an estimated one quarter (25%) are engaged in other economic activities (Taraba Agricultural Development Programme (TADP, 1998).

The population comprised the management staff as well as other personnel employed at the state and local government offices of Fadama III Project in the state. There are sixteen local government areas (LGAs) in Taraba State and all of them participated in Fadama III Project. Fifty per cent of the LGAs were selected through simple random sampling technique giving eight LGAs. All the management staff (9) at the State Fadama Coordinating Office, the 32 facilitators in the selected LGAs and the sixteen desk officers at the Local Fadama Desks Office was used giving 57 respondents for the study. The data were collected through use of questionnaire and secondary data from official documents.

The respondents were asked to indicate organizations that are linked to the implementation processes of the program and the strength of the linkages. (e.g linkage between the Fadama III staff in the state coordinating office and actors in federal government ministries/parastatals, donor agencies, state government ministries/parastatals, local governments, service providers etc). This was achieved using four point Likert-type scale of strong = 4, moderate = 3, weak = 2 and no linkages = 1. These values were added and divided by 4 to get the mean value of 2.5. Any value ≥ 2.50 was regarded as strong linkage while value <2.50 represented weak linkage.

The rules and regulations guiding the operation of Fadama III in areas of project management and preparation, staff recruitment, financial management, preparation of local development plans, environment compliance and friendliness and group formation were assessed. The respondents were asked to indicate the degree of adherence to these rules and regulations on 5-point Likert-type scale of very strong (5), strong (4), moderatel (3), weak (2), very weak (1). These values were added and divided by 5 to get a cut-off point of 3.0. Any value ≥ 3.0 meant strong adherence while value of <3.0 was considered weak adherence to the rules and regulations. Also information on the perceived institutional constraints (example lack of payment of counterpart fund by the government, delay in the payment of counterpart fund, late disbursement of funds by the Fadama office, bureaucratic bottlenecks) to the project was elicited from the respondents using 4-point Likert- type scale of very serious (4), serious (3), less serious (2) and not serious (1). The nominal values were added and divided by 4 to get a mean value of 2.50. Constraint items with mean scores ≥ 2.50 were considered as serious, while <2.50 was regarded as not a serious constraint.

Results and Discussion

Linkages of SFCO with parastatals and organizations

The level of linkages between the personnel of state Fadama coordinating office (SFCO) and other actors of Fadama III was discussed under the following headings:

State fadama coordinating office linkage/collaboration with federal ministries and parastatals

The SFCO had strong linkages with Federal Ministry of Agriculture and Water Resources (M=3.37), Federal Ministry of Environment and Federal Ministry of Finance (M=3.27), National Food Reserve Agency (M=3.25), and Federal Ministry of Education (M=2.58). However, the standard deviation shows wide variance in the respondents' opinion of the strength of the linkages. Relatively, the results suggest some level of partnership and cooperation between the staff of SFCO and those of the federal ministries and parastatals. Structurally, the project is designed such that relevant institutions partner at different implementation stage, field and around Fadama resources. Hence, the need for interaction, exchange and flow of information, among the

manpower and institutional elements in all the different collaborating ministries and agencies that are directly or indirectly involve in implementing the project. Where this is lacking it could undermine efforts in the implementation process. For instance , linkage according to Ani (2007) is very necessary among researchers, extension practitioners, farmers and policy makers because it enables each category of the actors to wake up to the realization that solution to problems related to linking knowledge generations, dissemination and utilization requires a system's approach. Linkages between the SFCO and the collaborating ministries/agencies ensure productive communication, efficient resource use and enhanced performance of the entire project implementation process.

Table 1. Mean score distribution on linkage between the staff of SFCO and other actors in Fadama III Project

Federal parastatals	Mean (M)	Standard deviation (SD)
Federal Ministry of Agric and Water Resources	3.37	0.850
National Food Reserve Agency	3.25	0.857
Federal Ministry of Environment	3.27	0.962
Federal Ministry of Finance	3.27	1.112
Federal Ministry of Education	2.58	1.270
State ministries/agencies		
State Ministry of Agric (SMA)	3.66	0.769
State Ministry of Finance/	3.61	0.803
State Ministry of Education	2.52	1.338
State Ministry of Women Affairs	2.69	1.168
State of Ministry Environment	3.07	1.016
Service providers		
ADPS	3.39	0.941
National Livestock Development Programmes	3.07	1.215
ARMTI	2.79	1.114
Universities	2.85	1.156
IITA	3.05	1.117
Individual consultants	3.19	0.981
Private entrepreneurs	3.00	1.000
Community groups	3.30	0.912
NGOs	3.19	1.111
Qualified Civil Organizations	3.07	1.035
Community level fadama organizations		
Local Fadama Community Desks	3.55	0.910
FCAs/FUGs	3.39	0.780

State fadama coordinating office (SFCO) linkages with state ministries and parastatals

The state Fadama coordinating office are strongly linked to the State Ministry of Agriculture and Natural Resources (M=3.66), State Ministry of Finance (M=3.61), State Ministry of Environment (M=3.07), State Ministry of Women affairs (M=2.69), and State Ministry of Education (M=2.52) (Table 1). The SFCO carries out day to day coordination, and using resources from the

line ministries approves engineering and other consultant services, reviews, screens, and provides clearance on the technical viability of all subprojects submitted for project funding. Besides, most of these organizations are involved in providing services in crop protection, horticulture, agro forestry, credit and environmental management, hence the need for strong linkage. This will boosts efficient resource use, and enhanced performance of the whole development project.

SFCO linkage with service providers

Table 1 shows that SFCO's had strong link with the ADP (M=3.39), community groups other than FUGs (M=3.30), individual consultants/privates (M=3.19), civil organizations/NGOs (M=3.1), National Livestock Development Programme (M=3.05), private entrepreneurs (M=3.00), universities (M=2.85), and ARMTI (M=2.79) (Table 1). This is not surprising because under Fadama III project, increased attention was placed on providing adequate training for the facilitators and Fadama community associations. In addition, much effort was placed on the involvement of civil society (e.g., NGOs and other community-level professional associations/service providers); particularly with respect to information, mobilizing the vulnerable groups, and supporting training activities at the community level (World Bank, 2008). These organizations played one or two of the above roles. Consequently, strong linkage with the organizations is apt to ensure adequate training of facilitators and subsequent provision of technical advice, efficient service delivery, community empowerment and improved quality of decision-making within FCAs. Many rural development interventions have failed because of neglect of interdependency in roles, particularly in integrated development approach.

SFCO linkage with Fadama Community Associations (FCAs)

The results show that both the officials of FCAs (M=3.59) and the staff of the Local Fadama Desk (M=3.55) had strong linkage with staff at SFCO (Table 1). Fadama community associations are apex organizations of economic interest groups (EIGs) which have a common interest; to derive their livelihood from the shared natural resources of Fadama. They identify, prepare, execute, supervise, operate, and maintain their subprojects (Nkonya et al., 2010). They are the direct beneficiaries of Fadama Project. The coordination function of the SFCO is not efficient without strong and functional grass root linkage. Strong linkage between staff of SFCO and the Fadama farmers' organizations at the community level guarantee adequate supervision and successful implementation of the project. Above all, strong linkage between the staff of the SFCO and the fadama organizations ensures cooperation, effective tracking and feedback on the project.

Adherence to rules and regulations in Fadama III Project

The project had clearly stated rules and regulations for the implementation process The staff adhered to all the rules regarding project management and preparation, staff recruitment requirement, financial management, preparation of local development plan, environmental compliance and screening and group formation (Table 2 & 3).

Table 2. Mean score on level of adherence to rules and regulations in Fadama 111 Project

Rules	(M)	(SD)
Rules on project management preparation		
Approval and periodicals review of work plan and budget	3.77	1.460
Compliance with sector policies during implementation	3.63	1.191
Approval of annual physical and financial reports, auditor's report and project accounts	4.05	1.163
Selection of contractors and award of contracts	3.92	1.392
Rules on staff recruitment		
Requirement on basic minimum qualifications	3.82	1.468
Requirement on post qualifications experience	3.72	1.273
Requirement on experience in a Community Driven Development project	3.56	1.331
Requirement on relevant skills and computer literacy	3.30	1.231
Requirement on adequate number of both technical and supportive staff	3.50	1.454
Rules on financial management		
Following due process in the approval and disbursement of funds	3.87	1.454
Deposit of 30% up front by the beneficiaries	4.08	1.075
Contribution of 70% by the project	4.08	1.217
Establishment of saving scheme by the FUGs	3.90	1.209
Release of counterpart funds on timely basis by the state government	3.05	1.413
Rules on preparation of Local Development Plans (LDPS)		
Identification of problems	3.85	1.182
Causes of problems	3.69	1.280
Proffering solutions/actions to be taken	3.66	1.236
Determining implementation period	3.59	1.322
Identifying resources for implementation	3.88	1.289

Table 3. Mean distribution of respondents based on adherence to rules on environmental compliance and group formation

Rules and regulations	Mean (M)	Standard deviation (SD)
Rules on environmental compliance and friendliness		
Rules on land clearing and cultivation e.g. bush should not be set on fire, forest should not be destroyed etc.	3.60	1.288
Rules on irrigation practices e.g. the project should not cause the loss of available surface water etc.	3.59	1.438
Rules on the use of agro-chemicals e.g. observing the rules on international pest management practices etc.	3.73	1.329
Rules on protecting beneficial animals and insects e.g. the project should not cause reduction on the number of species or reduce their habitat	3.59	1.328
Rules on fishing practices e.g. avoiding chemicals to kill fish in water	3.82	1.167
Rules on access roads and tracks construction e.g. road construction should not destroy animal habitat etc.	3.68	1.430
Rules on Group Formation		
Disabled persons (blind, deaf, dumb, cripples etc.)	4.05	1.158
Widows	3.81	1.330
The poor	3.79	1.301
Youths	3.74	1.245
Religious groups	3.38	1.426
Women	3.38	1.362

This is contrary to The World Bank appraisal report of Fadama II, which indicated low performance of existing manpower and inadequate adherence to institutional provisions (World Bank, 2008). Particularly, there was low performance in the preparation of local development plans by the officials of the FUGs/FCAs. Also disregard for laid down rules and regulations as cases of collusion between the facilitators and the service providers were reported (IFPRI, 2007). Generally, the results suggest that the project execution was carried out according to World Bank specification. This could be explained by many factors including the monitoring evaluation strategies employed, incorporation of lessons learnt from previous phases of the project; which together ensures rule compliance, enforcement, conformity to rules guarding implementation and minimizes the abuse and mismanagement of resources (Ghate and Nagendra, 2005). According to Tyler (2005), securing employee adherence to work place rules and company policies is one key antecedent of a successful coordination and functioning within organizations. This is quite very uncommon with most government agricultural programmes due to problems of corruption, administrative bottleneck and bureaucracy and similar man-made problems in public system. Adherence to rules and laws greatly account for the success or failure recorded in development interventions.

Perceived Institutional factors affecting the implementation of Fadama III Project

Table 4. Mean score distribution of institutional factors affecting the implementation of Fadama III project

Factors	Mean (M)	Standard Deviation
Lack of payment of counterpart fund by the government	3.04*	0.980
Delay in the payment of counterpart fund	3.39*	0.790
Late disbursement of fund by the fadama office	2.98*	1.083
Bureaucratic bottlenecks	2.75*	1.072
Embezzlement of funds	1.90	1.015
Lack of transport and other logistic support	3.06*	0.886
Insufficient office facilities (computers, furniture etc)	2.37	1.248
Shortage of funds	2.56*	1.037
Superstitious beliefs of project beneficiaries	2.00	1.066
Religious prohibition of the acceptance of some sub projects	1.83	1.080
Cultural taboos	1.71	1.054
Lack of effective monitoring and supervision of subprojects	2.19	1.093
Lack of capacity to enforce rules and regulations	2.35	1.153
Political differences	2.14	1.167
Bad leadership	1.78	1.083

Serious constraints factor

Table 4 shows that six out of the fifteen items were perceived by the Fadama staff as serious institutional constraints militating against the implementation of the project. These include delay in the payment of counterpart fund by the government (M=3.39), lack of transport and other logistic support (M=3.06), lack of payment of counterpart fund by the government (M=3.04), late disbursement of funds by the Fadama III office (M=2.98), bureaucratic bottlenecks (M=2.75) and shortage of funds (M=2.56) (Table 1). The findings reveals that most constraints affecting Fadama III project revolve around the way financial issues were handled by the government as well as the Fadama III project itself. In addition to several other criteria, the participation of states in the progamme was based on regular payment of counterpart fund, but as common with government interventions/programme, poor funding remains a formidable challenge. This confirms the finding of Agwu and Abah (2009) who observe poor funding and late disbursement of funds as problem in the second National Fadama Development Project. Timely and regular release of fund is expedient for financing and implementing main components and sub-components of project designed to transfer financial and technical resources to the beneficiary groups. Also, the problem of bureaucratic bottleneck is a reoccurring impediment in the execution of government programmes. It delays and forestalls timely implementation of activities of all the components of the project, leading to inefficiency and poor performance of the technical and administrative arrangements.

Conclusion

Institutional provisions are crucial for effective implementation of development programmes. Fadama 111 Development Project maintained strong linkages with relevant organizations, institutions and associations in the implementation and achievement of the project objectives. The staff strongly adhered to the rules and regulations guiding operation of different components of the project such as project management preparation, staff recruitment, financial management, preparation of Local Development Plans (LDPS) and others. However, the project is constrained by poor funding and bureaucratic bottleneck. In other words, the project had functional institutional arrangements for successful implementation and enhanced performance but lacked adequate funding to facilitate the whole process. Consequently, recording expected impact on the beneficiaries' income, livelihood and poverty level may be a mirage. The study therefore recommends government support through timely payment of counterpart fund and provision of logistic support to facilitate the entire implementation process. Policy makers and development planner should ensure functional mechanisms that could enhance and foster linkages; and promote adherence to rules and regulations prescribed for implementation of development programmes.

References

Agwu, A.E. and Abah, H.O. 2009. Attitude of farmers towards cost sharing in the second National Fadama Development Project: the case of Kogi State of Nigeria. *J. Agril. Ext.* 13 (2): 92-106.

Akinola, M.O. 2003. The performance of Fadama users association under the National Fadama Development Project phase One, Nigeria. PhD Thesis, Ahmadu Bello Univerty, Zaria. 73p.

Ani, A.O. 2007 Agricultural Extension: A Pathway for Sustainable Agricultural Development. Apani Publications, Kaduna, Nigeria.

FMA and WR. 2008. Third National Fadama Development Project (fadama III). Project Implementation Manuals. Federal Ministry of Agriculture and Water Resources.

Ghate, R. and Nagendrra, H. 2005. Role of monitoring in institutional performance. Forest management in Maharashtra, India. http://conservationandsociety.org/article.asp/ISSN 0972-4923. accessed 2nd August, 2010.

IFPRI. 2007. Beneficiary assessment/impact evaluation of Fadama II Project. A World Bank document on the third National Fadama Development Project (Fadama III). International Food Policy Research Institute. pp. 61-64.

Kanshahu, A.I. 2000. Planning and implementing sustainable projects in developing countries-Theory, practices and economics. 2nd revised and expanded edition. Singapore, AgBe publishing. p. 209.

Nkonya, E.H., Markel Ova, Kato, E., Alomolaron, A., Shetima, A.G., Ingawa, S., Madukwe, M.C., Olukosi, J., Phillip, D. and Park, D. 2010. Baseline Report of Fadama III Impact Assessment Study. International Food Policy Research Institute, Washington, DC, USA. pp. viii.

TADP. 1998. The adoption of technologies under the National Fadama Development Project in Taraba State. Report Prepared for planning, monitory and evaluation sub-programme of Taraba State Agricultural Development Programme.

Taraba State Official Diary. 2012. Official Diary of Taraba State government of Nigeria.

Tyler, T.R. 2005. Promoting employee policy adherence and role following work settings: the value of self regulatory approaches http/www.digitalcommonslaw.yale.edu/cgi/viewcontent.cgi, accessed 20th October, 2012.

World Bank. 2008. Project appraisal document on a proposed credit to the Federal Republic of Nigeria for third National Fadama Development (Fadama III) Project. p. 52.

15

FACTORS AFFECTING THE MILK PRODUCTION OF DAIRY CATTLE IN NORTHERN RURAL AREAS OF BANGLADESH

M.R. Begum[1*], M. Anaruzzaman[2], M.S.I. Khan[3] and M. Yousuf[4]

Abstract

A cross sectional study was conducted to observe the factors affecting the productive performance of dairy cattle from northern rural areas of Bangladesh during July and September 2013. Data of 105 cows, 85 (80.95%) from local and 20 (19.05%) cows from cross breed, were randomly selected for the study. A binary logistic regression, expressed by odds ratio with 95% confidence interval, was done to determine the association of daily milk production categorized into ≤ 2 and > 2 liters (L), based on median, with the significant explanatory variables of body weight, age at first calving, lactation period, vitamin use, type of floor and milking person. The result demonstrated that the probability of milk production of >2 L was 6.16, 4.5, 20.65 and 5.7 times higher from the with animal body weight of >140 kg, age at first calving of >36 m, lactation period of >8 m and vitamin use than that of body weight of ≤ 140 kg, age at first calving of ≤ 36 m, lactation period of ≤ 8 m, and not vitamin used respectively. The chance of milk production of > 2 L was 0.25 and 0.22 times lower for mud floor, and owner milking than that of brick floor and gowala (professional milking person) respectively.

Keywords: Breed, Factors, Milk Production, Odds Ratio, Logistic Regression

[1]Assistant Professor, Department of Agricultural Economics and Social Sciences, Chittagong Veterinary and Animal Sciences University, Khulshi, Chittagong, Bangladesh
[2]Doctor of Veterinary Medicine, Chittagong Veterinary and Animal Sciences University, Khulshi, Chittagong, Bangladesh
[3]Assistant Professor, Department of Food Microbiology, Patuakhali Science and Technology University (PSTU), Dumki, Patuakhali, Bangladesh
[4]Director (Program), SANGRAM (Sangathita Gramunnyan Karmasuchi), Bangladesh

*Corresponding author's email: rasustat@yahoo.com (M.R. Begum)

Introduction

An important component of national economy, Bangladesh has 47.51 million livestock of which 22.87 million are cattle, plays a vital role for economic development of Bangladesh (BBS, 2008). Better genotype and sound management are the major determinants of profitability of dairying at either farm or individual level (Djemali and Freeman, 1987; Rahman et al., 1987) although, performances of high yielding exotic crossbred cows may differ among different geographical areas (Jahan et al., 1990; Alam and Ghosh, 1994). Biological potential for milk production also depends on the age at puberty, first calving age, number of parity and calving interval (Djemali and Freeman, 1987; Rahman et al., 1987). Around 2 years age of first calving and less than 13 to 14 months calving interval are the indicators of better management index of a farm (Wiltbank, 1970; Sarder, 2001). Besides, some diseases also have influence on the milk production where mastitis is one of the most devastating and farmers in Bangladesh are not well aware of the best practices to control (Rehman et al., 1997). For good health and high milk yields, cows need to be fed the proper amounts of available minerals and vitamins (Weiss, 1998). The present study was therefore undertaken to detect the factors of milk production. To be specific, the goal was to quantify the significant association of milk production with other covariates and factors by implying binary logistic regression model.

Materials and Methods

A cross sectional study was conducted at Rangpur and Gaibandha districts of northern rural areas of Bangladesh. A total of 105 cows were randomly selected where 85 were local and 20 were cross breed between July and September 2013.

Statistical analysis

Logistic regression, a type of probabilistic statistical classification model can be used to predict a binary response from a binary predictor and for predicting the outcome of a categorical dependent variable based on one or more predictor variables.

The simple logistic model has the form

$$logit(y) = ln\left(\frac{\pi}{1-\pi}\right) = \alpha + \beta_1 x_1 \qquad (1)$$

After simplification,

$$\pi = Probability(Y = Outcome\ of\ interest|X = x) = \frac{e^{\alpha+\beta_1 x_1}}{1+e^{\alpha+\beta_1 x_1}} \qquad (2)$$

Where 'ln' denotes the natural logarithm, π is the probability that the dependent variable of milk production (>2 liters), α is the intercept from the linear regression, β_1 is the regression coefficient multiplied by predictor, x can be categorical or continuous but Y always categorical. According to equation 1, the relationship between logit (Y) and X is linear. Yet, according to equation 2, the relationship between the probability of Y and X is nonlinear. For this reason, the natural log transformation of the odds in equation 1 is necessary to make the relationship between a categorical outcome variable and its predictor(s) linear. A binary logistic regression model was fitted with least significant value <0.05 (Hosmer and Lemeshow, 2000) and result was interpreted by odds ratio. Data analysis was done by statistical software packages SAS 9.2 and RStudio.

Results

The average Body weight (Kg), Daily milk production (L), Lactation period (m), Age of first calving (m) and Calving interval (m) of local and cross breed were 131.49 and 253.3 kg, 1.66 and 6.63 liters, 7.41 and 10.8 months, 35.36 and 41.55 months, and, 14.96 and 15.2 months respectively presenting in Table 1.

Table 1. Descriptive statistics of different quantitative variables

Variables	Local		Cross	
	Mean	St. Dev	Mean	St. Dev
Body Weight (Kg)	131.49	31.29	253.3	77.12
Daily milk production (L)	1.66	0.68	6.63	2.8
Lactation period (m)	7.41	2.25	10.8	3.17
Age of first calving (m)	35.36	5.4	41.55	6.18
Calving interval (m)	14.96	4.92	15.2	4.79

The bar diagram shows in Fig. 1 represent the overall frequency percent distribution where 36 (34.3%) and 34 (32.4%) cows were in third and second parities respectively. The maximum frequency of milking daily once was found 87 (82.9%), and 29 (27.6%) owner collect milk by Gowala (Professional milkers) and remaining 76 (72.4%) by himself but the proportion of vitamin used during lactation period was only 29 (27.6%). It was observed that, 98 (93.3%) owner used regular anthelmintic and anthrax and FMD vaccine were used in 20 (19%) and 23 (21.9%) cases respectively and the remaining 62 (59%) reared their cows without any vaccination.

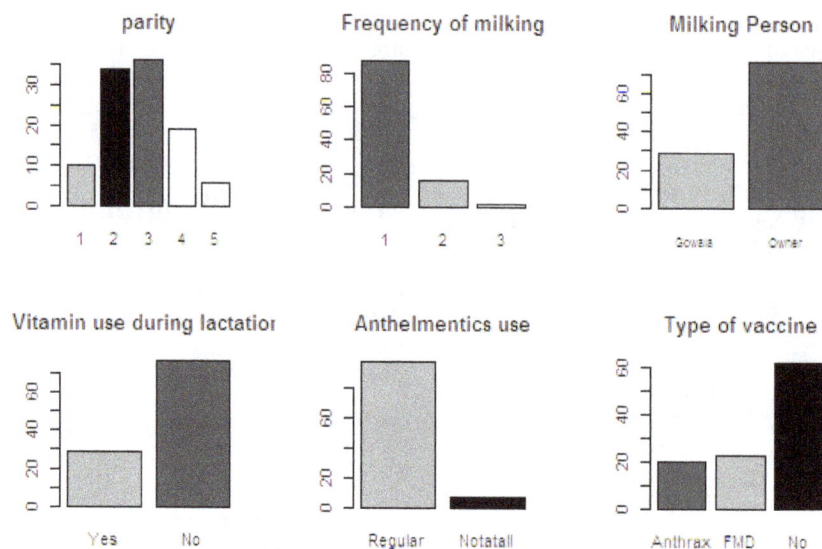

Fig. 1. Bar diagram of different categorical variables

The pie chart presents in Fig. 2 portrays the frequency and percentage distribution of different variables. Among the cows, local and cross breed were 85 (81%) and 20 (19%) where only 8 (8%) and 7 (7%) cows showed the clinical sign of mastitis and FMD respectively. The season of calving in summer, rainy and winter were respectively 50 (48%), 16 (15%) and 39 (37%). The number of vaccinated cows were only 42 makes up 41%. In case of cowsheds, the brick floor and mud floor were found 69 (56%) and 46 (44%) respectively.

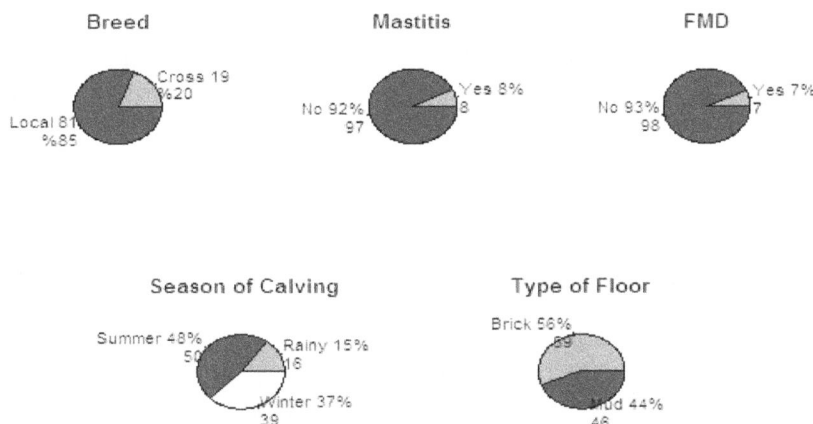

Fig. 2. Pie-chart of different categorical variables

A contingency analysis was conducted to observe the frequency and percentages among the different cells distribution. Table 2 depicts the frequency and percentage distribution among different categorical variables based on binary milk production categorized based on median (50%). Body weight, age of first calving, lactation period, calving interval were categorized based on median and only significant covariates were considered. In category of ≤2 L average milk production, maximum frequency and percentages were (44) 83.02%, (56) 73.68% and (59) 85.51% found within ≤140, ≤36 and ≤8 of body weight (kg), age of first calving (m) and lactation period(m), which was (29) 55.07%, (18) 62.77% and (28) 77.78% in > 140 m, > 36 m and > 8 for category of > 2 L milk production respectively. The maximum proportion was found (57) 75%, (37) 80.43% and (56) 73.68% for no vitamin use, mud floor and owner for the category of ≤2 L milk production, wherever, it was (19) 65.52%, (29) 49.15% and (18) 62.07% for vitamin use, brick floor and gowala in the category of > 2 L milk production.

From the binary logistic regression analysis, it was observed some variables were significantly associated with milk production. The odds of milk production > 2 liters for body weight >140 kg was 6.16 times than that of body weight ≤140 kg, 4.58 times higher for age of first calving of >36 m than ≤ 36 m, 20.65 times higher for lactation period >8 m than ≤ 8 m and, 5.7 times higher for vitamin use than no vitamin used cows. The probability of milk production > 2 liters for mud floor type was 0.25 times lower than that of brick floor type and 0.22 times lower in owner milking than gowala.

Table 2. Percent distribution of different categorical variables on the basis of milk production

Variables	Category	Milk production (liters)		Total	Odds Ratio(OR)	95% CI of OR
		≤2 (%)	>2 (%)			
Body Weight (Kg)*	≤140	44(83.02)	9(16.98)	53	6.16	2.50, 15.19
	>140	23(44.23)	29(55.77)	52		
Age of first calving(m)*	≤36	56(73.68)	20(26.32)	76	4.58	1.85, 11.35
	>36	11(37.93)	18(62.07)	47		
Lactation Period (m)*	≤8	59(85.51)	10(14.49)	69	20.65	7.35, 58.00
	>8	8(22.22)	28(77.78)	36		
Vitamin Use*	No	57(75.00)	19(25.00)	76	5.70	2.26, 14.38
	Yes	10(34.48)	19(65.52)	29		
Type of Floor*	Brick	30(50.85)	29(49.15)	59	0.25	0.10, 0.61
	Mud	37(80.43)	9(19.57)	46		
Milking Person*	Gowala	11(37.93)	18(62.07)	29	0.22	0.09, 0.54
	Owner	56(73.68)	20(26.32)	76		

*p-value <0.01

Discussion

From the study, it was revealed that, significant relation was pragmatic between milk production and age at first calving which was consistent with previous literature (Pirlo et al., 2000; Bayram et al., 2009; Bajwa et al., 2004). On the other hand, no significance relation of age of first calving with milk production in Brown Swiss Cattle investigated by Bayram et al. (2009). Lactation period was significantly associated with milk production, which was similarly demonstrated in the research result of Baul et al. (2012). In case of parity, no significant relation between milk yield and 4th gives the highest production. Dhumal et al. (1989) found no relation between milk yield and parity. Similarly, 4th parity was detected with highest production by Bajwa et al. (2004) whereas Tahir et al. (1989) registered 5th lactation. Similar significant effect of vitamin use on milk production of this study was observed by Bregsten et al. (2003) for high yielding cows whereas 20 mg of supplemental biotin per day in higher producing cows (>75 lbs day^{-1}) increase milk production 2 to 7 lbs day^{-1} but no response in lower producing cows (<45 lbs day^{-1}). In this study, floor type has significant effect on milk production, in opponent, no significant relation was observed by Kremer et al. (2007). In this study, there was no significant relation between anthelmentic use and milk production. However, a study by Gross et al. (1999) delineated that a median increase in milk production of 0.63 kg/cow per day might be expected after anthelmintic treatment. Another study of meta analysis done by Sanchez et al. (2004) showed that on average, an increase of milk production of 0.35 kg cow^{-1} day^{-1} might be expected after anthelmintic treatment of naturally infected lactating dairy cows. Finally it was observed that the milk production was 0.22 times lower for owner milkers than gowala, this may be due to owner's inefficiency or their tendency to give the adequate milk left for calf.

Acknowledgements

The author thanks to Veterinary Surgeon of Gaibandha Upazila Livestock Hospital and all farmers who provided information during conducting this study.

References

Alam, M.G.S. and Ghosh, A. 1994. Plasma and milk progesterone concentrations early pregnancy in Zebu cows. *Asian Australasian J. Animal Sci.* 7: 131-136.

Bajwa, I.R., Khan, M.S., Khan, M.A. and Gondal, K.Z. 2004. Environmental factors affecting milk yield and lactation length in Sahiwal cattle. *Pakistan Vet. J.* 24(1): 23-27.

Baul, S., Cziszter, L.T., Acatincǎi, S., Gavojdian, D., Tripon, I., Erina, S. and Rǎducan, G.G. 2012. Phenotypic correlations among milk yield and chemical composition per normal lactation in Romanian black and white Breed. *Animal Sci. Biotech.* 45 (2): 275-277.

Bayram, B., Yanar, M. and Akbulut, O. 2009. The effect of average daily gain and age at first calving on reproductive and milk production traits of brown swiss and holstein friesian cattle. *Bulgarian J. Agril. Sci.* 15 (5): 453-462.

BBS . 2008. Statistical Pocketbook of Bangladesh. Bangladesh Bureau of Statistics, Statistics Division, Ministry of Planning, Government of the People of Republic of Bangladesh, Dhaka. p. 256.

Bergsten, C., Greenough, P.R., Gay, J.M., Seymour, W.M. and Gay, C.C. 2003. Effects of biotin supplementation on performance and claw lesions on a commercial dairy farm. *J. Dairy Sci.* 86: 3953-3962.

Dhumal, M.V., Salhare, P.G. and Deshpande, K.S. 1989. Factors affecting lactation milk yield and lactation length in Red Kandhari and crossbred cows. *Indian J. Dairy Sci.* 42: 102-104.

Djemali, M. and Freeman, A.E. 1987. Reporting of dystocia scores and effects of dystocia on production days open, and days dry from dairy herd improvement data. *J. Dairy Sci.* 70: 2127-2131.

Gross, S.J., Ryan, W.G. and Ploeger, H.W. 1999. Anthelmintic treatment of dairy cows and its effect on milk production. *Vet. Rec.* 144: 581–587.

Hosmer, D.W. and Lemeshow, S. 2000. Applied Logistic Regression. New York: *John Wiley and Sons, 2nd edition.* pp. 147-156.

Jahan, M.K.I. Chowdhury, M.M.R., Nahar, S.M.Z.H. and Rahman, M.F. 1990. Performances of local and cross-bred dairy cows in farm condition. *The Bangladesh Veterinarian,* 7: 48-49.

Kremer, P.V., Nueske, S., Scholz, A.M..and Forester, M. 2007. Comparison of claw health and milk yield in dairy cows on elastic or concrete flooring. *J. Dairy Sci.* 90 (10): 4603-11.

Pirlo, G., Miglior, F. and Speroni, M. 2000. Effect of age at first calving on production traits and on difference between milk yield returns and rearing costs in Italian Holsteins. *J. Dairy Sci.* 83: 603–608.

Rahman, M.S. Ahmed, M. and Ahmed, A.R. 1987. A comparative study on some productive and reproductive performance of dairy cows at Savar Dairy cattle improvement farm. *Bangladesh Vet. J.* 21: 55-61.

Rehman, M.S., Nooruddin, M., and. Rahman, M.N. 1997. Prevalence and distribution of mastitis in crossbred and exotic dairy cows. *The Bangladesh Veterinarian,* 14: 104.

Sanchez, J., Dohoo, I., Carrier, J. and DesCoteaux, L. 2004. A meta-analysis of the milk-production response after anthelmintic treatment in naturally infected adult dairy cows. *Preventive Vet. Med.* 63: 237–256.

Sarder, M.J.U. 2001. Reproductive and productive performances of indigenous cows. *The Bangladesh Veterinarian,* 18: 123-129.

Tahir, M., Qureshi, M.R. and Ahmad, W. 1989. Some of the environmental factors influencing milk yield in Sahiwal cows. *Pakistan Vet. J.* 9: 173-175.

Weiss, W.P. 1998. Requirements of fat-soluble vitamins for dairy cows: A review. *J. Dairy Sci.* 81: 2493–2501

Wiltbank, J.B. 1970. Research needs in beef cattle reproduction. *J. Animal Sci.* 3: 755-762.

16

EFFECT OF INTEGRATED APPROACH OF PLANT NUTRIENTS ON YIELD AND YIELD ATTRIBUTES OF DIFFERENT CROPS IN WHEAT-SESAME-T. AMAN CROPPING PATTERN

M.A. Islam[1]*, Mst. A. Begum[2] and M.M. Jahangir[3]

Abstract

The experiment was carried out at FSRD site, Pushpopara, Pabna, during November, 2010 to December, 2011 to observe the comparative performance of integrated plant nutrients management (IPNS) system through the use of organic (cowdung, cowdung slurry) manure and inorganic fertilizer on wheat, sesame and T. Aman crops under wheat-sesame-T. Aman cropping pattern. The experiment was consisted with four treatments viz. T_1: Soil test based inorganic fertilizer dose for high yield goal, T_2: Cowdung @ 5 t ha^{-1} + IPNS basis inorganic fertilizer dose for high yield goal, T_3: Cowdung slurry @ 5 t ha^{-1} + IPNS basis inorganic fertilizer dose for high yield goal and T_4: Fertilizer dose usually practiced by the farmers. In case of wheat, the highest grain yield (3.80 t ha^{-1}) was obtained from bio-slurry treated plot that means T_3 treatment followed by T_2 and the lowest (3.31 t ha^{-1}) from T_4. Higher seed yield (1.31 t ha^{-1}) of sesame was obtained from T_3 that was statistically identical to T_2 and T_1 and the lower (1.01 t ha^{-1}) from T_4. For T. Aman rice, the highest grain yield (4.89 t ha^{-1}) was obtained from T_3 which was statistically indistinguishable from T_1 where as the lowest grain yield (4.1 t ha^{-1}) was recorded from T_4. Considering the whole pattern, it is observed that the highest gross return (271100 Tk ha^{-1}) was obtained from T_3 followed by T_2 and the lowest (225650 Tk ha^{-1}) from T_1 treatment. Total variable cost was recorded as the highest (100368 Tk ha^{-1}) in T_2 followed by T_3 and the lowest (86775 Tk ha^{-1}) in T_4 treatment. The highest marginal value of product (45450 Tk ha^{-1}) was recorded in T_3 followed by T_2 where as the minimum (28710 Tk ha^{-1}) was found in T_1 over the T_4 treatment. Marginal variable cost was observed as the highest (13593 Tk ha^{-1}) in T_2 treatment followed by T_3 and the minimum (8899 Tk ha^{-1}) was recorded in T_1 treatment. The highest MBCR (4.15) was recorded from T_3 followed by T_2 and the minimum (2.31) from T_2 treatment.

Keywords: IPNS System, Organic Manure, Rice Equivalent Yield, Cropping Pattern

[1]On Farm Research Division, Regional Agricultural Research Station, BARI, Pabna, Bangladesh
[2]Department of Agronomy, Bangladesh Agricultural University, Mymensingh, Bangladesh
[3]Department of Soil Science, Bangladesh Agricultural University, Mymensingh, Bangladesh

*Corresponding author's email: amin_bau@yahoo.com (M.A. Islam)

Introduction

The basic concept of Integrated Plant Nutrition System (IPNS) is the management of all available plant nutrient sources, organic and inorganic, to provide optimum and sustainable crop production conditions within the prevailing farming system. Therefore, in IPNS an appropriate combination of mineral fertilizers, organic manures, crop residues, compost, N-fixing crops and bio fertilizer is used according to the local ecological conditions, land use systems and the individual farmer's social and economic conditions.

The main aim of integrated plant nutrition system is to increase and sustain soil fertility to provide a sound basis for flexible food production systems that, within the constraints of soil and climate, can grow a wide range of crops to meet changing needs (FAO, 2001). Therefore, it is necessary to use inorganic and organic fertilizers in an integrated way so as to obtain economically profitable crop yield, without incurring loss to soil fertility (Haque et al., 2001). IPNS can produce comparable or higher crop yield compared to sole fertilizer use (BARC, 2005).

Soil fertility deterioration is a major constraint for higher crop production in Bangladesh. The increasing land use intensity without adequate and balanced use of chemical fertilizers and with little or no use of organic manure have caused severe fertility deterioration of our soils resulting in stagnating or even declining of crop productivity. The farmers of this country use on an average, 215 kg nutrients/ha annually (149 kg N + 37 kg P2O5 + 22 kg K2O + 7 kg S, Zn, B and others), while the crop removal is about 280-350 kg ha^{-1} (Islam, 2008). Since fertile soil is the fundamental resource for higher crop production, its maintenance is a prerequisite for long- term sustainable crop productivity. Soil organic matter is a key factor for sustainable soil fertility and crop productivity. Organic matter undergoes mineralization with the release of substantial quantities of N, P, and S and smaller amount of micronutrients. In Bangladesh, most of the cultivated soils have less than 1.5% organic matter and some soils even less than 1%, while a good agricultural soil should contain at least 2.5% organic matter (BARC, 2005). Moreover, this important component of soils is declining with time due to intensive cropping and use of higher doses of nitrogenous fertilizers with little or no addition of organic manure.

In Bangladesh, major food crops remove about 2.98 million tons of nutrients annually against a total addition of 0.72 million ton (Rahman *et al.*, 2008). According to an appraisal report of Bangladesh soil resources, soils of about 6.10 m ha contain very low (less than 1%) organic matter, 2.15 m ha contain low (1-2%) organic matter and the remaining 0.90 ha contain more than 2 % organic matter. Ali (1997) reported that during the years 1967-1995, the highest depletion of organic matter occurred in soils of Meghna River Floodplain (35%) followed by Madhupur Tract (29%), Brahmaputra Floodplain (21%), Old Himalayan Piedmont Plains (18%) and Gangetic Floodplain (15%).

The average organic matter content of top soils has decline by 20-46% over past 20 years due to intensive cropping without inclusion of legume crops, imbalance use of fertilizer, use of modern varieties and scanty use of organic manure. It is agreed that decrease in soil fertility is a major constraint for higher crop production in Bangladesh. The beneficial effect of organic manure in crops production has been demonstrated by many workers (Joshi *et al.*, 1994; Batsai *et al.*, 1979; Singh *et al.*, 1970; Subhan, 1991).

A suitable combination of organic and inorganic source of nutrients is necessary for sustainable agriculture that can ensure food production with high quality (Reganold *et al.*, 1990). Nambiar (1991) viewed that integrated use of organic manure and chemical fertilizers would be quite promising not only in providing greater stability in production, but also in maintaining better soil fertility. The long-term research of BRRI revealed that the application of cowdung 5 t/ha/yr improved rice productivity as well as prevented the soil resources from degradation (Bhuiyan, 1991). Thus, it is necessary to use fertilizer and manure in an integrated way in order to obtain sustainable crop yield without

affecting soil fertility. Based on the soil fertility problem as discussed above, the present study was undertaken to investigate the effect of combined use of chemical fertilizers and organic manures in wheat-sesame-T. Aman rice cropping pattern regarding yield and economic return.

Material and Methods

The experiment was carried out at FSRD site, Pushpopara, Pabna, during November, 2010 to December, 2011 to observe the comparative performance of integrated plant nutrients management (IPNS) system through the use of organic (cowdung, cowdung slurry) and inorganic fertilizer on wheat, sesame and T. Aman crops under wheat-sesame-T. Aman cropping pattern. The trial was conducted in 6 different locations at the farmer's field. Before conducting the experiment, initial composite soil samples (0-15 cm depth) were collected from the experimental plots and were analyzed. Nutrient packages was calculated on the basis of soil test value according to the instructions outlined by BARC fertilizer recommendation guide 2005.

$$F_r = U_f - X \frac{C_i}{C_s} (S_t - L_s)$$

Where

F_r = Fertilizer nutrient required for given soil test value

U_f = Upper limit of the recommended fertilizer nutrient for the respective STVI class

C_i = Units of class intervals used for fertilizer nutrient recommendation

C_s = Units of class intervals used for STVI class

S_t = Soil test value

L_s = Lower limit of the soil test value within STVI class

The experiment was laid out in Randomized Complete Block Design with a unit plot size of 5m×6m. Four fertilizer treatments viz. T_1: Soil test based inorganic basis fertilizer dose for high yield goal, T_2: Cowdung @ 5 t ha^{-1} + IPNS basis inorganic fertilizer dose for high yield goal, T_3: Cowdung slurry @ 5 t ha^{-1} + IPNS basis inorganic fertilizer dose for high yield goal and T_4: Farmers practice. Variety BARI Gom 24, BARI Til 4 and Swarna was selected for wheat, sesame and T. Aman respectively by the cooperator farmers and used in the trial. Wheat seeds were sown in the field on December 15-25, 2010 maintaining spacing of 20 cm × 5 cm. Fifty percent urea and entire amount of TSP, MoP, Gypsum, Zinc sulphate , Boric acid and organic manure were applied as basal as per treatment specification. Rest 50% urea was applied at 18-21 days after sowing (DAS). The seeds of sesame were broadcasted @ 7 kg ha^{-1} on April 6, 2011. All fertilizers with full amount except urea were applied during final land preparation. Fifty percent urea was applied as basal and the rest amount was applied as top dress during first irrigation. In case of T. Aman rice, thirty day's old seedling was transplanted on 16 and 17

August, 2011. Entire amount of all fertilizers including cowdung manure and cowdung slurry except urea were applied as basal. Urea was applied in two equal splits at 15 and 55 DAT as top dress. Intercultural operations viz. weeding, thinning, irrigation and spraying of pesticides were done as and when required in order to support normal plant growth and development. The crops were harvested after full maturity. Wheat, sesame and T. Aman were harvested on March 12-17, July 7 and 25-29 November, 2011 respectively.

The recorded data were statistically analyzed following Gomez and Gomez (1984). All types of variable production cost are recorded to find out the marginal benefit cost ratio (MBCR). Economic analysis with respect to net return was carried out to evaluate the profitability of different treatments.

Table 1. Soil test value (Nutrient status) of the initial soil of the experimental field at FSRD site, Pushpopara, Pabna

pH	OM (%)	Total N (%)	P	S	B	K (me/100g soil)
			(μg/g soil)			
8.23	1.91	0.11	17	21	0.39	0.23
Alkaline	M	L	M	M	M	O

Table 2. Analytical value of N, P and K nutrient from different manure (Cowdung and bioslurry)

Kind of Manure	Nutrient supply (kg) from 1 ton material		
	N	P	K
Cowdung	3	1	3
Cowdung slurry	4.6	1.6	5

Table 3. Calculated amount of different nutrients as per treatment specification applied for different crops under wheat – sesame – T. Aman cropping pattern (Soil test based)

Treatment	Manure	Wheat						Sesame						T. aman					
		N	P	K	S	B	Zn	N	P	K	S	B	Zn	N	P	K	S	B	Zn
T_1	-	112	18	47	6	1	1.5	73	16	25	9	1	1.5	84	9	27	5	1	1
T_2	CD @ 5 t ha⁻¹	97	13	32	6	1	1.5	58	11	10	9	1	1.5	69	4.0	12	5	1	1
T_3	CDS @ 5 t ha⁻¹	89	10.5	22	6	1	1.5	50	8.5	-	9	1	1.5	61	1.5	2	5	1	1
T_4	-	69	22	19	-	-	-	35	12	18	-	-	-	69	20	37	10	-	-

Results and Discussion

Yield and yield attributes of wheat

Significantly higher plant height (98.84 cm) was obtained from T_3 that was statistically similar to T_2 and lower from T_4 treatment. The spike m⁻² (333.55) and grain spike⁻¹ (44.05) were observed as the highest from T_3 followed by T_2 and the lowest from T_4 treatment. Significantly the highest1000 grain weight (42.61 g) was obtained from T_3 and the lowest (3.31 g) from T_4. The highest grain yield (3.80 t ha⁻¹) was obtained from bio-slurry treated plot that means T_3 treatment followed by T_2 and the lowest (3.31 t ha⁻¹) from T_4 treatment. The highest straw yield (5.92 t ha⁻¹) was also obtained from bio-slurry

treated plot i.e. from T_3 and the lowest (5.38 t ha⁻¹) from T_4. Significant variation of grain yield indicated that nutrient management packages had significant influence on wheat production. It was observed that IPNS systems with organic manure (cowdung, bio-slurry) based nutrient packages showed better performance over T_1 and T_4 packages as those dealt with chemical fertilizers only. This might be due to the positive effect of organic manure (bio-slurry, cowdung) on yield and yield contributing characters of wheat. The result of yield increment of wheat was supported by the findings of Bodruzzaman et al. (2002), Karki et al. (1996), Kologi et al. (1993), Maskey (1978).

Table 4. Yield and yield contributing characters of wheat at FSRD site pushpapara, Pabna

Treatment	Plant height (cm)	Spikem⁻²	Filled grainspike⁻¹	1000 grain weight (g)	Grain yield (t ha⁻¹)	Straw yield (t ha⁻¹)
T_1	95.63 b	312.27 c	38.58 c	42.21 b	3.51 c	5.50 c
T_2	98.38 a	320.05 b	39.75 b	42.28 b	3.60 b	5.80 b
T_3	98.84 a	333.55 a	44.05 a	42.61 a	3.80 a	5.92 a
T_4	95.91 b	308.00 c	38.00 c	42.10 c	3.31 d	5.38 d
CV (%)	8.49	11.77	5.90	2.24	6.22	4.77

Within column values followed by same letter(s) did not differ significantly by DMRT

Yield and yield attributes of sesame

Data presented in the Table 5 revealed that plant height, plant population, nos. of siliqua plant^{-1} and nos. of seeds siliqua^{-1} of sesame were statistically non significant but varied numerically among the treatments. The higher plant height (136.8 cm) was recorded from T_3 followed by T_2 and the lower (130.1 cm) from T_4 treatment. Similar trend was found in case of plant populations. The maximum nos. of siliqua plant^{-1} (36.16) was obtained from T_3 followed by T_2 and the minimum (33.42) from T_4 treatment. Numerically higher nos. of seeds siliqua^{-1} (47.00) was found in T_3 followed by T_2 and the lower (44.47) from T_4 treatment. Higher seed yield (1.31 t ha^{-1}) of sesame was obtained from T_3 that was statistically identical to T_2 and T_1 and the lower (1.01 t ha^{-1}) from T_4. The cumulative positive effect of nos. of siliqua plant^{-1}, nos. of seeds siliqua^{-1} and weight of 1000 seeds might be contributed to higher seed yield in T_3. Lower seed yield was attained from farmers practice treatment probably the poor performance of yield contributing characters. Maximum stover yield (3.53 t ha^{-1}) was also recorded in T_3 was statistically indistinguishable from T_2 and T_1 and the minimum (3.32 t ha^{-1}) from T_4. The availability of nutrients and balanced uptake of nutrient might be enhanced optimum plant growth and finally maximized grain yield in T_3 treatment as compared to other treatments. Haruna and Abimiku, (2011) reported about higher yield of sesame from application of organic manure.

Table 5. Yield and yield contributing characters of sesame at FSRD site, Pushpapara, Pabna

Treatment	Plant height (cm)	Plant population	Nos. of siliqua plant^{-1}	Nos. of seeds siliqua^{-1}	1000 seeds weight (g)	Seed yield (t ha^{-1})	Stover yield (t ha^{-1})
T_1	134.0	34.00	34.45	44.52	2.62 b	1.21 ab	3.41 ab
T_2	135.9	35.16	35.42	45.00	2.64 ab	1.25 ab	3.44 ab
T_3	136.8	35.42	36.16	47.00	2.68 a	1.31 a	3.53 a
T_4	130.1	33.42	33.42	44.47	2.62 b	1.01 b	3.32 b
CV (%)	8.04	14.71	11.49	8.52	4.32	12.07	12.42

Within column values followed by same letter(s) did not differ significantly by DMRT.

Yield and yield attributes of T. Aman rice

The yield and yield contributing characters were statistically significant except panicle length (Table 6). Yield attributes like filled grain panicle^{-1}, 1000 grain weight were found as the highest in T_3 and the minimum in T_4. The results revealed that the yield contributing characters exhibited better performance due to IPNS with CDS based fertilizer management. The highest grain yield (5.05 t ha^{-1}) was obtained from T_3 which was statistically indistinguishable from T_2 where as the lowest grain yield (4.26 t ha^{-1}) was recorded from T_4 treatment. The maximum yield in T_3 treatment might be due to the cumulative positive effect of yield contributing parameters. Probably integrated fertilizer management using both inorganic and organic sources improved the availability of nutrients and their balanced uptake facilitated optimum growth and development of the crop which ultimately increased grain yield. The poor performance of yield attributes might be attributed to lower grain yield in T_4. The highest straw yield (7.90 t ha^{-1}) was obtained from T_3 followed by T_2 and the lowest (6.6 t ha^{-1}) from T_4 treatment. Similar trend of yield increment in rice with the application of bioslurry was also found by Bodruzzaman et al. (2002), Kanthaswamy (1993), Tripathi (1993), Gupta (1991), Singh et al. (1995), Maskey (1978).

Table 6. Yield and yield contributing characters of T. aman rice at FSRD site, Pushpopara, Pabna

Treatment	Plant height (cm)	Nos. of tiller hill^{-1}	Nos. of panicle m^{-2}	Nos. of filled grain panicle^{-1}	1000 grain wt (g)	Grain yield (t ha^{-1})	Straw yield (t ha^{-1})
T_1	108.4 b	8.96 b	276.1 b	117.8 c	20.28 ab	4.81 b	7.06 c
T_2	103.1 d	8.57 c	285.7 ab	123.4 b	20.17 bc	4.91 ab	7.35 b
T_3	105.8 c	9.38 a	293.7 a	128.7 a	20.35 a	5.05 a	7.90 a
T_4	111.0 a	9.31 a	297.3 a	108.4 d	20.14 c	4.26 c	6.6 d
CV (%)	6.54	8.07	10.17	7.35	4.66	5.59	6.06

Within column values followed by same letter(s) did not differ significantly by DMRT

Economics

Gross return and total variable cost for wheat, sesame and T. Aman are presented against different treatments in the Table 7. It is observed that the highest gross return (276130 Tk ha^{-1}) was obtained from T_3 followed by T_2 and the lowest (230080 Tk ha^{-1}) from T_1 treatment. Many research work related to integrated use of fertilizer have found profitable (Haque *et al.*, 2001; Gupta, 1991). Total variable cost for purchasing different inputs i.e. fertilizers, pesticides and carrying out various intercultural operations recorded as the highest (97674 Tk ha^{-1}) in T_1 followed by T_2 and the lowest (86775 Tk ha^{-1}) in T_4 treatment. Comparatively high variable cost in T_2 than T_3 treatment might be due to higher labour cost for intercultural operations i.e. weed management. Fresh cowdung contains viable weed seeds that compete with the crops and requires farmers to put extra labours for weeding where as bioslurry is also reported to be free from weed seeds (Tripathi, 1993; Van Brake, 1980). The lowest variable cost found in T_4 treatment due to less input cost for purchasing fertilizers as farmers do not use optimum dose of fertilizers for crop production. This statement agrees to the findings reported by Jahiruddin *et al.* (2009). This has created unbalanced use of fertilizers which produces negative impact on soil fertility and crop yield and there by ultimately on economic return. The highest marginal value of product (46050 Tk ha^{-1}) was recorded in T_3 followed by T_2 where as the minimum (27440 Tk ha^{-1}) was found in T_1 over the T_4 treatment. The highest marginal value of product in T_4 treatment is mainly of higher yield and economic return facilitated by IPNS with cowdung slurry. Marginal variable cost was observed as the highest (10899 Tk ha^{-1}) in T_2 treatment followed by T_3 and the minimum (7949 Tk ha^{-1}) was recorded in T_1 treatment. Comparatively low marginal variable cost in T_2 and T_3 over T_1 treatment probably due to less input cost for purchasing chemical fertilizers. The highest MBCR (5.79) was recorded from T_3 followed by T_2 and the minimum (2.52) from T_1 treatment.

Table 7. Profitability of the IPNS system of fertilizer application over the farmers' practice obtained from wheat-sesame-T. Aman rice cropping pattern

Treatment	Rice equivalent yield (t ha^{-1})	Gross return from Wheat- Sesame- T. Aman Pattern (Tk ha^{-1})	Total variable cost for Wheat-Sesame-T. Aman Pattern (Tk ha^{-1})	Marginal value of product (Tk ha^{-1})	Marginal variable cost (Tk ha^{-1})	MBCR
T_1	12.88	257520	97674	27440	10899	2.52
T_2	13.22	264365	95425	34285	8650	3.96
T_3	13.81	276130	94724	46050	7949	5.79
T_4	11.50	230080	86775	-	-	-

Item of Input	Unit price (Tk kg^{-1})	Item of Output	Unit price (Tk kg^{-1})
Cowdung / Cowdung slurry	1.0	Wheat grain	20.0
Urea	20.0	Wheat straw	1.00
TSP	24.0	Rice grain	20
MoP	15.0	Rice straw	2.5
Gypsum	10.0	Sesame seed	55
ZnSO$_4$	130.0	Sesame stover	1.0
Boric acid	130.0		

Conclusion

Considering yield and return it can be concluded that treatment T_3 that means IPNS with cowdung slurry is the most profitable as compared to other treatments. Many researchers claim that IPNS based fertilizer management systems have been found promising for maintaining soil health. In order to maintain sustainable agriculture we should give priority for creating awareness and using IPNS based fertilizer package for successful crop production keeping sustenance of soil health.

References

Ali, M.M., Saheed, S.M. and Kubota, D. 1997. Soil degradation during the period 1967-1995 in Bangladesh. II. Selected chemical characters. *Soil Sci. Plant Nutr.* 43 (4): 879-890.

BARC. 2005. Fertilizer Recommendation Guide. Bangladesh Agricultural Research Council, Farmgate, Dhaka-1215. 260 p.

Batsai, S.T., Polyakev, A.A. and Nedbal, R.F. 1979. Effect of organic and mineral fertilizers on the yield and quality of irrigated late white cabbage in the steppe region of the Crimea. *Hort. Abst.* 49 (11): 730.

Bhuiyan, N.I. 1991. Issues concerning declining/stagnating productivity in Bangladesh Agriculture. A paper presented at the National Workshop on Risk Management in Bangladesh Agriculture, held at BARC, Dhaka. August, 1991. pp. 24-27.

Bodruzzaman, M., Sadat, M.A., Meisner, C.A., Hossain, A.B.S. and Khan, H.H. 2002. Direct and residual effects of applied organic manures on yield in a wheat-rice cropping pattern. Paper presented in 17th WCSS Symposium held in Thailand. 14-21 August, 2002. pp. 7-8.

Brady, N.C. 1974. Organic matter of mineral soils. *In:* The nature and properties of soils. (Buckman, H.O. and Brady N.C. ed.) Macmillan Publishing Co., New York, pp. 137-163.

FAO. 2001. Report of the 2nd Research Co-ordination Meeting of the FAO/IAEA Co-ordinated Research Project on The Use of Nuclear Techniques for developing Integrated Nutrient and Water Management Practices for Agroforesty Systems. 7-11 May, Kuala Lumpur. IAEA - 311-D1-RC-735.2.

Gomez, K.A. and Gomez, A.A. 1984. Statistical Procedure for Agricultural Research (2nd edition), Jhon Willey and Sons, New York. pp. 202-215.

Gupta, D.R. 1991. Bio-Fertilizer from Biogas Plants/ In Changing Villages, Vol. 10, No. 1. Jan- Mar., 1991. pp. 25-26.

Haque, M.Q., Rahman, M.H., Islam, F., Rijpma, J. and Kadir, M.M. 2001. Integrated nutrient management in relation to soil fertility and yield sustainability under Wheat-Mung-T.Aman cropping pattern. *J. Biol. Sci.* 1 (8): 731-734.

Haruna, I.M. and Abimiku, M.S. 2011. Yield of Sesame (*Sesamum indicum* L.) as Influenced by Organic Fertilizers in the Southern Guinea Savanna of Nigeria. *African J. Biotech.* 10 (66): 14881-14887.

Islam, M.S. 2008. Soil fertility history, present status and future scenario in Bangladesh. Paper presented at the IPI-BFA-BRRI International workshopon Balanced Fertilization for Increasing and Sustaining Crop Productivity held at Hotel Rajmoni Ishakha, Dhaka, Bangladesh. 30 Mar.-01 Apr. 2008. 78 p.

Jahiruddin, M., Islam, M.R. and Miah, M.A.M. 2009. Constraints of farmers' access to fertilizer for food production. Final Report.

National Food Policy Capacity Strengthening Programme. FAO. Dhaka. 52 p.

Joshi, J.R., Moncrief, J.F., Swan, J.B. and Malzer, G.L. 1994. Long-term conservation tillage and liquid dairy manure effects on corn. II. Nitrate concentration in soil water. *Soil & Till. Res.* 31 (2-3): 225-233.

Kanthaswamy, V. 1993. Effect of bio-digested slurry in Rice. In Biogas Slurry Utilsation. New Deihi: CORT. 46 p.

Karki, Krishna, B. and Gurung, B. 1996. Evaluation of Biogas Slurry Extension Pilot Programme. Kathmandu: BSP, SNV-Nepal. 2 p.

Kologi, S.D., Nagalikar, S.S. and Hirevendkanagoudar, L.V. 1993. Effect of Biogas Slurry in Crop Yield/ In Biogas Slurry Utilsation. New DeIhi:CORT. pp. 20-22.

Maskey, S.L. 1978. Manurial Value of Biogas Slurry: Some Observations. (Paper presented at International Workshop on Microbiologial aspects of Biogas production. May 31- June 3, Kathmandu). 2 p.

Nambiar, K.K.M. 1991. Long-term fertility effects on wheat productivity. Wheat for the nontraditional warm areas. Proceedings International Conference, Mexico, DF (Mexico) CIMMYT. pp. 516-521.

Rahman, M.A., Ullah, M.M., Sen, R., Hasan, M.K., Islam, M.B. and Khan, M.S. 2008. Bio-Slurry management and its Effect on Soil Fertility and Crop Production. Project Report, Sept. 2008. 5 p.

Reganold, J.P., Robert, I.P. and Parr, J.F. 1990. Sustainability of agriculture in the United States- An overview. *Sustainable Agric. Sci. Am.* 262: 112-120.

Singh, K., Gill, J.S. and Verma, O.P. 1970. Studies on poultry manure in relation to vegetable production. *Indian. J. Hort.* 27: 42-47.

Singh, S.P., Verma, H.N., Vatsa, D.K. and Kalia, A.K. 1995. Effect of Biogas Digested Slurry on Pea, Okra, Soybean and Maize/ In Biogas Forum Vol. IV, No. 63. 14 p.

Subhan, 1991. Effect of organic materials on growth and production of cabbage (*Barssica oleracca* L.) *Soils & Fert.* 54 (4): 587.

Tripathi, A.K. 1993. Biogas Slurry - A Boon for Agriculture Crops/ In Biogas Slurry Utilisation. New Delhi: CORT. pp. 11-14.

Van Brake. 1980. The Ignis Fatuus of Biogas. Small Scale Anaerobic digesters (biogas plants): a critical review of the pre-1970 literature Delft, the Netherlands. Delft University Press. 106 p.

17

GRAIN-FILLING PATTERN OF SUPER HYBRID RICE LIANGYOUPEIJIU UNDER DIRECT SEEDING AND TRANSPLANTING CONDITION

M.A. Badshah and Tu Nai Mei*

Abstract

To evaluate the grain-filling pattern, Chinese first super hybrid rice, Liangyoupeijiu was grown under tillage and establishment methods at a spacing of 20 cm × 20 cm with one seedling hill^{-1} and at a seeding rate of 22.5 kg ha^{-1} in Changsha, Hunan Province, China in 2012. Our results showed that, superior grain weight in TP had always higher than DS up to 24 DAH but at 36 DAH, grain weight had similar in both TP and DS. Middle grain weight was higher in TP than DS up to 18DAH but it was higher in DS than TP at 24 – 36 DAH and at 36 DAH, grain weight of DS had significantly higher than TP. Inferior grain weight was higher in TP than DS up to 12 DAH but it was higher in DS than TP at 24 -36 DAH and at 36 DAH, grain weight of DS had significantly higher than TP. Grain-filling rate of superior grain had higher in TP than DS up to 18 DAH but it was higher in DS than TP at 30 DAH. In middle grain, it was higher in TP at 6DAH but in DS, it was higher at 30 DAH. In inferior grain, it was higher in TP at 36 DAH but in DS, it was higher at 30 DAH. The heavier grain was found in TP only in superior grain but DS had heavier grain both in middle and inferior grain. Grain-filling rate of superior grain was higher in TP than DS and it was similar in both TP and DS in middle grain. But in inferior grain, it was significantly higher in DS than TP. Transplanting method produced slightly higher grain yield due to higher sink size (more number of spikelet's caused by longer panicle and more number of spikelet per cm of panicle) but it was statistically similar with DS.

Keywords: Direct Seeding, Grain-filling Pattern, Super Hybrid Rice, Transplanting, Tillage

College of Agronomy, Hunan Agriculture University, Hunan, China

*Corresponding author's email: tnm505@163.com (Tu Nai Mei)

Introduction

The degree and rate of grain-filling in rice spikelet's differ largely with their positions on a panicle. In general, earlier flowering superior spikelet's has usually located on apical primary branches, fill fast and produce larger and heavier grains. While later-flowering inferior spikelet's, usually located on proximal secondary branches, are either sterile or fill slowly and poorly to produce grains unsuitable for human consumption (Mohapatra *et al.*, 1993; Yang *et al.*, 2000; Yang *et al.*, 2006). The slow grain-filling problem in inferior spikelet's is more aggravated in the newly bred 'super' rice cultivars, although they generally show a yield potential of 8–20% more than other conventional rice cultivars (Cheng *et al.*, 2007; Zhang, 2007; Nakamura *et al.*, 1989). Grain-filling is actually a process of starch accumulation and it has been reported that there are 33 major enzymes involved in the metabolism of carbohydrates in developing rice endosperm (Nakamura *et al.*, 1989). Grain weight is determined by grain size and degree of grain-filling. The grain size is a stable cultivar characteristic determined before anthesis and it is rigidly controlled by size of the hull (Samonte *et al.*, 1998). Although both grain-filling rate and duration are associated with the degree of grain-filling, some studies reported that variation in grain-filling duration was responsible for the difference in grain weight between hybrid and inbred rice (Cheng *et al.*, 2007; Wang *et al.*, 2004). However, there has been contradictory statement that poor grain-filling of super hybrid rice was attributed not to source limitation but to poor partitioning of assimilates to grains (i.e. low harvest index) (Yang *et al.*, 2002). Rice yield depends upon not only the genetic characteristics but also the agronomic practices (Zou *et al.*, 2003). Sink size (spikelet number per unit land area), spikelet filling percentage and grain weight are the determinants of rice yield. Sink size is considered as the primary determinant of the rice yield (Kropff *et al.*, 1994). Sink size can be increased by increasing panicle number per unit land area or spikelet number per panicle or both (Ying *et al.*, 1998). There is limited information currently available on grain-filling pattern of Liangyoupeijiu but no information is available about the effect of transplanting and direct seeding on grain-filling pattern. Therefore, the objective of this study was to evaluate grain-filling pattern of super hybrid rice, Liangyoupeijiu under transplanting and direct seeding condition.

Materials and Methods

Experiment location and soil

A field experiment was conducted under moist sub-tropical monsoon climate during 2012 (May to September). Mean annual air temperature was about 17.0°C, rainfall was about 1355 mm and sunshine hour was about 1677. The soil properties of the experimental field were presented in Table 1.

Table 1. The soil properties of the experimental field, Changsha, Hunan, China

Treatments	Bulk density (g cm^{-3})	pH	Active organic carbon (mg g^{-1})	NaOH hydrolysable N (mg kg^{-1})	Double acid P (mg kg^{-1})	NH$_4$OAc extractable K (mg kg^{-1})
0-5 cm soil depth						
CTTP	1.06	5.94	3.01	198	27.7	44.5
NTTP	1.07	5.83	3.45	197	27.1	46.1
NTDS	1.01	5.91	4.42	239	28.7	52.8
CTDS	1.04	5.81	4.02	227	29.1	52.0
5-10 cm soil depth						
CTTP	1.08	6.01	2.62	160	30.8	33.7
NTTP	1.26	5.91	2.90	160	28.3	33.3
NTDS	1.27	6.18	2.07	136	31.6	29.7
CTDS	1.06	5.99	2.11	133	33.0	30.9

Experiment design and fertilizer management

The field experiment was conducted in a factorial randomized complete block design with four replications. The unit plot size was 30 m². Factor A was tillage system, viz. conventional tillage (CT) and no tillage (NT), factor B was crop establishment method, viz. transplanting (TP) and direct seeding (DS). The treatment combinations were: conventional tillage and transplanting (CTTP), no-tillage and transplanting (NTTP), conventional tillage and direct seeding (CTDS), and no tillage and direct seeding (NTDS). For CT, land was prepared by animal-drawn plough followed by harrowing, and for the plots of NT, by using non-selective herbicide and soaking. For TP, twenty five-days-old seedlings were manually transplanted at a spacing of 20 cm x 20 cm with one seedling per hill on 8th June. For DS, pre-germinated seeds were manually broadcasted on the soil surface at a seed rate of 22.5 kg ha^{-1} on 24th May. Fertilizer (ha^{-1}) was used as 150 kg N, 90 kg P$_2$O$_5$ and 180 kg K$_2$O. Fertilizer N was split as 90, 45 and 15 kg ha^{-1} at basal, mid-tillering and panicle initiation, respectively. Fertilizer P$_2$O$_5$ was applied at basal. K$_2$O was split equally at basal and panicle initiation (PI). Weeds, insects and diseases were controlled by recommended methods.

Sampling and method

Ten (10) representative panicles were sampled for each replication starting from 6-day after heading (DAH) to 36 DAH. Grains on upper (superior), middle and basal (inferior) branches (Fig. 1) of each panicle were separated and oven-dried at

70°C to constant weight to determine grain weight. At MA, 5 m² areas were harvested for grain yield and it was adjusted at 14% moisture level.

Heading was determined at when around 80% of the stem had more than 50% of panicle was exerted.

Data analysis

Statistical analyses were performed using Statistic 8, Analytical software, Tallahassee, FL, USA. Means of cultivation methods were compared according to the least significant difference test (LSD) at the 0.05 probability level. Figures were performed using MS Excel 2003.

Fig. 1. Upper, middle and basal branches in a rice panicle

Results and Discussion

Weather condition

Weather (flowering to maturity) data during 2012 was presented in Fig. 1. Average highest temperature decreased sharply at 6 DAH and then increased up to 18 DAH then again sharply decreased at 24 DAH and then decreased up to 36 DAH. Average lowest temperature sharply decreased at 6 DAH, then increased up to 18 DAH and then decreased up to 36 DAH. There was a continuous fluctuating tendency in sunshine hour during whole sampling period and at 36 DAH there was no sun hour. Average rainfall had very small up to 24 DAH and then drastically increased at 30 DAH and then no rainfall at 36 DAH (Fig. 2).

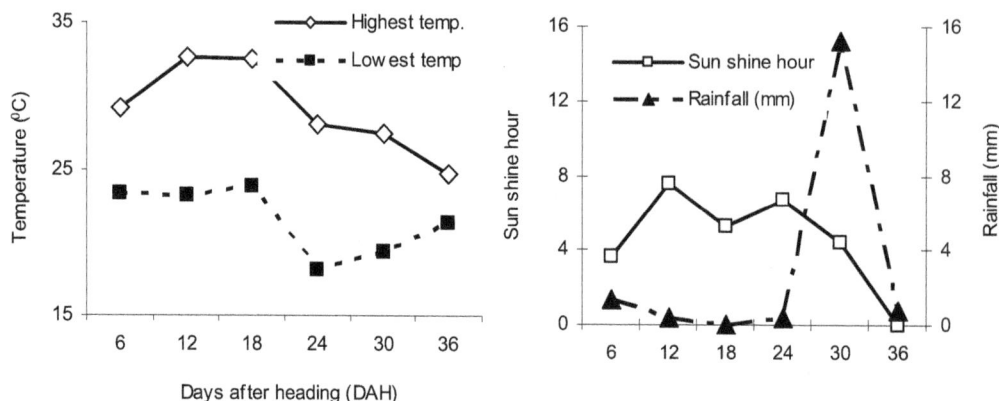

Fig. 2. Temperature, Sunshine hour and Rainfall at different days after heading (DAH)

Grain-filling pattern

Pattern of grain weight (mg): Superior grain weight in TP had always higher than DS up to 24 DAH and DS had higher grain weight at 30 DAH. At 36 DAH, both in TP and DS had similar weight of grain. In TP, grain weight gradually increased up to 30 DAH and then sharply increased at 30 – 36 DAH but in DS, it increased gradually up to 24 DAH and sharply increased at 24-30 DAH. Grain weight of superior grain was higher in TP than DS up to 24 DAH due to higher source capacity (more functional leaf/stem) but it was higher in DS than TP at 30 DAH due to higher source- sink ratio in DS. Grain weight is determined not only by the grain capacity to receive assimilates, but also by the source capacity (photosynthetic leaves) to supply assimilates (Ntanos and Koutroubas, 2002) and by the partitioning of assimilates to grains (Yang et al., 2002).

Middle grain weight was higher in TP than DS up to 18 DAH but it was higher in DS than TP at 24 – 36 DAH. Grain weight of TP gradually increased up to 24 DAH and then slightly decreased at 30 and 36 DAH but DS increased gradually up to 30DAH and then slightly decreased at 36 DAH. At 36 DAH, grain weight of DS had significantly higher than TP.

Inferior grain weight was higher in TP than DS up to 12 DAH but it was higher in DS than TP at 24 - 36 DAH. Grain weight of TP gradually increased up to 24 DAH and then slightly decreased at 30 DAH and then sharply increased at 36 DAH but grain weight in DS, increased gradually up to 30 DAH and then slightly decreased at 36 DAH. At 36 DAH, grain weight of DS had significantly higher than TP. Mohapatra et al. (1993) reported that inferior spikelet's accumulated higher concentrations of soluble assimilates than superior spikelet's during the grain-filling period. These results suggest that assimilate supply is not the main factor that leads to poor grain-filling and that there are other, unknown, factors resulting in slow or aborted grain-filling in inferior spikelet's. Inferior grain weight increased slowly at early grain-filling stage but grain-filling increased sharply at later grain-filling period and was closed to superior grain in DS. The slow grain-filling rate and low grain weight of inferior spikelet's have often been attributed to a limitation in carbohydrate supply (Sikder and Das Gupta, 1976).

Grain weight of superior grain was higher than middle and inferior grain and it was 27.7 mg in TP and 27.5 mg in DS. For middle grain weight, it was 20.1 mg in TP and 22.3 mg in DS. However, in inferior grain, it was 24.3 mg in TP and 27.6 mg in DS (Fig 3).

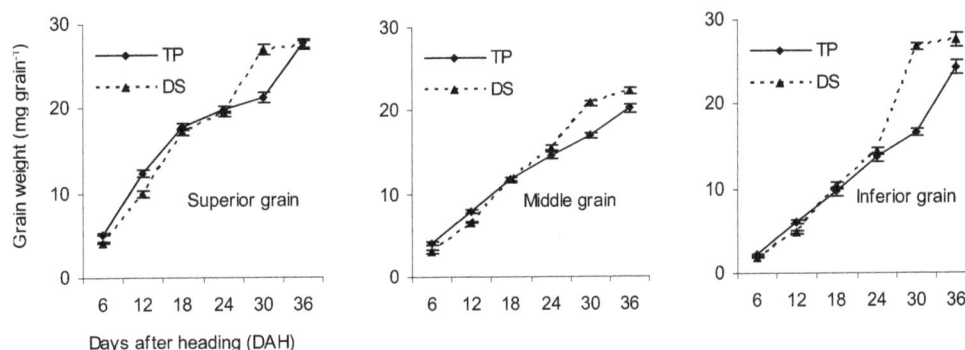

Fig. 3. Grain weight of Liangyoupeijiu (superior, middle and inferior grains) at different DAH

Grain-filling rate (mg grain^{-1} day^{-1})

Grain-filling rate of superior grain had higher in TP than DS up to 18 DAH but it was higher in DS than TP at 30 DAH. Grain-filling rate of TP increased up to 12 DAH, then decreased up to 30 DAH and again slightly increased at 36 DAH. In DS, grain-filling rate increased up to 18 DAH, then decreased at 24 DAH and again slightly increased at 30 DAH.

In middle grain, it was higher in TP at 6 DAH but in DS, it was higher at 30 DAH. In TP, grain-filling rate was higher at 6 DAH and then gradually decreased up to 36 DAH. However, in DS, it increased up to 30 DAH and then decreased at 36 DAH.

In inferior grain, it was higher in TP at 36 DAH but in DS, it was higher at 30 DAH. In TP, grain-filling rate increased up to 24 DAH, then decreased at 30 DAH and again increased at 36 DAH. However, in DS, it increased up to 30 DAH and then decreased at 36 DAH.

In superior grain, grain-filling rate in TP had 1.03 mg grain^{-1} day^{-1} and in DS, it was 0.96 mg grain^{-1} day^{-1}. In middle grain, it was 0.68 mg grain^{-1} day^{-1} in TP and 0.70 mg grain^{-1} day^{-1} in DS. However, in inferior grain, it was 0.68 mg grain^{-1} day^{-1} in TP and 0.90 mg grain^{-1} day^{-1} in DS (Fig. 4).

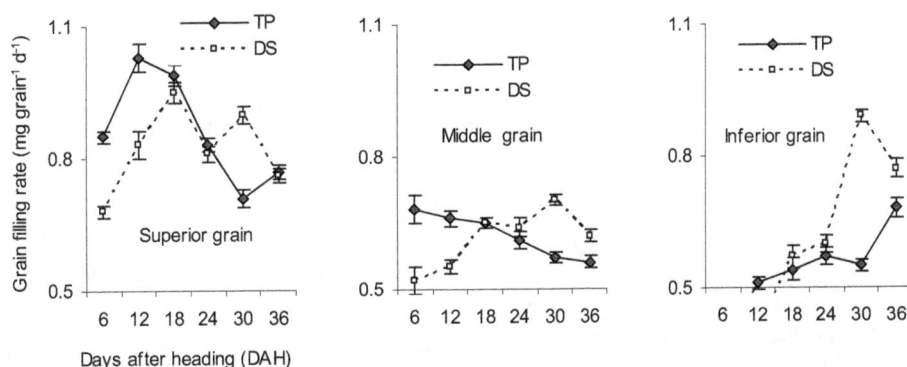

Fig. 4. Grain-filling rate of Liangyoupeijiu (superior, middle and inferior grains) at different DAH

Relationship between yield and yield components

TP had significantly higher number of spikelet's (per m^2) than DS. There was no significant difference in spikelet filling rate between TP and DS. TP produced significantly longer panicle than DS also higher number of spikelet per cm of panicle was observed in TP than DS. Although, average grain weight (weight of superior, middle and inferior grain) had higher in DS but TP produced slightly higher grain yield due to higher sink size (more number of spikelet's caused by longer panicle and more number of spikelet per cm of panicle) but it was statistically similar with DS. Sink size (spikelet number per unit land area), spikelet filling percentage and grain weight are the determinants of rice yield. However, sink size is considered as the primary determinant of the rice yield (Kropff et al., 1994) (Table 2).

Table. 2 Yield and yield components of super hybrid rice Liangyoupeijiu during 2012

Treatments	Total spikelet (m⁻²)	Spikelet filling rate (%)	Panicle length (cm)	spikelet per cm of panicle	Grain yield (t/ha)
TP	48869 a	72.5	24.9 a	9 a	9.7
DS	42713 b	74.3	22.1 b	6 b	9.4
Analysis of variance					
Establishment method (A)	*	NS	*	*	NS
Tillage (B)	NS	NS	NS	NS	NS
A X B	*	NS	*	*	*
CV (%)	4.28	3.79	1.9	8.6	4.49
SE	1387.3	1.96	0.32	0.44	0.30

Conclusion

The heavier grain was found in TP only in superior grain but DS had heavier grain both in middle and inferior grain. Grain-filling rate of superior grain was higher in TP than DS and it was similar in both TP and DS in middle grain. However, in inferior grain, it was significantly higher in DS than TP. Transplanting method produced slightly higher grain yield due to higher sink size (more number of spikelet is caused by longer panicle and more number of spikelet per cm of panicle) but it was statistically similar with DS.

Acknowledgments

This study was a part of the PhD thesis research of the first author. The author greatly appreciates the financial support provided by NSFC project, code no. 311712494, China. The author also appreciates the financial support provided by NATP, BARC, Dhaka, Bangladesh.

References

Cheng, S., Zhuang, J., Fan, Y., Du, J. and Cao, L. 2007. Progress in research and development on hybrid rice: A super-domesticate in China. *Ann. Bot.* 100: 959–966.

Kropff, M.J., Cassman, K.G., Peng, S., Matthews, R.B. and Setter, T.L. 1994. Quantitative understanding of yield potential. In: K.G. Cassman, ed., Breaking the Yield Barrier. International Rice Research Institute, Los Baños. pp. 21-38.

Mohapatra, P.K., Patel, R. and Sahu, S.K. 1993. Time of flowering affects grain quality and spikelet partitioning within the rice panicle. *Australian J. Plant Physiol.* 20: 231–242.

Nakamura, Y. Yuki, K. and Park, S.Y. 1989. Carbohydrate metabolism in the developing endosperm of rice grains. *Plant and Cell Physiol.* 30: 833–839.

Ntanos, D.A. and Koutroubas, S.D. 2002. Dry matter and N accumulation and translocation for Indica and Japonica rice under Mediterranean conditions. *Field Crops Res.* 74 : 93-101.

Samonte, S.O.P.B., Wilson, L.T. and McClung, A.M. 1998. Path analyses of yield and yield-related traits of fifteen diverse rice genotypes. *Crop Sci.* 38: 1130-1136.

Sikder, H.P. and Das Gupta, D.K. 1976. Physiology of grain-filling in rice. III. Effects of defoliation and removal of spikelet's. *Indian Agriculturalist* 20: 133-141.

Wang, Z., Gu, Y.J., Hirasawa, T., Ookawa, T. and Yanahara, S. 2004. Comparison of caryopsis development between two rice varieties with remarkable difference in grain weights. *Acta Bot. Sin.* 46: 698-710.

Yang, J., Peng, S., Visperas, R.M., Sanico, A.L., Zhu, Q. and Gu, S. 2000. Grain-filling pattern and cytokinin content in the grains and roots of rice plants. *Plant Growth Regulation* 30: 261–270.

Yang, J., Peng, S., Zhang, Z., Wang, Z., Visperas, R.M. and Zhu, Q. 2002. Grain and dry matter yields and partitioning of assimilates in japonica/indica hybrid rice. *Crop Sci.* 42 : 766-772.

Yang, J., Zhang, J., Wang, Z., Liu, K. and Wang, P. 2006. Post-anthesis development of inferior and superior spikelet's in rice in relation to abscisic acid and ethylene. *J. Expt. Bot.* 57: 149–160.

Ying, J., Peng, S., He, Q., Yang, H., Yang, C., Visperas, R.M. and Cassman, K.G. 1998. Comparison of high-yield rice in tropical and subtropical environments: I. Determinants of grain and dry matter yields. *Field Crops Res.* 57: 71-84.

Zhang, Q. 2007. Strategies for developing green super rice. *In:* Proceedings of the National Academy of Sciences, USA, 104: 16402–16409.

Zou, Y.B., Zhou, S.Y. and Tang, Q.Y. 2003. Status and prospect of high yielding cultivation researches on China super hybrid rice. *J. Hunan Agric. Univ.* 29: 78–84.

18

PHYSICAL, CHEMICAL AND MICROBIOLOGICAL ANALYSIS OF THE WATER QUALITY OF RAWAL LAKE, PAKISTAN

Mehreen Hassan* and Sana Hanif

Abstract

What better gift of nature would be than good quality water? In order to assess the quality of water of Rawal Lake, following research was carried out. Rawal lake is a source of drinking water supplied to many areas of Rawalpindi and Islamabad' the capital city of Pakistan. Water of this lake is being highly polluted by the local communities alongside the lake through solid waste dumping. Samples of surface water were collected, tested and analyzed in the laboratory on the basis of physical, chemical and microbiological parameters. The results showed uncertainties in many of the selected parameters. Microbiological analysis revealed high contamination of *E. coli,* fecal coliform and total coliform in the samples proving it unfit for drinking. It was found that the concentration of all physical parameters such as nitrates, chloride, pH and conductivity were within the normal limits. The level of heavy metals like lead, iron, chromium etc. was also found low. Turbidity at some points exceeded the maximum acceptable limit as per WHO statement.

Keywords: Water Quality, Rawal Lake, Water Contamination

Institute of Agricultural Sciences, University of the Punjab, Lahore, Pakistan

*Corresponding author's email: mehrinabbas@yahoo.com (Mehreen Hassan)

Introduction

The surface water and ground water quality is deteriorating day by day throughout the country. The discharge of industrial and domestic wastewater into open water bodies and groundwater is the main threat to the country's water reserves. The non implementation of legislative measures and standards is aggravating the situation resulting in further deterioration in water quality observed over the year. The issue is becoming very serious as open water-bodies; like lakes, rivers, and streams are being increasingly contaminated (Au, 1995; Jaffer and Saleem, 1987).

Rawal lake is an important source of water supply; surface water is pumped through shallow wells or deep tube wells/turbines. The total withdrawal of water from surface is estimated at 712 MAF (Mateen and Garstang, 2008). Water, during its passage through rivers and through the ground acquires various types of dissolved and suspended impurities and heavy metals.

Water quality is characterized in terms of its physical, chemical and biological composition. It is important to note that many of the physical properties, chemical and biological characteristics are inter-related (Marletta, 1986). The biological characteristics of water quality are of fundamental importance in the control of diseases caused by pathogenic organisms of human origin. The microorganisms found in surface and wastewater includes viruses, bacteria, fungi, algae, protozoans, plants and animals. The chemical constituents of wastewater are typically classified as inorganic and organic. Inorganic constituents of concern include nutrients, nonmetallic constituent, metals and gases (Marletta, 1986).

Rawal Lake is the main source of water supply for Rawalpindi city. It is situated in the capital territory of Islamabad. It is situated at N 33° 41' 47.18", E 73° 08' 07.64" and at an Elevation of 1728 ft above sea level. The lake is located on Korang river and has a catchment area of 106 sq miles, which generates 84,000 cubic feet of water in an average rainfall year. There are four major streams and 43 small streams contributing to its storage. The total storage capacity is 47,500 acre feet. Live storage is 43,000 acre feet (Mateen and Garstang, 2008). There are many people living around Rawal lake and its contributing streams who dump a large amount of solid waste, sewage waste, organic and inorganic wastes in it. This sewage waste generates different types of bacteria and contaminates the water of the lake (Mateen and Garstang, 2008). The people who use this water for drinking purposes are extremely vulnerable to fatal diseases. Therefore, it is necessary to monitor and analyse the water quality on all parameters and prepare a management plan for ensuring the water quality up to WHO standards. This report is about the pollution and drinking water quality of Rawal lake compared with its sources and their possible prevention management plan. Major objectives of the study were:

- To assess the variation in drinking water quality of Rawal lake and its contribution to water sources.
- Creating awareness among the country about importance of potable water.
- To make recommendations for prevention of water pollution.

Materials and Methods

Selected sampling sites

Prior to field sampling and testing, a sampling plan for water and wastewater monitoring in the area was made. A reconnaissance survey for identification of sampling points was made as a part of environmental monitoring program. Following sites were selected for sample collection

Natural stream with sewage pollution 1: This is the stream coming from the Quaid-E-Azam University Islamabad which eventually enters in the lake. It has large amounts of solid waste and chemicals added from the laboratory of the university. Therefore, it was a point worth selecting for sampling.

Natural stream with sewage pollution 2: This stream carries a large amount of solid wastes, which is added to the lake water.

Lake resort area: This area has a large human interference. As hundreds of people visit daily. Solid waste is added in the lake. The oils discharged from the boats there also pollute the water.

Inlet of Korang River: It has a large population near it adding solid waste in it.

Drain/ sewage of Shahdara: A large number of livestock of the near town pollutes the water.

Drain/ sewage of Banigala: This point was selected for sampling because it has a large human and livestock interference.

Drain/ sewage of Satrameel/ Chattar: It carries a large amount of solid waste and livestock excreta.

Centre of the lake: This point was selected because it has least human and livestock interference and can be used for comparison of source and lake water.

The above mentioned points were then marked for coordinates with the help of GPS. Samples were collected manually according to method developed by the Chief Technical Advisor of PWP for examination of wastewater. GIS based map was developed and average coordinates were taken by using GPS (Garmin) receiver over the bridges at downstream positions and marked. In addition fixed points photographs were taken for the purpose of analysis. Sampling protocols were finalized and undertaken and were analysed in PCRWR laboratory, Islamabad. A detailed and thorough examination of the given map was made for the identification of suitable sampling points for surface water. Collected samples were kept in an ice box to prevent bacterial growth and were sent to the laboratory within 3-4 hours. Following parameters were tested

Table 1. Physical, chemical and microbiological parameters studied in the experiment

Physical Parameters	Chemical Parameters	Microbiological Parameters
Temperature	Chloride	Total coliform
pH	Fluoride	Fecal coliform
BOD	Iron	E.coli
COD	Chlorine	
TSS	Lead	
TDS	Mercury	
Conductivity	Manganese	
Turbidity	Nitrate	
Colour	Calcium	
Hardness	Carbonate	
	Sodium	
	Potassium	

Results and Discussion

Results showed that the physical parameters were within the permissible limits except in the values of turbidity, which is due to the presence of suspended matter such as silt clay, finely divided organic and inorganic matter. Some heavy metals like lead, cadmium, potassium etc. were detected but they were in very low concentrations. A few of them were below detectable limit. However, most of the trace elements lie within the permissible limit prescribed by WHO. Coliform bacteria were

also detected in very high amounts. The source of coliform bacteria is the sewage water dumping into the Rawal lake. There is a high concentration of sedimentation transported into the lake through Korang river because there is very less forest and vegetation, which promotes the weathering and erosion process. Locations of sampling points in Rawal lake are given in Table 2. All the samples were collected during April and May 2008.

Table 2. Physical parameters for surface water analysis

S. No.	Locations	Colour	Odour	Turbidity (NTU)	pH	EC (U/cm)	Alkalinity (ppm)	Hardness (ppm)
1	Permissible limits/WHO limits	Colorless	UO	5	6.5 -8.5	NGVS	NGVS	500
2	Natural stream with sewage pollution 1	Colorless	UO	49.8	7.56	476	162	197
3	Natural stream with sewage pollution 2	Colorless	UO	76	7.54	477	167	192
4	Inlet of Korang river	Colorless	U.O	4.4	8.5	442	209	226
5	Lake resort area	Colorless	U.O	4.3	8.4	442	209	225
6	Drain/ sewage of shahdara	Colorless	UO	14.7	7.70	650	267	302
7	Drain/ sewage of banigala	Colorless	UO	13.9	8.60	442	147	167
8	Drain/ sewage of satrameel	Colorless	UO	7.4	8.08	454	157	177
9	Centre of the lake	Colorless	U.O	4.1	8.2	440	205	225

U.O: Unobjectionable
NGVS: No Guideline Value Set

Table 3. Chemical parameters for surface water analysis

Sl. No.	Locations	TSS	Cl⁻	SO₄⁻	Ca	Mg++	NO3⁻	COD	BOD
1	Natural stream with sewage pollution 1	88	12	-	45	21	0.7	07	18
2	Natural stream with sewage pollution 2	95	12	-	45	19	0.8	10	21
3	Inlet of Korang river	8.7	16	67	56	20	10	29	4
4	Lake resort area	-	9.0	40	50	18	9	27	3
5	Drain/ sewage of shahdara	-	9	82	89	19	2	-	-
6	Drain/ sewage of banigala	-	14	72	37	18	0.6	-	-
7	Drain/ sewage of satrameel	BDL	10	-	41	18	0.7	BDL	7
8	Centre of the lake	-	12	63	54.3	18	9	26	2

(All concentration is expressed in ppm.
BDL : below detectable limit
TSS: Total suspended solids
COD: Chemical Oxygen Demand
BOD: Biochemical Oxygen Demand)

Table 4. Heavy metals parameters for surface water analysis

Sl. No.	Locations	Pb	Ar	Cr	Cu++	Fe	Mg++	Cd	K++
1	Natural stream with sewage pollution 1	0.56	-	-	-	BDL	21	-	3
2	Natural stream with sewage pollution 2	0.61	-	-	-	0.03	19	-	2.8
3	Inlet of Korang river	0.09	-	-	-	-	0.02	0.03	2.99
4	Lake resort area	0.08	-	-	-	-	-	0.03	2.4
5	Drain/ sewage of shahdara	-	-	-	-	-	19	-	4.3
6	Drain/ sewage of banigala	-	-	-	-	-	18	-	3.6
7	Drain/ sewage of satrameel	0.69	-	-	-	0.04	18	-	2.4
8	Centre of the lake	0.07	-	-	-	0.15	-	-	2.1

All concentrations are expressed in ppm.

Table 5. Microbiological parameters for surface water analysis

S.No.	Locations	Fecal coliform MPN/100ml	Total coliform MPN/100ml	*E. coli* +ve/-ve	Total plate count at 35°C CFU/ml
1	Natural stream with sewage pollution 1	>1600	>1600	+ve	-
2	Natural stream with sewage pollution 2	>1600	>1600	+ve	-
3	Inlet of Korang river	>1600	>1600	+ve	-
4	Lake resort area	>1600	>1600	+ve	-
5	Drain/ sewage of shahdara	240	240	+ve	-
6	Drain/ sewage of banigala	>1600	>1600	+ve	-
7	Drain/ sewage of satrameel	>1600	>1600	+ve	356.5
8	Centre of the lake	>1600	>1600	+ve	-

Water samples, which were collected from different locations of Rawal lake and its source inlets contain a high amount of bacterial contamination, which makes the water unhygienic for the drinking purposes and can cause water borne disease. So by following the pollution prevention management plan water could be made safer for drinking and other purposes.

Pollution level in the sources of lake water is slightly larger in case of heavy metals and chemical contamination. However, in the centre of the lake almost all of the physical parameters were within the permissible limits, which on the other hand are very high in the source streams. Therefore, it can be concluded that there is less contamination in the centre of the lake as compared to the sources where human interference is more.

So, after the complete monitoring and analysis of Rawal Lake water a management plan for the prevention of pollution can be made. By following this plan, we can minimize water pollution to a non hazardous level.

Recommendations

- Identify and plan for means to effectively protect and improve surface water quality.
- Prevent erosion and sedimentation of soil into surface water systems by increasing vegetation and growing more trees in the waterways.
- Environmental education should be necessary at grass root level. e.g curriculum & syllabus.
- The government should enforce Pakistan environmental protection act properly.

- Stop deforestation by enforcing law and proper monitoring of forestry.
- There should be an alternative way to put sewage water, rather to put into the Rawal lake water. e.g. there should be seperate sewage treatment plants.
- A focal point organization must be identified to monitor the progress of the implementation of recommendations and their effect on overall water availability for drinking and other purposes. So, the progress can be monitored and any desired precautionary steps could be taken to protect our water resources.

References

Au, E. 1995. *EIA Follow Up and Monitoring.* EIA Process Strengthening Workshop, Canberra. pp. 222-225.

Jaffer, M. and Saleem, M. 1987. Concentration of selected toxic trace metals in some vegetables and fruits of local origin. *Pak. J. Agric. Sci.* 24: 140-146.

Marletta, G.P. 1986. Determination of cadmium in roadside grapes. *J. Sci. food Agric.* 37: 1090-1096.

Mateen, H.A. and Garstang, R. 2008. Environmental Monitoring of Natural Streams/Nullahs in Sector F-7, Islamabad: Detailed analytical baseline report of surface water a research report submitted to the Ministry of Environment's Pakistan Wetlands Programme. p. 16.

19

TOP SOIL SALINITY PREDICTION IN SOUTH-WESTERN PART OF URMIA LAKE WITH GROUND WATER DATA

N. Hamzehpour[1*], M.K. Eghbal[2], P. Bogaert[3] and N. Toomanian[4]

Abstract

Drying of Urmia Lake in the north-west of Iran threatens all the agricultural lands around the Lake. Therefore, soil salinity appears to be the major threat to the agricultural lands in the area. The aim of the present study was to investigate the spatial variation of top soil salinity by taking into account of underground water quality data as secondary information. The research was performed on a grid of 500 m in an area of 5000 ha. Soil samples were gathered during the autumn of 2009 and were repeated in the spring of 2010. Electrical conductivity of soil samples was measured in a 1:2.5 soil to water suspension. Then covariance functions were build for each data set and soil salinity prediction were done on a grid of 100 m using kriging estimator with taking into account the mean variation. Afterwards sodium activity ratio derived from underground water quality database was used as covariate to develop cross-semivarograms in prediction of top soil salinity using co-kriging method. Results demonstrated that soil salinity varied from values lower than 0.5 to more than 35 dSm^{-1} as a function of distance to the Lake. Cross-validating the results from salinity predictions using only kriging estimator to that of cokriging with sodium activity ratio data revealed that kriging offered better estimations with ME of 0.04 for autumn 2009 and -0.12 for spring 2010. Cokriging estimator had more smoother and diffused boundaries than that of kriging and resulted in more bias estimations (ME= -0.11 and -0.21 for first and second data sets). Although kriging method had better performance in top soil salinity prediction, but cokring method resulted in smoother boundaries and reduced the negative effects of mean variation in the area.

Keywords: Urmia Lake, Electrical Conductivity, Sodium Activity Ratio, Cross-semi variogram

[1]Soil Science Department, Maragheh University, Maragheh, East Azerbaijan, Iran
[2]Soil Science Department, Tarbiat Modares University, Tehran, Iran
[3]Faculty of Bio-engineering, Agronomy and Environment, Catholic University of Leuven, Louvain-la-Neuve, Belgium
[4]Isfahan Research Centre for Agriculture and Natural Resources, Isfahan, Iran

*Corresponding author's email: nikooh62@yahoo.com (N. Hamzehpour)

Introduction

Urmia Lake is one of the biggest hyper saline lakes in the world and the biggest one in the Middle East, which is located in the north-west of Iran (Zarghami, 2011; Hassanzadeh *et al.*, 2011). It is estimated that the surface area of the Lake was more than 6100 km in 1995, which has started to decrease rapidly since then (Eimanifar and Mohebbi, 2007).

Salinity of soils around Urmia Lake can be divided to two categories, lands, which are inherently and primarily salty, and areas affected by secondary salinization. Secondary salinization can occur due to the use of low quality saline irrigation water.

In order to manage salt affected soils, first it is required to monitor spatial variation of soil salinity to recognize regions with potential salinity and to prioritize the regions for temporal monitoring. Recent improvements in the field of geostatistics and advances in calculating complex problems have made the analysis of variables with spatial correlation possible. Kriging methods have widespread use in geostatistical methods and in soil salinity prediction models, which have been discussed in detail in several papers (Li and Heap, 2008). There have been numerous attempts in mapping spatial variability of soil salinity using kriging methods (Peck and Hatton, 2003; Triantafilis *et al.*, 2004; Malins and Metternicht, 2006; De Clercq *et al.*, 2009; Giordano *et al.*, 2010; Acosta *et al.*, 2011; Li *et al.*, 2011).

Stein and Corsten (1991) discussed the relationship between universal kriging and cokriging with regression kriging. Mondal *et al.* (2001) used linear and non-linear methods to predict top soil salinity in Bangladesh. De Clercq *et al.* (2009) utilized a first order polynomial equation for mapping spatial and temporal variation of soil salinity. Juan *et al.* (2011) took advantage of a spatial Gaussian linear mixed model to calculate soil salinity using soil electrical conductivity and Na content.

The aim of this research was to predict spatial variability of soil salinity in Urmia Plain, west of Urmia Lake, and to investigate possible effects of Urmia Lake dehydration on agricultural lands using underground water sodium activity ratio (SAR) as secondary information.

Materials and Methods

Study site

The region under investigation is located in the western part of Urmia Lake, north-west of Iran. It is located between 45° 5′ to 45° 20′ E and 37° 15′ to 37° 35′ N. The mean annual precipitation is 367 mm. The mean annual temperature for the coldest month is -5.2°C and for the warmest one is 32°C. Potential evaporation in the area is between 900-1170 mm. In terms of geology, the study area is composed of two different deposits: saline playa deposits and young alluvial terraces and alluvial fans with very low salinity.

Soil salinity data set

The location of the study area and sampling points are shown in Fig. 1. Soil samples were taken from depth 0-20 cm on a grid of 500 meter, once during autumn 2009 (first data set) and were repeated during spring 2010 (second data set). All soil samples were sieved and analyzed for their electrical conductivity (EC) in a 1:2.5 soil to water suspension.

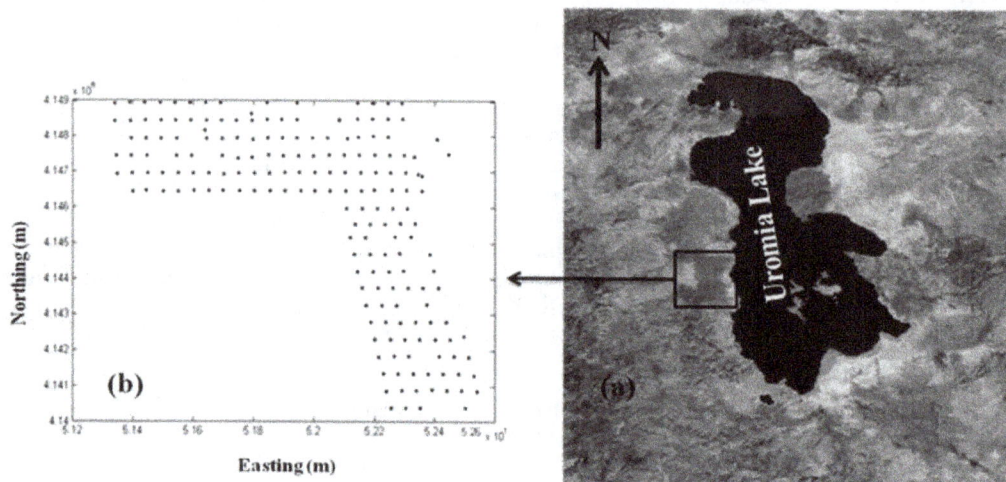

Fig. 1. Study area (a) and soil sampling locations (b) in west of Urmia Lake, North-west of Iran

Underground water quality dataset

In this study, underground water (UGW) quality analysis, which was available for seven years from 2001 to 2007, was utilized as secondary information in soil salinity prediction. These analyses included UGW electrical conductivity (EC), total dissolved salts (TDS), pH, all common anions, and Cations (e.g. CO_3^{2-}, HCO_3^-, Cl, K^+, Na^+). Some other quality parameters like sodium activity ratio (SAR) were also calculated using equation 1.

$$SAR = \frac{\sqrt{(Ca^{2+} + Mg^{2+})}}{Na^+} \qquad (1)$$

Among aforementioned parameters, SAR values of UGW were spatially correlated with top soil salinity data set and consequently were chosen as covariate in soil salinity prediction.

Covariance function

Assuming random fields with spatial homogeneity, the mean function is constant:

$$\mu_x(p) = \mu_x \qquad (2)$$

Hence, the covariance function can be written as follows:

$$c_x(h) = E\{[Z(x) - \mu_x][Z(x+h) - \mu_x]\} = E[Z(x)Z \qquad (3)$$

If the mean function is known, the moment's estimator of covariance function is:

$$\hat{c}_x(h) = \frac{1}{N(h)} \sum_{i=1}^{N(h)} \{[z(x \qquad (4)$$

Where $N(h)$ is the number of pairs of data separated by the spatial lag.

Cross-semivariance functions

Kriging estimators' basic equation is defined as follows: (Li and Heap, 2008) (5)

$$Z'(x_.) - \mu = \sum_{i-1}^{n} \lambda_i [Z(x_.) - \mu(x_.)]$$

Equation 4 can be extended to incorporate the additional information to derive equation 5 as follows:

$$\hat{Z}_1(x_0) - \mu_1 = \sum_{i_1=1}^{n_1} \lambda_i \left[Z_1(x_{i_1}) - \mu_1(x_{i_1}) \right] + \sum_{j=2}^{n_v} \sum_{i_j=1}^{n_j} \lambda_{ij} \left[Z_j(x_{ij}) \right] \qquad (6)$$

Where μ_1 is an acknowledged stationary mean of the primary variable, $Z_1(x_{i_1})$ is the data at point i_1, $\mu_1(x_{i_1})$ is the mean of samples within the search window, n_1 is the number of sample points within the search window for point x_o used to make the estimation, (λ_{i_1}) is the weight selected to minimize the estimation variance of the primary variable, n_v is the number of secondary variables, n_j is the number of jth secondary variable within the search window, λ_{ij} is the weight assigned to i_j^{th} point of jth secondary variable, $Z_j(x_{ij})$ is the data at i_j^{th} point of jth secondary variable, and $\mu_j(x_{ij})$ is the mean of samples of j^t secondary variable within the search window.

The cross-semi variance can be estimated from data using the following equation:

$$\hat{\gamma}_{12}(h) = \frac{1}{2n} \sum_{i=1}^{n} [z_1(x_i) - z_1(x_i) - z_1(x_i + h)][z_2(x_i) - \qquad (7)$$

In the case of this research, Z_1 refers to the soil salinity and Z_2 refers to the underground water SAR values, which can be possibly used as an indicator of soil salinity.

Validation and comparison criteria

In order to compare kriging with cokring method (SAR of ground water as covariate), two thirds of available data were used for modeling and the rest for comparing the two different models. Hence, three global performance criteria were computed: r, which is the Pearson correlation coefficient, the mean error (ME), and the mean squared error (MSE). Accurate predictions are thus characterized by a ME value that should be close to zero and a MSE that should be as small as possible.

All the analyses were done using the BMElib toolbox (Christakos et al., 2002) written using Matlab (MathWorks, 1999).

Results and Discussion

Soil salinity data analyses

Pole plots for top soil $EC_{2.5}$ measurments during autumn 2009 and spring 2010 are presented in Fig. 2. The mean values for first and second data sets were 3.68 dSm^{-1} and 4.73 dSm^{-1}, respectively. The increase of soil salinity mean value during spring 2010 was due to the seasonal variation of soil salinity, which is caused by rainfalls and water table fluctuations.

Soil salinity had a wide range of variation in the study area, from very low values (less than 0.5 dSm^{-1}) in agricultural lands distant from Urmia Lake to very high values (more than 35 dSm^{-1}) in lands adjacent to the lake (Fig. 2).

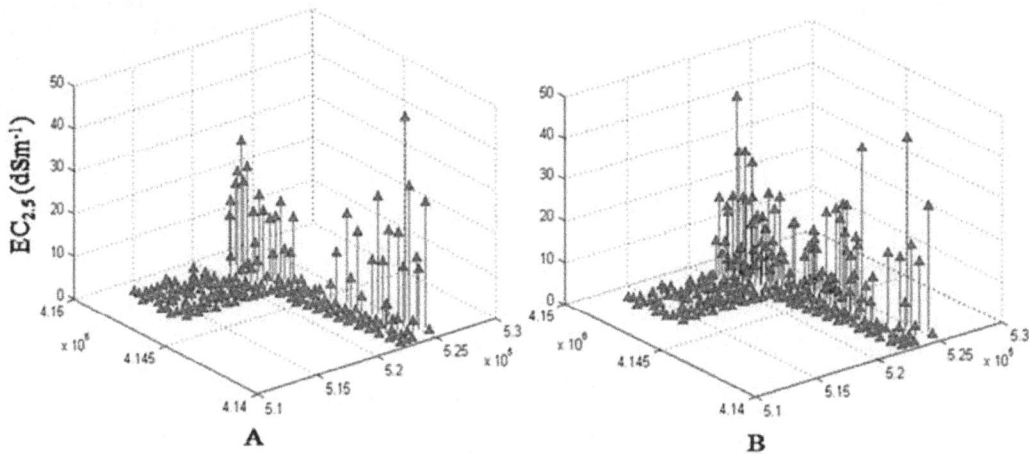

Fig. 2. Pole plots for top soil (0-20 cm) $EC_{2.5}$ laboratory measurements for autumn 2009 (A) and spring 2010 (B)

On the basis of the obtained results, soil salinity variation was not gradual all over the study area, however there were sudden variations in an exact distance from Urmia Lake. This phenomina resulted in a big variation of soil salinity mean values around a boundary, which will be called soil salinity boundary afterwards.

Based on the geological information, the study area consists of two main deposits: saline playa deposits and young alluvial terraces and alluvial fans with very low salinity. The soil salinity boundary, where the main variation in soil salinity mean occurs, had a reasonable match with the geological boundaries. This means that spatial variation of soil salinity is also affected by geology

of the study area as well as the distance between samples.

Underground water data analysis

Color plots of UGW SAR values for seven years are displayed in Fig. 3. Temporal studies of these data showed that there was no significant variation in the SAR values of UGW from 2001 to 2007. This means that the quality of the UGW, which is being used for irrigation of the agricultural lands around the Urmia Lake, was not the main factor in secondary salinization of the lands through time. As the UGW SAR did not show significant temporal variation, mean of SAR data at each point of observation over years was taken to develop the covariance and cross-semivariance functions.

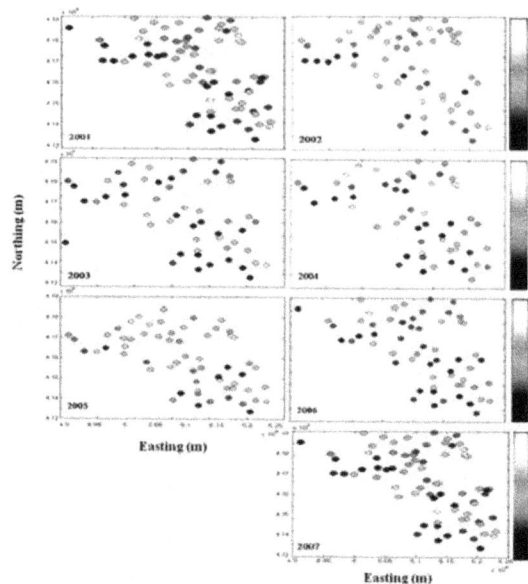

Fig.3. Color plots for the seven years underground water sodium activity ratio (SAR). The level of color reflects the SAR values in log scale.

Covariance and Cross-semi variance functions

As it was mentioned in previous sections, soil salinity had a variable mean throughout the study area. Therefore in order to consider the mean variation among datasets, the spatial components of the mean trend were computed and subtracted from measured $EC_{2.5}$ values of both first and second datasets, which resulted in residuals. Then the covariance function for each dataset was calculated and modeled separately based on the residuals (Fig. 4 and 5, parts B).

Parameters of fitted covariance functions on soil salinity datasets during autumn 2009 and spring

2010 are presented in Table 1. The fitted covariance model for autumn 2009 had three parts: a nugget effect equal to 0.145 (dS m^{-1})^2 and two spherical parts, consisting of a small scale component with sill of 2.31 (dS m^{-1})^2 and spatial range of 2.5 km, along with a larger scale component with sill of 0.01 (dS m^{-1})^2 and spatial range of 5 km (Table 1). These results for spring 2010 were a nugget effect equal to 0.025 (dS m^{-1})^2 a small-scale component with sill of 2.57 (dS m^{-1})^2 and spatial range of 2.5 km and a larger scale component with sill of 0.01 (dS m^{-1})^2 and spatial range of 5 km (Table 1).

Table 1. Covariance functions parameters for first and second data sets

Covariance function parameters	Nugget	Small scale component sill (2.5 km)	Large scale component sill (5 km)
Autumn 2009	0.145	2.31	0.01
Spring 2010	0.025	2.57	0.01

In order to use UGW SAR in prediction of top soil salinity, the mean covariance function of available data for seven years was calculated and is shown in Fig. 4 and 5, parts A. Then cross-variograms were calculated using UGW SAR as covariate. The cross-variograms for first and second datasets are presented in Fig. 4 and 5, parts C, respectively.

The calculated cross-semivariograms for both datasets have only one sill. This means that the use of UGW SAR as covariate has smoothened the soil salinity mean variation and has reduced the geological discontinuity effect on spatial variation of soil salinity through the study area.

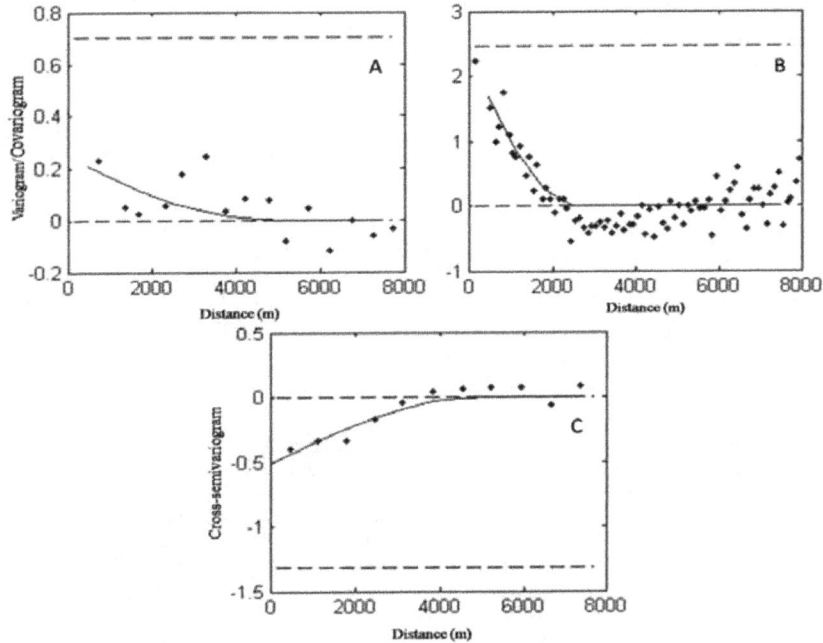

Fig. 4. Spatial covariance and cross-covariance functions for top soil salinity prediction during autumn 2009. A: covariance function for mean UGW SAR, B: covariance function for top soil $EC_{2.5}$, and C: cross-covariance function for top soil $EC_{2.5}$ using UGW SAR as covariate. Dots corresponds to estimated values, solid line is the corresponding fitted model.

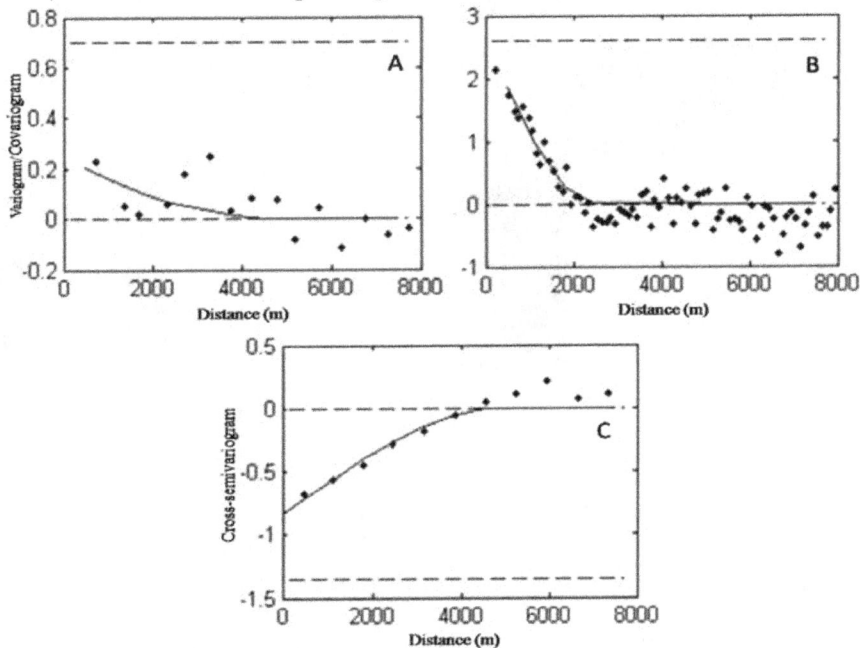

Fig. 5. Spatial covariance and cross-covariance functions for top soil salinity prediction during spring 2010. A: covariance function for mean UGW SAR, B: Covariance function for top soil $EC_{2.5}$, and C: cross-covariance function for top soil $EC_{2.5}$ using UGW SAR as covariate. Dots corresponds to estimated values, solid line is the corresponding fitted model.

Soil salinity prediction

Soil salinity prediction maps for autumn 2009 and spring 2010 with only top soil $EC_{2.5}$ data and UGW SAR data as covariate are shown in Figure 6. Comparing the cross-validation results from soil salinity predictions using only top soil $EC_{2.5}$ data with those of UGW SAR as covariate

indicated that the use of UGW SAR data as secondary information in top soil salinity prediction resulted in lower r and higher ME and MSE both for autumn 2009 and spring 2010. Higher negative ME values shows that the application of UGW SAR as covariate resulted in more under estimations of top soil salinity.

Table 2. Quantitative criteria to compare soil salinity prediction methods during autumn 2009 and spring 2010

Criterion	Autumn 2009		Spring 2010	
	EC2.5	$EC_{2.5}$ & SAR	EC2.5	$EC_{2.5}$ & SAR
r	0.94	0.87	0.92	0.85
ME (dS m^{-1})	0.04	-0.11	-0.12	-0.21
MSE (dS m^{-1})^2	0.33	0.51	0.55	0.63

In Fig. 6, parts B and D have smoother and more defuse bounders which, show that use of UGW SAR has homogenized the salinity variation with

removing the effect of mean variation in the study area within the range of 2.5 km.

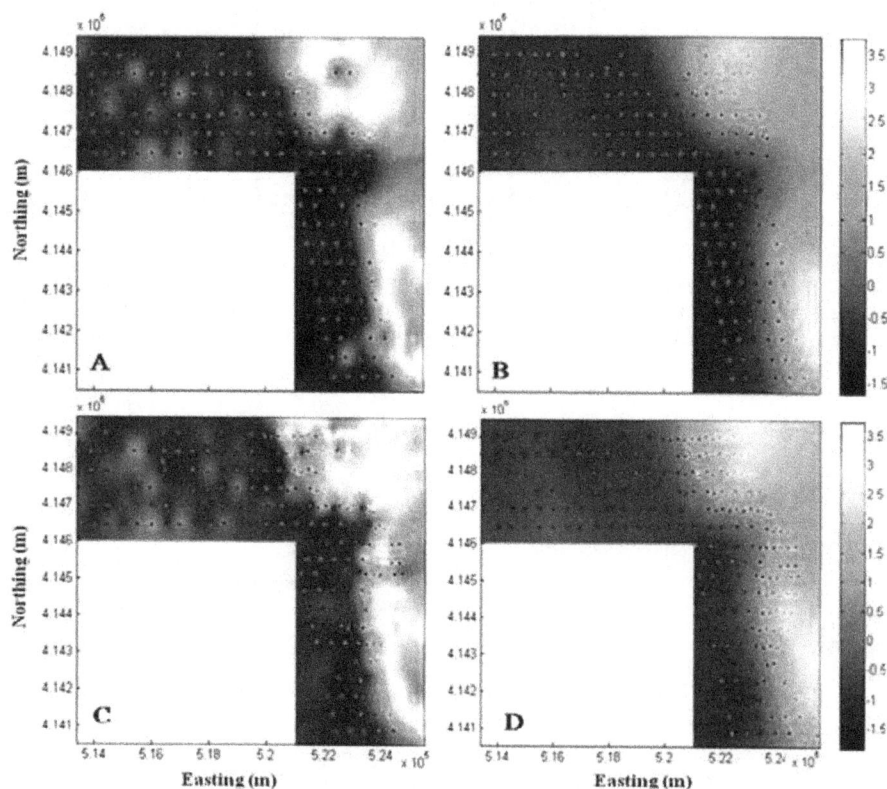

Fig. 6. Soil salinity prediction maps. A: soil salinity predictions for autumn 2009 with only top soil $EC_{2.5}$ values; B: soil salinity predictions for autumn 2009 with UGW SAR as covariate; C: soil salinity predictions for spring 2010 with only top soil EC2.5 values, and D: soil salinity predictions for autumn 2009 with UGW SAR as covariate. Red points represent the places with top soil $EC_{2.5}$ measurements.

Conclusion

This research was conducted to investigate the spatial variation of top soil salinity in Urmia Plain as a function of distance from Urmia Lake. Top soil $EC_{2.5}$ measurements during autumn 2009 and spring 2010 and underground water SAR data for

seven years were used to develop covariance and cross-semi variance functions. Results revealed that top soil salinity increased as distance from the Lake decreased. The increase in top soil salinity was not gradual and there was an instant increase in salinity data all around a boundary parallel to the Lake. To take into account the mean variation in kriging equations, some assumptions were made and soil salinity was predicted on a 100 m grid. Afterwards as there was no significant temporal variation in UGW SAR data, the mean SAR data over time were used to develop the cross-semivariograms. Salinity predictions maps using cross-semivariograms and cokriging method showed that the use of UGW SAR data as covariate had a smoothing effect on $EC_{2.5}$ covariance functions and produced more diffused and gradual soil salinity boundaries, which resulted in more underestimations (higher ME) than that of kriging.

References

Acosta, J.A., Faz, A., Jansen, B., Kalbitz, K. and Martinez-Martinez, S. 2011. Assessment of salinity status in intensively cultivated soils under semiarid climate, Murcia, SE Spain. *J. Arid Environ.* 75: 1056-1066.

Christakos, G., Bogaert, P. and Serre, M.L. 2002. Temporal GIS. Advanced Functions for Field-Based Applications. Springer-Verlag, New York.

De Clercq, W.P., Van Meirvenn, M. and Fey, M.V. 2009. Prediction of the soil-depth salinity-trend in a vineyard after sustained irrigation with saline water. *Agril. Water Manag.* 96: 395-404.

Eimanifar, A. and Mohebbi, F. 2007. Uromia Lake (Northwest Iran): a brief review. *Saline Systems* 3: 5. doi:10.1186/1746-1448-3-5.

Giordano, R., Liersch, S., Vurro, M. and Hirsch, D. 2010. Integrating local and technical knowledge to support soil salinity monitoring in the Amudarya river basin. *J. Environ. Manag.* 91: 1718-1729.

Hassanzadeh, E., Zarghami, M. and Hassanzadeh, Y. 2011. Determining the main factor in declining the Uromia Lake level by using System Dynamics Modeling. *Water Resources Manag.* 26 (1): 129-145.

Juan, P., Mateu, J., Jordan, M.M., Mataix-Solera, J., Melendez-Pastor, I. and Navarro-Pedreno, J. 2011. Geostatistical methods to identify and map spatial variations od soil salinity. *J. Geochem. Explor.* 108: 62-72.

Li, J. and Heap, A.D. 2008. A Review of Spatial Interpolation Methods for Environmental Scientists. Geoscience Australia, Canberra, Australia.

Li, K.L., Chen, J., Tan, M.Z., Zhao, B.Z., Mi, S.X. and Shi, X.Z. 2011. Spatio-Temporal variability of soil salinity in Alluvial Plain of the lower reaches of the Yellow River- a case study. *Pedosphere* 21 (6): 793-801.

Malins, D. and Metternicht, G. 2006. Assessing the spatial extent of dry land salinity through fuzzy modeling. *Ecological Modelling* 193: 387-411.

Math Works Inc. 1998. MatLab, the language of technical computing, using MATLAB version 5. the Mathwork Inc. http://www.mathworks.com, Natick.

Mondal, M.K., Bhuiyan, S.I. and Franco, D.T. 2001. Soil salinity reduction and prediction of salt dynamics in the coastal rice lands of Bangladesh. *Agril. Water Manag.* 47: 9-23.

Peck, A.J. and Hatton, T. 2003. Salinity and the discharge of salts from catchments in *Aus. J. Hydrol.* 272: 191-202.

Stein, A. and Corsten, L.C.A. 1991. Universal kriging and cokriging as a regression procedure. *Biometrics* 47: 575–587.

Triantafilis, J., Odeh, I.O.A., Warr, B. and Ahmed, M.F. 2004. Mapping of salinity risk in the lower Namoi valley using non-linear kriging methods. *Agril. Water Manag.* 69: 203-231.

Zarghami, M. 2011. Effective watershed management; Case study of Urmia Lake, Iran. *Lake and Reservoir Manag.* 27 (1): 87-94.

20

PRELIMINARY STUDY ON THE PRODUCTION OF COMMON CARP CULTURED IN FRESHWATER RIVER CAGES

A.H. M. Kohinoor* and M.M. Rahman

Abstract

A preliminary study was conducted to assess the performance of the common carp, cultured in freshwater river of Brahmaputra cages at different stocking densities during November 2011 to March 2012. The stocking densities tested were 80, 100 and 120 fish/m³. Fish were fed a 28% protein diet at the rate of 15-5% of body weight. The result of the study showed that fish in the T1 stocked at the rate of 80 fish/m³ resulted the best individual weight followed by T2 and T3. The productions of fish in T1, T2 and T3 were 22.33 ± 1.20, 19.00 ± 0.58 and 18.00 ± 1.15 kg/m³, respectively. The results of the present study indicated that the best individual growth and production of common carp was obtained at a density of 80 fish/m³. The results also showed that the individual mean harvesting weights were negatively correlated with stocking density. Therefore, the stocking density of 80 fish/m³ is considered optimum for the rearing phase.

Keywords: Production, Common Carp, Cage

Freshwater Station, Bangladesh Fisheries Research Institute, Mymensingh, Bangladesh

*Corresponding author's email: kohinoor41@gmail.com (A.H. M. Kohinoor)

Introduction

Culture of fish in cages is comparatively a new method of aquaculture, which has gained much popularity throughout the world due to a number of advantages over the conventional methods of fish farming. The scope for increasing fish production in inland waters through cage culture is highly expected in Bangladesh and it would be a very profitable industry like Japan, Thailand, Cambodia, Philippines, Malaysia, USA, and UK (Hasan *et al.,* 1982). But, unfortunately cage culture on commercial basis are yet to be popularized in Bangladesh due to many reasons such as, lack of knowledge about proper management, selection of species, determination of appropriate stocking densities, unavailability of cage materials and socio-economic constraints. For cage culture or any other intensive culture system, selection of species is also important since not all species are suitable for all culture system. In Bangladesh, *C. carpio* is an exotic fish, which has been introduced in 1960. Among the various culturable species *C. carpio* are particularly important for their fast growth, lucrative size, good taste, and high market demand in Bangladesh. However, there is published information on cage culture of *C. carpio* in Bangladesh (Haque *et al.,* 1994). Stocking density is an important factor for the production of all fishes as well as *C. carpio* in cages. High stocking density decreases the production of fish and low stocking density is not commercially profitable. Using appropriate stocking densities and feeding strategies are two of the keys to success in aquaculture management. Sodikin (1977) stated that the fish culture in cages could be developed by improving stocking density, feeding methods, selection of species and regulating the culture cycle for maximum profitability. Lack of nutritionally adequate and low-cost feed has always been one of the constrains to the successful practice of cage fish culture in many developing countries (Otubusin, 1987). Coche (1979) observed that cost of feeding in intensive cage culture could be account for more than 50% of the production costs.

There are number of literature on the culture of fish in cages with different commercial fish species conducted in different parts of the world. Even in Bangladesh some information's are available on the culture of carps and other catfishes (Ahmed, 1982; Rahman *et al.,* 1992; Hannan *et al.,* 1988; Haque *et al.,* 1984; Haque *et al.,* 1988; Haque *et al.,* 1994) but not much works have been done on the culture of *C. Carpio* in cages. Thus, the present study was carried to optimize the suitable stocking density of common carp (*C. carpio*) in net cages.

Materials and Methods

Experimental site

The trial was conducted in the Brhamaputra River at Khagdohar Ghat, Mymensingh district for five months during November to April 2012.

Construction of net cages

The cages were rectangular shaped, made of iron framework and covered by high-density polyethylene net. The size of each cage was 3m³. The mesh size of the net was 1 cm. Iron rods were welded to construct a rectangular shape frame and nets were attached to the rod with the help of nylon twine. One edge of the side of each cage was kept open which was tied with nylon threads in such a way that it could be opened to deliver feed in the cage.

Installation of the cages

The cages were installed by several bamboo bars. The cage was fixed with bamboo poles inserted into the river bottom. The cages were tied with the frame by nylon ropes at the time of suspension; about 1 feet of the upper portion of the cages were always kept above the water level. For easy management, the cages were numbered as 1 to 9 and were divided into three treatment groups viz. T1, T2 and T3.

Experimental fish

Common carp fingerlings were used in this experiment. Fish fry were collected from hatchery of Bangladesh Fisheries Research Institute (BFRI), Mymensingh. All the fishes were of the same age group having mean weight of 11.92±1.32g. Prior to start of the experiment, the fingerlings were acclimatized to the new environment in floating cages for seven days. Then the fishes were stocked in the cages according to experimental design.

Experimental design

The designs of experiments are as follows:

Table 1. Stocking density of common carp in net cages in three different treatments

Species	Treatment	No of replication	Stocking density (m³)
	T1	3	80
Common carp	T2	3	100
	T3	3	120

Feeding

Fish in each cage were fed a 28% crude protein (Saudi-Bangla Fish Feed Company Ltd.) ration 2 times/day. Feeding rate was initially 15% of total body weight per day and was subsequently reduced to 13, 11, 9, 7 and 5% on days 30, 60, 90, 120 and 150, respectively. Feed ration was maintained based on monthly samples of fish from each cage. At monthly intervals, 100 fish from each cage were sampled.

Monitoring of water quality parameters

The water quality parameter such as temperature (°C), pH dissolved oxygen (mg/l), transparency (inches) and ammonia (mg/l) were recorded weekly interval throughout the experimental period at 9 am to 10 am by using a water testing kit. The samples were always collected from the sub-surface of with minimum disturbance of water.

Harvesting

After 150 days of culture, all fish from each cage were harvested. Then total number and weight of fishes were recorded and survival, production were also calculated.

Data analysis

A statistical test on the treatment means was done using the procedure for randomized block design. When means were significantly different, the Duncan Multiple Range Test at 5% level was used.

Results

The physico-chemical parameters of Brahmaputra River water viz., temperature, transparency, pH, dissolved oxygen and total ammonia of are presented in Table 2. The values of temperature, transparency, dissolved oxygen, pH and total ammonia were 18.60-29.30°C, 1.20-1.40 cm, 6.72-8.55 mg/l, pH 7.21-8.66 and 1.1-1.80 mg/l, respectively.

Table 2. Water quality parameters of Brahmaputra River during experimental period

Parameters	Value
Water Temperature (°C)	18.6 – 29.30
Transparency (cm)	1.20-1.40
pH	6.72 - 8.55
DO (mg/l)	6.72 – 8.55
Total ammonia (mg/l)	1.10 – 1.80

On the basis of final growth attained, it was observed that the highest average weight was found in treatment-T1. At harvest, the mean weights were 282.37±12.17, 268.55±14.87 and 246±15.62g, in treatments-T1, T2, and T3, respectively. The harvesting weight of treatment-1 was significantly higher (P<0.05) than those of other treatments.

The survival rate of carpio varied between 66 to 78%. There were significant difference (P>0.05) of survival of Carpio among the treatments. The highest survival was recorded in T1 and lowest in T3.

The mean food conversion Ratio (FCR) values in different treatments varied between 2.4 to 3.5.

The highest FCR values (3.5) were recorded in T3 and the lowest (2.4) in T1. The values of FCR in three different treatments were significantly (P>0.05) different from each other.

The productions obtained in cages were 22.33±1.20, 19.00±0.58 and 18.00±1.15 kg/m3 from treatments-T1, T2 and T3, respectively. The highest production was obtained from treatment-T1, which differed significantly (P<0.05) from other two treatments, when analyzed statistically. In higher stocking densities, the harvesting weight of Common carp was found linearly. However, the production of fish was not varied significantly from the stocking density of 100 and 120/m3.

Table 3. Harvesting wt., survival and production of common carp under different treatments

Treatments	Harvesting Wt. (g)	SGR (%)	FCR	Survival (%)	Production/m3 (kg)
T1 (80/m3)	282±6.11[a]	1.76	2.4±.11[a]	78±2.88[a]	22.33±1.20[a]
T2 (100/m3)	268±5.29[a]	1.73	3.0±.05[b]	74±3.46[ab]	19.00±0.58[ab]
T3 (120/m3)	238±4.61[b]	1.66	3.5±.15[c]	66±3.20[b]	18.00±1.15[b]

*Dissimilar superscript indicates significant difference at 5% level of probability

Correlation matrix among stocking density, harvesting weight, survival and production of fish is shown in Table 4. Stocking density showed an inverse relationship with survival. It means that if stocking density increased, then survival of fish decreased. While, harvesting weight citied positive correlation with production and survival. Whereas, production also showed positive correlation with survival.

Table 4. Correlation matrix among stocking density, harvesting weight, survival, and production of fish

Parameters	Density	Harvesting Wt. (g)	Survival (%)	Production/m3 (kg)
Density	1			
Harvesting Wt. (g)	-0.97741164	1		
Survival (%)	-0.98198051	0.999739528	1	
Production/m3 (kg)	-0.96671061	0.890797029	0.900935671	1

*Significant difference at 5% level of probability

Discussion

The water quality parameters such as temperature, dissolve oxygen, pH and alkalinity studied during the experimental period were found suitable for fish farming and did not hamper the normal fish growth (Jhingran, 1983).

The results indicated that the growth rate of *C. carpio* varied in different stocking densities. Among the treatments, T$_1$ (40 fishes/m3) showed the best result in terms of growth and feed utilization. Ahmed (1982) reported that the stocking rate of *Labeo rohita*, 10 fishes /m3 in floating cages in ponds gave best result in terms of individual growth followed by 20 and 30 fishes respectively. While, Haque *et al.* (1994) found that in case of *Cyprinus carpio* cultured in floating ponds, the best growth was recorded at lowest density (5 fishes/m3) and least growth was recorded at highest stocking density (20 fishes/m3).

While, Dimitrov (1976) observed that low stocking densities 20 fish/m3 gave the highest production of carps in net cages compare to the high densities of 80 and 150 fishes/m3.

In another study, Chaitiamvong (1977) found higher production in low stocking densities and lower production in higher stocking densities of carp. Shiremen *et al.* (1977) and Ahmed *et al.* (1983) reported that the growth and production of fish are to a certain extent, dependent on the population density. Whereas, Powell (1972) reported that the harmful effects of higher stocking density on the culture of fish were the reduction of growth rate, increase of food conversion ratio and lowering of survival rate.

The similar type of observation was reported by Azimuddin (1998) who found that in cage culture of pangus, the stocking density of 40 fishes/m³ gave the best growth in comparison with 50 and 60 fishes/m³.

The results of the study indicated that the best individual growth and production of carpio was obtained at a density of 80 fishes/m³.

References

Ahmed, G. 1982. Intensive culture of *Labeo rohita* (Hamilton) in floating ponds with special reference to different stocking densities. M.Sc. Thesis. Department of Fisheries Biology and Limnology. Bangladesh Agricultural University, Mymensingh. 38p.

Ahmed, G.U., Aminul Haque, A.K.M., Aminul Islam, M. and Mahfuzul Haque, M. 1983, Intensive culture of *Labeo rohita* (Hamilton) in floating ponds with special reference to different stocking densities. *Bangladesh J. Fish.* 6 (1-2): 11-17.

Azimuddin, M.K. 1998. Effect of stocking density on the growth of Thai Pangus, *Pangasius sutchi* (Fowler) in net cage by using formulated diet. M.Sc. Thesis. Department of Aquaculture. Bangladesh Agricultural University, Mymensingh. 62p.

Chaitiamvong, C. 1977. Review of Fresh Water Fisheries Division, Department of Fisheries, Bangkok, Thailand. 32p.

Coche, A.G. 1979. A review of cage fish culture and its application in Africa. *In:* T.V.R. Pillay and W.A. Dill (Editors), Advances in Aquaculture. Fishing News Books, Farham, Survey, pp. 428-441.

Dimitrov, M. 1976. Carp culture in net cages. *FAO Aquaculture Bull.* 8 (1): 8-16.

Hannan, M.A., Alam, A.K.M.N., Mazid, M.A. and Humayn, N.M. 1988. Preliminary study of the culture of *Pangasius pangasius*. *Bangladesh J. Fish.* 11 (1): 19-22.

Haque, M.M., Alam, A.K.M.N and Aminul Islam, M. 1988. Culture prospects of Shingi *Heteropneustes fossilis* (Bloch) in floating net cages. *Bangladesh J. Fish.* 11 (2): 15-19.

Haque, M.M., Haque, A.K.M., Aminul Islam, M. and Ahmed, G.U. 1994. Effect of stocking density on the survival and growth of *Cyprinus Carpio* culture in floating ponds. *Bangladesh J. Agril. Sci.* 11 (2): 175-179.

Haque, M.M., Islam, M.A., Ahmed, G.U. and Haque, M.S. 1984. Intensive culture of Java Tilapia (*Oreochromis mossambica*) in floating ponds at different stocking densities. *Bangladesh J. Fish.* 7 (1-2): 55-59.

Hasan, M.R., Aminul Haque, A.K.M., Aminul Islam, M. and Kaiser Khan, E.U. 1982. Studies on the effects of stocking density on the growth of Nile tilapia, *Sarotherodon nilotica* (Linnaeus) in floating ponds. *Bangladesh J. Fish.* 2-5 (1-2): 73-81.

Jhingran, V.G. 1983. Fish and Fisheries of India. Hindustan Publishing Corporation, Delhi. 666p.

Otubusin, S.O. 1987. Effect of different levels of blood meal in pelleted feeds on Tilapia, *Oreochomis niloticus*, production in floating bamboo net-cages. *Aquaculture* 65:263-266.

Powell, M.R. 1972. Cage and raceway culture of stripped bass in brackish water in Alabama. Proc. 26 Ann. Conf. South east. Assoc. Game Fish Comm. pp. 553-565.

Rahman, M.K., Akter, J.N., Mazid, M.A. and Halder, G.C. 1992. Comparison of fingerling growth rates and survival of two *Pangasius* species. *J. Inland Fish. Soc. India* 24 (2): 40-44.

Shiremen, J.V., Colle, D.E. and Rattman, R.W. 1977. Intensive culture of grass carp, *Ctenopharyngodon idella*, in circular tanks. *J. Fish Biol.* 2: 267-272.

Sodikin, D. 1977. Fish cage culture in Indonesia: construction and management. Technical report of workshop on Aquaculture engineering, Vol. II. held at SEAFDEC Aquaculture Department Facilities, Tigbauan, lloili, Philippines. Nov. 27-Dec. 3. pp. 351-357.

21

EFFECTS OF DIFFERENT KINDS OF FERTILIZERS ON PRODUCTION OF FISHES IN POLYCULTURE SYSTEM

M.J. Alam[1,2], M. Shahjahan[1*], M.S. Rahman[1], H. Rashid[1], M.A. Hosen[1,2]

Abstract

A study was conducted to assess the effects of different kinds of inorganic fertilizers on the production of fishes in six ponds during October to December 2011. There were three treatments with two replications under each treatment and each of the ponds was stocked with 80 fish fry. In treatments I, II and III, ponds were fertilized fortnightly @ urea 100 g decimal[-1], T.S.P. 100 g decimal[-1] and urea 50 g decimal[-1] + T.S.P. 50 g decimal[-1], respectively. Selected water-quality parameters of ponds under study were more or less similar and within the productive range. Mean phytoplankton and zooplankton densities under treatments I, II and III were 57.08 ± 1.35, 8.80 ± 0.09 and 77.29 ± 3.72, 12.88 ± 0.74 and 98.93 ± 1.61, 16.16 ± 1.75 (x10^3) cells L^{-1}, respectively. The net and gross fish productions of the ponds under treatments I, II and III were 0.85 and 3.11 t ha^{-1} yr^{-1} and 1.32 and 3.58 t ha^{-1} yr^{-1} and 1.85 and 4.11 t ha^{-1} yr^{-1}, respectively. Fish production under treatment III was better than those under treatments I and II because plankton population densities under treatment III was the highest. Therefore, the mixed fertilization is suitable for production of plankton that enhance growth and production of fishes.

Keywords: Fish Culture, Pond, Limnological Factors, Fertilizer, Fish Production

[1]Department of Fisheries Management, Bangladesh Agricultural University, Mymensingh-2202, Bangladesh
[2]Department of Fisheries, Ministry of Fisheries and Livestock, Dhaka, Bangladesh
*Corresponding author's email: mdshahjahan@bau.edu.bd (M. Shahjahan)

Introduction

Optimizing production in pond fish culture by the use of fertilizers is an important task. Fertilizer is helpful for the increase of natural food of fish i.e. plankton, benthos and periphyton. Plankton is the basic food of all the organisms living in the water. Fishes and other aquatic organisms depend on this basic food directly and indirectly. Extensive work on water quality and pond fertilization has been conducted elsewhere (Boyd, 1982) but very few of them have relevance to the Asian carp culture. Both over and under fertilization may cause adverse effects on fish production, water quality and economic returns. It is therefore necessary to evaluate fertilization regimes and recommend appropriate fertilization strategies to farmers in order to maximize fish production, maintain good water quality, reduce environmental bad impact and maximize economic returns.

Successful fisheries management and scientific fish culture depends on the various limnological factors of the water bodies. According to Hickling (1968) fish farming is a practical application of limnology and fresh-water biology, According to Reid (1971) the chemical analysis for dissolved gases and solids are highly important for the study of natural waters. He also reported that fish culture can be enhanced by the improvement of substratum by the use of fertilizers along with other pond management measures. The physico-chemical characteristics of pond water are of great importance and essential in case of fish culture and fisheries management. The physico-chemical properties play the most important role in governing the production of phytoplankton i.e. primary production in fishponds (Banerjee, 1967).

For successful aquaculture, knowledge on several factors is very important among which fertilization is one of them. The necessities and principles of fertilization of ponds for the increase of production are similar to those of crop production. The amount and proportions of various fertilizers needed vary from country to country and even from area to area within a country. The use of fertilizers in proper doses is also very important to reduce the unit cost of production. As for example, Hepher *et al.* (1971) found in Israel that if there applied no fertilizers in fishpond, the cost of production per ton was 935 dollars and after applying fertilizers, the cost of production of fish per ton was 691 dollars. Therefore, the present study was conducted to know the effects of different kinds of fertilizers on production of fishes in polyculture system along with some limnological parameters.

Materials and Methods

The experiment was conducted for a period of three months from October to December 2011 in six experimental ponds at Bangladesh Agricultural University, Mymensingh, Bangladesh. The ponds were rectangular in size and similar in area (about 40 m²). They were numbered arbitrarily as P_1, P_2, P_3, P_4, P_5 and P_6 for the convenience of experimental work and data analysis. The ponds were treated by lime before two weeks of starting the experiment.

Table 1. Experimental layout

Treatments	Replication	Fish species & ratio	Fertilization
I	2 (P1 & P2)	Silver carp : tilapia : mrigal (2 : 2 : 1)	Urea 100 g decimal⁻¹
II	2 (P3 & P4)	Do	TSP 100 g decimal⁻¹
III	2 (P5 & P6)	Do	Urea 50 g+TSP 50 g decimal⁻¹

Stocking the pond

The ponds were stocked with fingerlings of silver carp, tilapia and mrigal with the stocking density of 80 (32+32+16) fingerlings per decimal. The average lengths of fingerlings of silver carp, tilapia and mrigal were 12.14 cm, 10.15 cm and 8.12 cm, respectively.

Fertilization

Inorganic fertilizers such as only urea, only TSP and urea and TSP mixed were used for the experiment. Fertilizers were applied at fifteen days interval. In the ponds of treatment 1 (T_1) only urea was applied at the rate of 100 g decimal⁻¹. In the ponds of treatment 2 (T_2) only TSP was applied at the rate of 100 g decimal⁻¹. In the ponds of treatment III, urea and T.S.P. were applied at the rate of 50 g urea and 50 g TSP decimal⁻¹.

Study of physico-chemical factors

The water samples were collected twice a month in the morning at about 10 a.m. The water samples were collected by dipping the bottle just below the surface water. All the bottles containing water samples were carried to the Laboratory of Limnology for analysis. The Physico-chemical factors which were studied are water temperature (°C), water depth (m), transparency (cm), dissolved oxygen (mg L⁻¹), Free CO_2 (mg L⁻¹), P^H and total alkalinity (mg L⁻¹). Water temperature was recorded with a Celsius thermometer and transparency was measured with a Secchi disc of 30 cm diameter. Dissolved oxygen was measured directly with a DO meter (Lutron, DO-5509) and a portable digital pH meter was used to measure pH. Free CO_2 and total alkalinity were determined by titrimetric method (APHA, 1992).

Study of plankton

For the quantitative and qualitative study of phytoplankton and zooplankton of water, samples were collected from the different spots of water of the ponds with the help of bottle. Fifteen to thirty liters of water was passed through the plankton net (mesh size 55 μ) for each sample and the volume of sample collected was 30 ml. The concentrated samples were poured into vials and preserved by 5% formalin. Plankton samples were collected fortnightly. The study of plankton was done by a haemocytometer under a compound microscope and calculating by using following formula (Rahman, 1992):

$$N = \frac{A \times 1000 \times C}{V \times F \times L}$$

Where,

N= No. of plankton cells per L,
A= Total no. of plankton counted,
C= Volume of final concentrate of samples in ml,
V= Volume of a field in cubic mm,
F= Number of the fields counted,
L= Volume of original water in liter,

Study of growth of fishes

Before releasing the fingerlings in the ponds average initial length (cm) and average weight (g) were recorded with the help of a meter scale and a balance. At the end of the experiment all the fishes were caught by a cast net and then by using rotenone in the ponds. Fish mortality, gross and net productions have been calculated using the following formulas:

(i) The survival rate was estimated by the following formula

$$\text{Survival rate (\%)} = \frac{\text{No. of harvested fishes}}{\text{Initial no. of fish}} \times 100$$

(ii) Calculation of gross fish production (t ha⁻¹ yr⁻¹)

$$= \frac{\text{Gross weight (kg) of fish per decimal per month} \times 250 \times 12}{1000}$$

(iii) Calculation of net fish production (t ha⁻¹ yr⁻¹)

$$= \frac{\text{Net weight (kg) of fish per decimal per month} \times 250 \times 12}{1000}$$

Statistical analysis

T-test of net fish production of the ponds under three treatments was done by a computer using SPSS package programme.

Results

Physico-chemical parameters

The results of physico-chemical parameters are shown in Table 2. All physical and chemical parameters of the ponds were found to be within the acceptable ranges for fish culture in all treatments.

Table 2. Physico-chemical parameters (Means ± SD; n = 3) of the ponds during the experimental period

Parameters	Treatment II	Treatment II	Treatment III
Average water depth (m)	0.89±0.01	0.85±0.04	0.91±0.02
Water temperature (°C)	20.53±4.34	20.53±4.34	20.53±4.34
Transparency (cm)	35.41±2.89	31.91±0.75	31.35±1.71
pH	7.38±0.20	7.33±0.10	7.48±0.37
Dissolved oxygen (mg L^{-1})	9.50±0.58	9.79±0.52	10.12±0.36
Free CO_2 (mg L^{-1})	1.21±0.16	1.18±0.11	1.22±0.12
Total alkalinity (mg L^{-1})	106.08±5.02	104.67±3.11	112.17±5.56

Plankton

Mean phytoplankton and zooplankton densities under treatments I, II and III were 57.08 ± 1.35, 8.80 ± 0.09 and 77.29 ± 3.72, 12.88 ± 0.74 and 98.93 ± 1.61, 16.16 ± 1.75 (x10^3) cells L^{-1}, respectively (Fig. 1 and 2). It was observed that both the phytoplankton and zooplankton densities were significantly higher in Treatment III. The generic status of phytoplankton and zooplankton found during the tenure of experiment are shown in Table 3. During the study period, 45 genera of phytoplankton belonging to five groups and 10 genera of zooplankton belonging to three groups were found in all the experimental ponds.

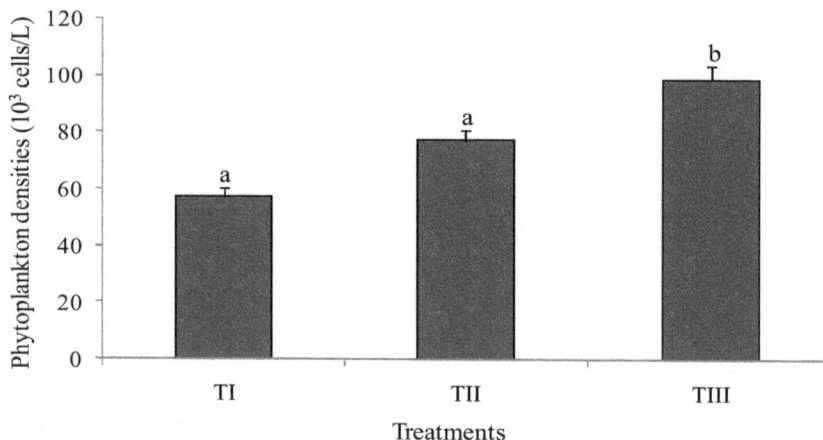

Fig. 1. Mean phytoplankton densities. Values accompanied by different letters are statistically significantly different (p < 0.05).

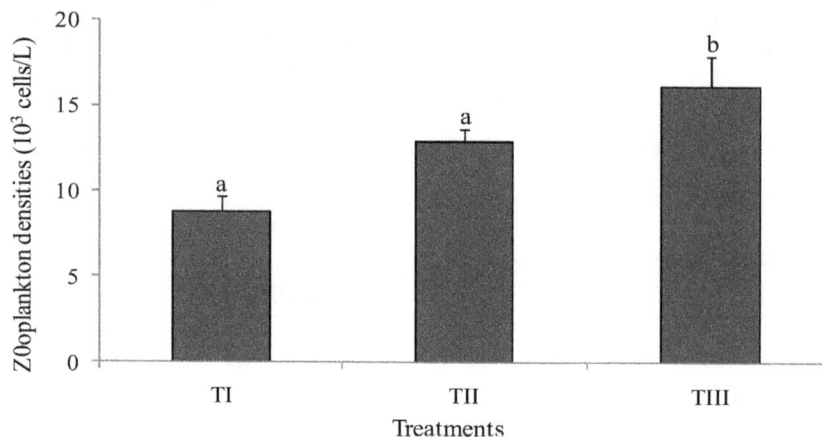

Fig. 2. Mean zooplankton densities. Values accompanied by different letters are statistically significantly different (p < 0.05).

Table 3. Generic status of phytoplankton and zooplankton found in the experimental ponds

Phytoplankton	Zooplankton
Bacillariophyceae: *Navicula , Diatoma,,Cyclotella, Melosira, Frustulia, Asterionella, Stauronesis, Fragilaria, Synedra, Frustulia, Surirella* **Chlorophyceae:** *Ulothrix, Ganatozygon, Gloeocystis, Spirogyra, Euastrum,Chroococcus, Ophiocytium, Protococcus, Volvox,Coelastrum, Cladophora, Tetraedron,Histococcus, Actinastrum,Closterium,Melosira, Scenedesmus, Pediastru, Pediastrum, Ankistrodesmus, Selenastrum, Actinastrum, Stistococcus* **Cyanophyceae:** *Anabaena, Oscillatoria, Microcystis, Nostoc, Spirulina, Aphanocapsa* **Euglenophyceae:** *Euglena, Phacus, Trachelomonas* **Dinophyceae:** *Ceratium, Peridinium*	**Rotifera:** *Brachionus, Keratella, Filinia, Polyarthra* **Cladocera:** *Daphnia, Diaphanosoma, Moina* **Copepoda:** *Nauplius, Cyclops, Diaptomus*

Growth and production of fish

The net productions (t ha^{-1} yr^{-1}) of silver carp, tilapia and mrigal under three treatments have been presented in Fig. 3. The net productions of fishes are shown in Figure 4. Significantly highest net productions of fishes were obtained in Treatment III.

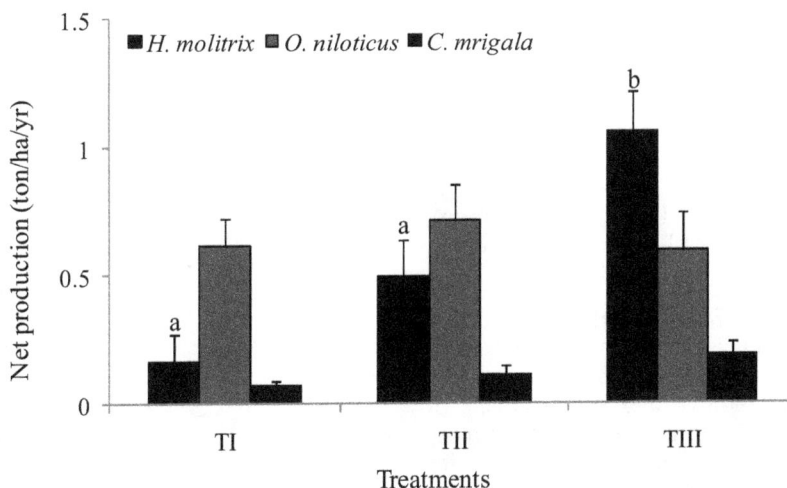

Fig. 3. Species wise net production of fish. Values accompanied by different letters are statistically significantly different (p < 0.05).

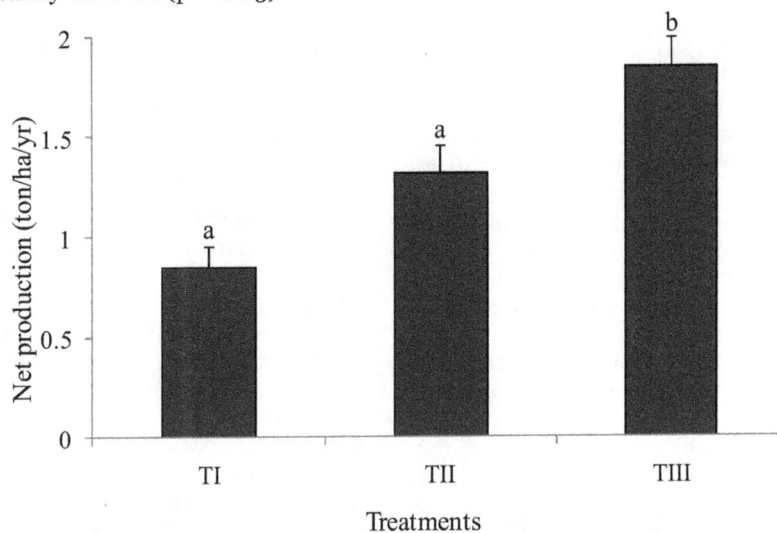

Fig. 4. Net production of fish. Values accompanied by different letters are statistically significantly different (p < 0.05).

Discussion

The present study was conducted to evaluate the effects of fertilization on growth and production of fishes in polyculture of tilapia, silver carp and mrigal. The fishes showed better growth and production performance in supply of mixed fertilizer as both phytoplankton and zooplankton production were highest in this condition.

The physico-chemical parameters of the experimental ponds were within the productive ranges for the growth of plankton and benthos during the tenure of experiment (Table 2). Within limit productive ranges of such water quality parameters have also been observed by a number of authors (Uddin *et al.*, 2007; Chowdhury *et al.*, 2008; Uddin *et al.*, 2012; Rahman *et al.*, 2012; Talukdar *et al.*, 2012; Siddika *et al.*, 2012; Nupur *et al.*, 2013) in the aquaculture ponds of BAU area which are in conformity with those of the present study.

In the present study, phytoplankton and zooplankton population densities were significantly higher in Treatment III (Figures 1 & 2) indicating that both urea and TSP fertilizer needed for their production. More or less similar results also observed by Amin *et al.* (2005), Ferdousi *et al.* (2005), Uddin *et al.* (2007), Chowdhury *et al.* (2008) and Talukdar *et al.* (2012).

The net production of fishes of the ponds under treatment III was higher than those of treatments I and II. Higher net fish production of treatment III indicates the positive and better effects of mixed fertilization of urea and TSP on primary productivity as well as fish growth. All the species of fishes showed the highest rate of growth in all respects in treatment III, which is probably due to high production of both phytoplankton and zooplankton. More or less similar results recorded by Rabanal (1967). Saha *et al.* (1974) recorded that fertilizers enhanced growth of phytoplankton and zooplankton, which in turn induced better growth of fish. The yields of fish were higher in fertilized ponds compare to unfertilized ponds (Shahjahan *et al.*, 2003). Donbrovskij *et al.* (1975) reported that introduction of mineral nitrogenous and phosphoric fertilizers increased the productivity of ponds by 1.4 times at the same stocking rate in fertilized ponds and by 2.3 times at higher stocking rates in fertilized ponds. Therefore, it can be concluded that the higher fish productions of treatment III was due to application of mixed fertilization of urea and TSP in the experimental ponds.

In conclusion, effects of fertilizer on the growth and production of fishes along with some limnological conditions were conducted in polyculture system under three treatments. In treatments I, II and III, ponds were fertilized fortnightly at the rates of urea 100 g decimal⁻¹, T.S.P. 100 g decimal⁻¹ and urea 50 g decimal⁻¹ + T.S.P. 50 g decimal⁻¹, respectively. Physico-chemical parameters were more or less similar in the ponds under three treatments and were within suitable ranges. The net fish production of fish was significantly higher in Treatment III compared to Treatments I and II indicated that mixed fertilization might play a vital role in pond fish culture to increase production of fishes in polyculture system.

References

Amin, A.K.M.R., Bapary, M.A.J., Islam, M.S., Shahjahan, M. and Hossain, M.A.R. 2005. The Impacts of Compensatory Growth on Food Intake, Growth rate and Efficiency of Feed Utilization in Thai Pangas. *Pakistan J. Biol. Sci.* 8: 766-770.

APHA. 1992. Standard Methods for the Examination of Water and Wastewater. American Public Health Association, Washington DC. p. 874.

Banerjee, S.M. 1967. Water quality and soil condition of fish ponds in some states of India in relation to fish production. *Indian J. Fish.* 14: 115-144.

Boyd, C.E. 1982. Water Quality Management For Pond Fish Culture. Elsevier Science Publishers B.V., 1000 AH Amsterdam, The Netherlands. 318p.

Chowdhury, M.M.R., Shahjahan, M., Rahman, M.S. and Sadiqul Islam, M. 2008. Duckweed (*Lemna minor*) as supplementary feed in monoculture of nile tilapia, *Oreochromis niloticus*. *J. Fish & Aqu. Sci.* 3: 54–59.

Donbrovskij, V.K., Lyakhnovich, V.P., Evodokimova, L.V. and Levchekova, N.I. 1975. Effects of mineral fertilizer and stoking rate on the fish proctivity of ponds and on the growth, feeding and energy balance of carps. *Tr. Beintirkh,* 10: 208.

Ferdoushi, Z., Shahjahan, M. and Haque, F. 2005. Impacts of different aquatic macrophytes (duckweed) on the growth and production of different fish. *Prog. Agric.* 16: 149-155.

Hepher, B., Chervinski, T. and Tagari, H. 1971. Studies on carp and silver carp nutrition III. Experiments on the effect on fish of dietary protein source and concentration. *Bamidgeh,* 23 (1): 11-37.

Hickling, C.F. 1968. The Farming of Fish, 1st ed., Pergamon Press Ltd., Oxford, 88p.

Nupur, N., Shahjahan, M., Rahman, M.S. and Fatema, M.K. 2013. Abundance of macrozoobenthos in relation to bottom soil textural types and water depth in aquaculture ponds. *Int. J. Agril. Res. Innov. & Tech.* 3 (2): 1-6.

Rabanal, H.R. 1967. Inorganic fertilizer for pond fish culture. *FAO Fish. Rep.* 44: 164-173.

Rahman, M.S. 1992. Water Quality Management in Aquaculture. BRAC Prokashana, Mohakhali, Dhaka, Bangladesh, 84p.

Rahman, M.S., Shahjahan, M., Haque, M.M. and Khan, S. 2012. Control of euglenophyte bloom and fish production enhancement using duckweed and lime. *Iranian J. Fish. Sci.* 11: 358-371.

Reid, G.K. 1971. Ecology of Inland Waters and Estuaries. Reinhold Publishing Corporation, New York, Amsterdam. 767p.

Saha, G.N., Chatterjee, D.K. and Raman, K. 1974. Observations on manuring nursery ponds with only chemical nitrogenous fertilizers for successful rearing of carp spawn. *J. Inland fish. Soc. India,* 6: 162-166.

Shahjahan, M., Islam, M.S., Bapary, M.A.J. and Miah M.I. 2003. Socioeconomic conditions of fisherman of the jamuna River. *Bangladesh J. Fish.* 26: 47-52.

Siddika, F., Shahjahan, M. and Rahman, M.S. 2012. Abundance of plankton population densities in relation to bottom soil textural types in aquaculture ponds. *Int. J. Agril. Res. Innov. & Tech.* 2 (1): 56-61.

Talukdar, M.Z.H., Shahjahan, M. and Rahman, M.S. 2012. Suitability of duckweed (*Lemna minor*) as feed for fish in polyculture system. *Int. J. Agril. Res. Innov. & Tech.* 2 (1): 42-46.

Uddin, M.N., Rahman, M.S. and Shahjahan, M. 2007. Effects of duckweed (*Lemna minor*) as supplementary feed on monoculture of GIFT strain of tilapia (*Oreochromis niloticus*). *Prog. Agric.* 18: 183-188.

Uddin, M.N., Shahjahan, M. and Haque, M.M. 2012. Manipulation of species composition in small scale carp polyculture to enhance fish production. *Bangladesh J. Prog. Sci. & Tech.* 10: 9-12.

22

EVALUATION OF GROWTH OF *Chlorella ellipsoidea* IN DIFFERENT CULTURE MEDIA

F. Jahan[1], M.S. Rahman[1] and M.A. Hossain[2*]

Abstract

An experiment was conducted to evaluate the growth of *Chlorella ellipsoidea* in three different media viz,. medium I (pulse bran), medium II (soil extract) and medium III (inorganic) under the natural environmental conditions. The alga, *C. ellipsoidea*, reached maximum cell density of 56.32×10^6 cells ml^{-1} in 10 days in medium I (pulse bran), maximum cell density of 102.99×10^6 cells ml^{-1} in 11 days in medium II (soil extract) and maximum cell density of 64.23×10^6 cells ml^{-1} in 12 days in medium III (inorganic medium). The ranges of water temperature, air temperature and light intensity were 22 to 32°C, 22 to 34°C and 2.11 to 4.31 ($\times 10^3$) lux, respectively during the culture period. The average sunshine period was 7.65 ± 1.57 hours. Total alkalinity, free CO_2, pH, NO_3-N, PO_4-P of algal culture medium I, medium II and medium III were 220, 200 and 150 mg L^{-1} ; 26, 9 and 19 mg L^{-1}; 7.9, 7.6 and 7.5; 45, 45 and 133.33 mg L^{-1}; 10.9, 15.1 and 37.06 mg L^{-1}, respectively. Cell densities of cultures of *C. ellipsoidea* under three treatments I, II and III, it can be concluded that cell densities under 3 treatments are significantly different (F=39.78) and treatment II (soil extract medium) is the best for algal (*C. ellipsoidea*) culture among three treatments.

Keywords: Soil Extract, *Chlorella* , Culture Media, Envioronmental Factors, Chemical Quality of Media

[1]Department of Fisheries Management, Bangladesh Agricultural University, Mymensingh, Bangladesh
[2]Department of Aquaculture, Bangabandhu Sheikh Mujibur Rahman Agricultural University, Gazipur, Bangladesh

*Corresponding author's email: amzad@bsmrau.edu.bd (M.A. Hossain)

Introduction

Microalgae, such as *Chlorella* are rich in nutrients especially protein, lipid and minerals (Becker, 1994a & b; Soeder, 1980; Geldenhuys *et al.*, 1988). A dried sample of pure *Chlorella* has nutrients which contain moisture 4.5%, crude protein 51.8%, crude fat 13.6%, carbohydrate 25.2%, crude fibre 2.3%, crude ash 4.9%, and chlorophyll 1.52%. Available energy from *Chlorella* was 430 cal g^{-1} (Tang and Suter, 2011). *Chlorella* contains all other essential nutrients that are needed for growth and development of both aquatic and terrestrial animals. So, people of some countries such as China and Japan have been using seaweeds and certain other algae as a source of food. Having perceived the miraculous nutritional quality of *Chlorella*, Japanese manufactured *Chlorella* tablets in the late 1950s; and from that time, numerous endeavors have emerged at specialized industries worldwide with a view to producing health food, food additives, animal feed, biofertilizers, biofuel and an assortment of natural products (Sasson, 1997; Christi, 2007). Two algae, *Scenedesmus* sp. and *Chlorella* sp. are of commercial importance due to their commercial significance of *Chlorella* sp. (Rydlo, 1973) which are (i) food, protein supplement/fortification in diets for malnourished children and adults; (ii) feed,

protein/vitamin supplement in feed for poultry, cattle, pigs, and bivalves; (iii) health food, *Chlorella* sp. powder as ingredients and supplement in health food recipes and products; (iv) therapeutics, β-carotene as possible anti skin-cancer treatment is used for variety of skin diseases; (v) pigments, β-carotene as food color and food supplement (provitamin A); and (vi) other uses, biofertilizers, soil conditioner, and waste treatment.

So, considering very high importance of microalgae culture, *Chlorella ellipsoidea,* a very important microalga, has been selected to culture in inexpensive culture media, especially soil extract, to reduce the cost of culture and to introduce easy technique of culture.

Materials and Methods

Preparation of three algal culture media

1. *Preparation of inexpensive culture medium using pulse bran (masakalai bran, Vigna mungo) according to the method of Rahman (2000):*

This medium was prepared by mixing 1 kg pulse bran (*Vigna mungo*) in 20-litre tap water. After a week, 15 g urea [$(NH_2)_2 CO$] was added into the mixture. After three weeks, pulse bran mixture was filtered through thin markin cloth to discard solid materials and then after several days the clear supernatant was siphoned to another bucket. To clear the medium, lime (CaO) was added to the medium at the rate of 2 g per litre and after a week supernatant was siphoned to another bucket. After adding lime, pH of the medium increased to about 10. Then to lower the pH to about 7.9, conc. H_2SO_4 was added to the medium at the rate of 0.325 mlL^{-1} and after one week, the medium was ready to be used for algae culture.

2. Preparation of soil extract medium

Soil (textural class, silty clay loam, pH, 7.3; organic carbon (%), 0.95; organic matter (%), 0.64; available phosphorous (mg L^{-1}), 17.12; total nitrogen (%), 0.087 was dried for 2 weeks. After drying, soil was crushed into powder to facilitate sieving. Soil was sieved through a small mesh sieve usually used to sieve rice powder for making cake. Then 2 kg soil was mixed with 5L tap water in a plastic bucket. Soil-water mixture was kept for 5 days and during this period, mixture was stirred everyday for half an hour. Then soil-water mixture was kept in this condition for several days untill the settling of soil particles at the bottom of the bucket. Then supernatant (soil extract) was siphoned with a plastic pipe to a plastic container. This supernatant was sterilized in an autoclave at 121°C temperature and 15 lb/inch^2 pressure for 20 minutes. Then soil extract was treated with commercial urea (5 g per litre) and T.S.P (2.5 g per litre) fertilizers, this soil extract were used as algal culture medium.

3. Preparation of inorganic medium

Inorganic medium was prepared with the inoculation of stock solutions of 8 major nutrients (NaNO$_3$, MgSO$_4$.7H$_2$O, K$_2$HPO$_4$, CaCl$_2$.2H$_2$O, Na$_2$CO$_3$, EDTA, Ferric ammonium citrate, Citric acid) and 6 minor nutrients (H$_3$BO$_3$, MnCl$_2$.4H$_2$O, Na$_2$MoO$_4$.H$_2$O, ZnSO$_4$.7H$_2$O CuSO$_4$.7H$_2$O, CaSO$_4$.7H$_2$O). Ten liter distilled water was taken in a 30 litre plastic bucket and stock solutions were added and mixed well in the bucket and stored in a 15 litre plastic container. Stock solutions were prepared in distilled water using different chemical compounds as major nutrients and trace elements soil extract were used as algal culture medium.

Study of the environmental factors

Water temperature, air temperature was determined by thermometer and light intensity were determined by a lux-meter (LUX-101). Data of sunshine period and rainfall were collected from the "Weather Yard" of Bangladesh Agricultural University, Mymensingh.

Results

Cell densities of C. ellipsoidea in medium I (pulse bran)

The growth of C. ellipsoidea calculated in medium I (pulse bran) under treatment I in 4 replications in natural light and temperature conditions. Culture of these green algae, started with 0.152 × 10^6 cells ml^{-1} (inoculum was 5%) which attained the maximum cell density of 56.32 × 10^6 cells ml^{-1} in 10 days. The average cell density of C. ellipsoidea was 28.31 × 10^6 cells ml^{-1} after the culture period of 17 days (Table 1, Fig. 1).

Table 1. Daily variation of mean cell density of C. ellipsoidea (mean of 4 replications under the treatment) cultured in medium-I (pulse bran), medium-II (soil extract), medium-III (inorganic) and environmental conditions

Culture time (days)	Cell density (×10^6, cells ml^{-1})			Water temp. (°C) (Av.)	Air temp. (°C) (Av.)	Light intensity (×10^3 lux)	Sunshine Period* (hrs.)	Rainfall* (mm)
	Treatment I	Treatment II	Treatment III					
1	3.68	6.4	2.56	27.00	30.00	2.50	6.8	00.0
2	8.89	13.21	10.87	27.75	27.00	2.11	9.3	00.0
3	15.09	18.99	18.32	27.50	32.50	2.93	9.2	00.0
4	29.16	44.19	23.44	28.50	31.00	2.90	8.8	00.0
5	32.86	57.76	27.50	27.50	29.50	2.53	8.1	00.0
6	33.22	58.97	30.71	27.00	26.75	2.30	7.5	00.0
7	41.7	63.06	35.87	26.75	27.50	4.31	6.5	0.60
8	43.99	80.26	40.22	26.50	29.50	3.93	8.3	00.0
9	46.29	85.02	46.28	25.00	28.50	2.90	9.3	00.0
10	**56.32**	90.54	48.60	25.75	28.75	2.26	7.4	00.0
11	41.29	**102.99**	50.41	25.50	29.00	3.00	6.8	00.0
12	35.19	98.00	**64.23**	25.50	28.50	2.58	8.6	00.0
13	20.79	92.40	60.23	26.50	29.00	2.63	5.9	00.0
14	19.77	83.78	55.03	25.50	29.00	3.52	5.8	00.0
15	19.33	75.78	44.98	24.35	28.00	3.80	3.7	11.0
16	18.26	39.20	38.78	24.00	27.00	3.50	8.8	00.0
17	15.31	22.44	33.89	24.50	27.50	2.45	9.2	00.0
Mean± SD	28.31±15.08	60.76±30.56	36.98±15.65	26.18±1.31	28.76±1.49	2.95±0.64	7.65±1.57	0.68±0.97

N.B Starting date of the algal culture was 3.10.11. Initial cell density of culture on zero day was 0.152 × 10^6 cells ml^{-1}
* Data of sunshine period and rainfall were collected from "Weather Yard", Bangladesh Agricultural University, Mymensingh.

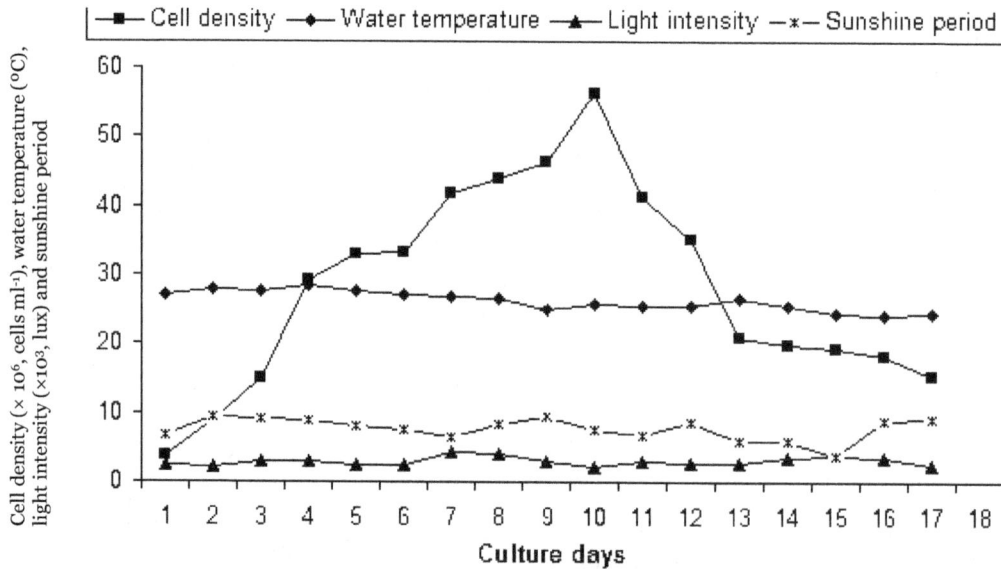

Fig. 1. Daily variations of mean cell density (×10⁶, cells ml⁻¹) of *C. ellipsoidea* cultured in medium-I (pulse bran extract, treatment-I) and average water temperature (°C), light intensity (×10³, lux) and sunshine period (hrs.)

Cell densities of *C. ellipsoidea* in medium II (soil extract medium)

The growth of *C. ellipsoidea* cultured in medium II (soil extract) attained the maximum cell density of 102.99 × 10⁶ cells ml⁻¹ in 11 days. The average density of *C. ellipsoidea* was 60.76 × 10⁶ cells ml⁻¹ after the culture period of 17 days (Table 1, Fig. 2).

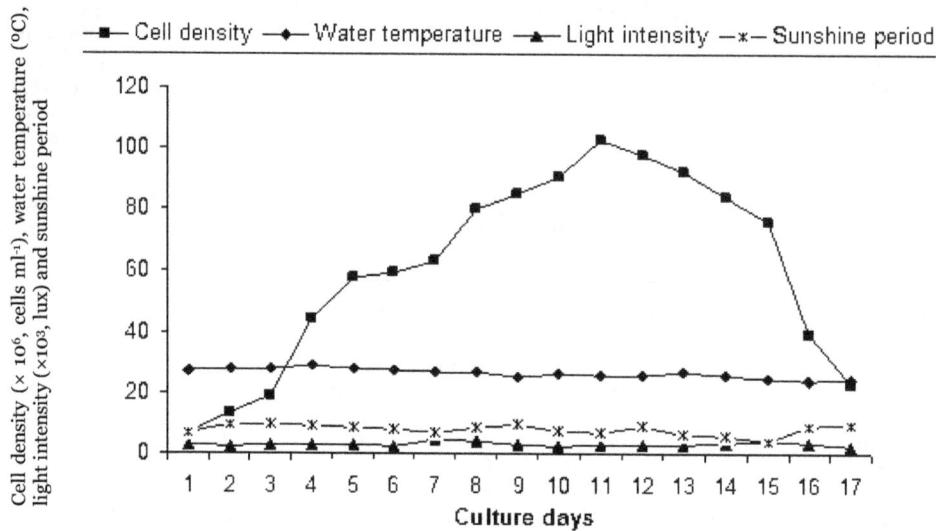

Fig. 2. Daily variation of mean cell density (×10⁶, cells ml⁻¹) of *C. ellipsoidea* cultured in medium-II (soil extract, treatment-II) and average water temperature (°C), light intensity (×10³, lux) and sunshine period (hrs.)

Cell densities of *C. ellipsoidea* in medium III (inorganic medium)

The growth of *C. ellipsoidea* cultured in medium III (inorganic) attained the maximum cell density of 64.23×10⁶ cells ml⁻¹ in 17 days. The average cell density of *C. ellipsoidea* was 37.17×10⁶ cells ml⁻¹ after the culture period of 17 days. (Table 1, Fig. 3). Comparison of daily variations of mean cell density (x10⁶ cell ml⁻¹) of *C. ellipsoidea* of media I , II, and III have been presented in Fig. 4.

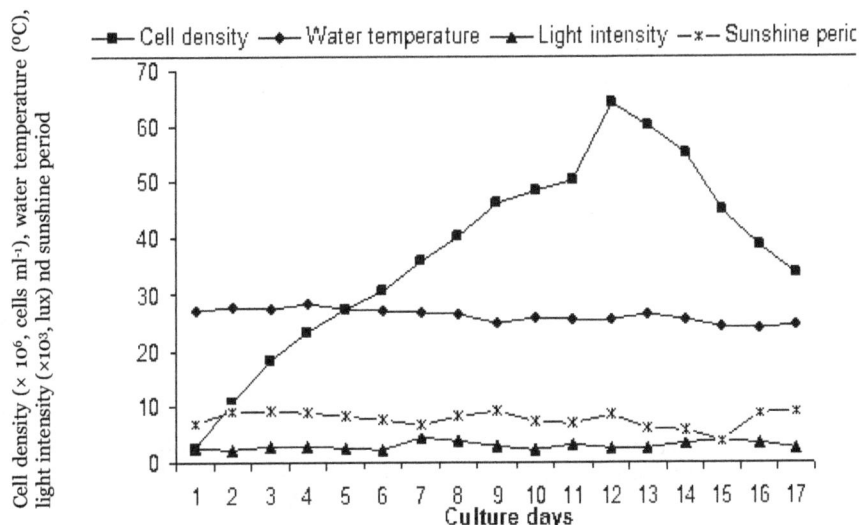

Fig. 3. Daily variation of mean cell density ($\times 10^6$ cells ml^{-1}) of *C. ellipsoidea* cultured in medium-III (Inorganic, treatment-III) and average water temperature (°C), light intensity ($\times 10^3$, lux) and sunshine period (hrs.)

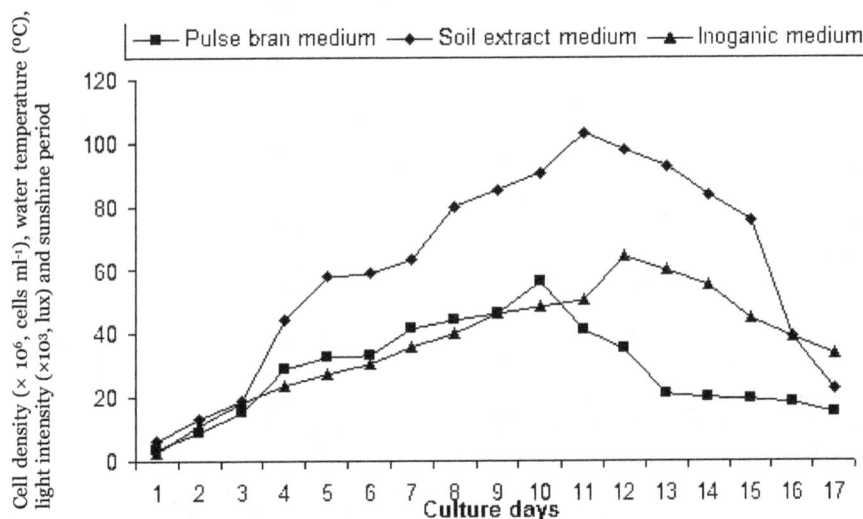

Fig. 4. Comparison of daily variation of mean cell density ($\times 10^6$ cells ml^{-1}) of *C. ellipsoidea* cultured in medium-I (pulse bran extract, treatment-I), medium-II (soil extract, treatment-II) and medium-III (inorganic, treatment-III)

Chemical properties of the three cultural media have been presented in Table 2 .

Table 2. Chemical properties of the culture media

Culture medium	pH	Free CO_2 (ppm)	Total alkalinity (ppm)	NO_3-N (ppm)	PO_4-P (ppm)
Pulse bran medium	7.9	26	220	45	10.9
Soil extract medium	7.6	9	200	45	15.1
Inorganic medium	7.5	19	150	133.33	37.06

Discussion

The range of cell density of *C. ellipsoidea* in medium I (pulse bran extract medium) was 3.68 to 56.32 $\times 10^6$ cells ml^{-1} during the culture period. The average cell density was 28.31 ± 15.084 $\times 10^6$ cells ml^{-1}. During the culture period of *C. ellipsoidea* in medium I exponential phase was up to 10th day from the beginning and then from stationary phase cell density started to decline towards death phase.

The range of cell density of *C. ellipsoidea* in medium II (soil extract medium) was 6.4 to 102.99 ×10^6 during the culture period. The average cell density was 60.76 ± 30.56 × 10^6 cells ml^{-1}. During the culture period of *C. ellipsoidea* in medium II, exponential phase was found up to 11th day from the beginning and then from stationary phase cell density started to decline towards death phase.

The range of cell density of *C. ellipsoidea* in medium III (inorganic medium) was 2.56 to 64.23×10^6 cells ml^{-1} during the culture period. The average cell density was 37.17 ± 16.43 × 10^6 cells ml^{-1}. During the culture period of *C. ellipsoidea* in medium III, exponential phase was found up to 12th day from the beginning and then from stationary phase cell density started to decline towards death phase.

Wongsnansilp *et al.* (2007) found, in an experiment of culture of an alga, *Chlorosarcinopsis* sp. (PSU/CHL20), the highest cell density of 14.8 × 10^6 cells ml^{-1} at 30°C on 14 day of culture which have little similarities with those of the present experiment but the highest densities found in the present experiment are much higher than those of Wongsnansilp *et al.* (2007). Singha (2001) performed an experiment on growth performance of *C. vulgaris* in different concentrations of sugarcane mill effluent medium (SMEM), press mud medium (PMM) and bold basal medium (BBM). The highest standing crop of 159.00 × 10^5 cell ml^{-1} resulted from the treatment of 50% SMEM, 163.25 × 10^5 cell ml^{-1} from PMM (1.0 g L^{-1}) and 213.043 × 10^5 cell ml^{-1} for all the media. The results of growth performance revealed that the growth of *C. vulgaris* was significantly higher (P≤0.01) in PPM at the concentration of 1.0 g L^{-1} than other concentrations of PPM and SMEM.

According to ANOVA and DMRT of cell densities of cultures of *C. ellipsoidea* under treatments I, II and III, it can be concluded that cell densities under 3 treatments (i.e. 3 media) are significantly different (F=39.78) and among these 3 treatments, treatment II (soil extract medium) is the best for algal culture than treatment I and treatment III. Preparations of soil extract medium and pulse bran medium are simple and the materials are inexpensive and easily available in Bangladesh. So, the use of inexpensive soil extract and pulse bran as algal culture medium might be used commercially and economically to culture algae, especially *Chlorella* sp., which can be used as feed for fish fry, poultry, livestock and as human food as live feed or as dried powder form.

References

Becker, E.W. 1994a. Microalgae, Biotechnology and Microbiology. Published by the Press Syndicate of the University of Cambridge. The Pitt Building, Trumpington, 293p.

Becker, E.W. 1994b. Production and Utilization of the blue-green algae *Spirulina* in India. *Biomass*, 4: 105-125.

Christi, Y. 2007. Biodiesel from microalgae. *Biotech. Adv.* Elesvier, Amsterdam, Netherlands, 25 (3): 294-306.

Geldenhuys, D.J., Walmsley, D.J. and Tofrien, D.J. 1988. Quality of algal material produced on a fertilizer-tap water medium in outdoor plastic enclosed system. *Aquaculture*, 68: 157-164.

Rahman, M.S. 2000. Mass culture of phytoplankton in inexpensive medium. Final Project Report, BAURES. Bangladesh Agricultural University, Mymensingh, pp. 15-25.

Rydlo, O. 1973. Verwendung einiger Milcroalgen in der praktischen Pharmazie. *Pharmazie*, 6: 147-149.

Sasson, A. 1997. Microalgal Biotechnologies: Recent developments and prospects for developing countries. Biotech Publication 1/25, pp. 42-76.

Singha, S.K. 2001. Study on the culture of *Chlorella vulgaris* in various concentrations of pressmud and sugarcane mill effluent. M.S. thesis, Department of Fisheries Management, Faculty of Fisheries. Bangladesh Agricultural University, Mymensingh. 73p.

Soeder, C.I. 1980. Massive cultivation of microalgae: results and prospects. *Hydrobiol.* 72: 197-209.

Tang, G. and Suter, P.M. 2011. Vitamin A, nutrition and health values of algae: Spirulina, Chlorella, and Dunaliella. *J. Pharm. Nutri. Sci.* 1: 111-118.

Wongsnansilp, T., Tansakul, P., Arunyanart, M. 2007. Factors affecting growth and beta-carotene content of *Chlorosarcinopsis* sp. (PSU/CHL20) in batch culture. *Kasetsart J. Natural Sci.* 41 (1): 153-157.

23

MORPHOLOGICAL CHARACTERIZATION AND GENETIC DIVERSITY IN LENTIL (*Lens culinaris* Medikus ssp. *culinaris*) GERMPLASM

K.U. Ahamed[1], B. Akhter[2], M.R. Islam[3], M.R. Humaun[4] and M.J. Alam[5]

Abstract

Genetic divergence of 110 lentil germplasm with checks was assessed based on morphological traits using multivariate analysis. Mahalanobis generalized distance (D2) analysis was used to group the lentil genotypes. Significant variations among lentil genotypes were observed in respect of days to 1st flowering, days to 50% flowering, days to maturity, plant height, and number of pods per peduncle, number of pods per plant, number of seeds per plant, 100 seed weight and yield per plant. Considering the mean values, the germplasm were grouped into ten clusters. The highest number of genotypes (17) was in cluster X and lowest (5) both in cluster II and IV. Cluster IV had the highest cluster mean for number of pods per plant (297.08), number of seeds per plant (594.16), 100 seed weight (1.44 g) and yield per plant (8.53 g). Among them, the highest inter-cluster distance was obtained between the cluster IV and I (24.61) followed by IV and III (22.33), while the lowest was between IX and II (1.63). The maximum value of inter-cluster distance indicated that genotypes belonging to cluster IV were far diverged from those of cluster I. The first female flower initiation was earlier in BD-3812 (49 days) in cluster I and cluster IV had highest grain yield per plant (8.53). BD-3807 produced significant maximum number of pods per plant (298.40) in cluster IV.

Keywords: Lentil, Morphological Characterization, Genetic Diversity, Germplasm

[1]Scientific Officer, PGRC, Regional Agricultural Research Station Ishurdi, Pabna, Bangladesh
[2]Scientific Officer, Plant Pathology Division, Regional Agricultural Research Station Ishurdi, Pabna, Bangladesh
[3]Scientific Officer, Agronomy Division, Regional Agricultural Research Station Ishurdi, Pabna, Bangladesh
[4]Scientific Officer, Plant Pathology Division, Regional Agricultural Research Station Ishurdi, Pabna, Bangladesh
[5]Scientific Officer, Pulses Research Centre, BARI, Ishurdi, Pabna, Bangladesh

*Corresponding author's email: kuahamed70@yahoo.com (K.U. Ahamed)

Introduction

Lentil (*Lens culinaris* Medikus ssp. *culinaris*) is a major and important food legume in Bangladesh. In Bangladesh, all the indigenous landraces and cultivars are *microsperma* with red cotyledons. Lentil is the early domesticated among crops. It plays an important role in human, animal and soil health improvement. Nutritionally, lentil is very rich and complementary to any cereal crop including rice. It is versatile source of nutrients for man, animals and soil containing, on an average, 25.1% protein, 59% carbohydrate,0.5% fat,2.1% minerals and sufficient amount of vitamins, viz. vitamin A 16 IU ; thiamine 0.23 mg and vitamin C 2.5 mg per gram lentil (Anonymous, 2003; Frederick *et al.*, 2006). Lentil fixes atmospheric nitrogen association with *Rhizobium* sp for this its cultivation improves soil nitrogen, carbon and organic matter balance in soil. Lentil ranks second in respect of area and production but consumer's preference its rank first among pulses in Bangladesh condition. Farmers are cultivated with sole crop or mixed crop with mustard in rabi season. Production of lentil has long lagged behind domestic demand in Bangladesh, where it is preferred pulse crop for human consumption. Therefore, it needs to increase its production with high yielding variety of lentil. Effort should

be made to develop such disease resistant varieties. Local germplasm of crop plants is an excellent source of economically useful plant characters (Pecetti *et al.*, 1996). The breeders must have a mean of choosing the accession most likely to posses the trait of interest. Quantitative traits provide an estimate of genetic diversity and various numerical taxonomic techniques have been successfully used to classify and measure the pattern of phenotypic diversity in the relationship of germplasm collections in a variety of crops by many scientists as in lentil (Ahmad *et al.*, 1997; Fratini *et al.*, 2007; Tullu *et al.*, 2008), Pea (Amurrio *et al.*, 1995) and Alfalfa (Smith *et al.*, 1991; Smith *et al.*, 1995; Warburton and Smith, 1993). Worldwide, lentil is grown on a total of 1.8 million hectares, of which 60% is in the South Asian region, which includes the lentil producing countries of Bangladesh, Burma, India, Nepal and Pakistan (Nazir *et al.*, 1994). However, lentil in general, does not respond to high inputs such as fertilizer and irrigation. Genetic diversity has been considered an important factor in any crop improvement program. However, the experiment of lentil was conducted to assess the genetic diversity of lentil genotypes in respect of agro-morphological traits.

Materials and Methods

The investigation was carried out at Regional Plant Genetic Resources Center, RARS, Bangladesh Agricultural Research Institute, Ishurdi, Pabna during the rabi season of 2010-11. The experiment was laid out non-replication. Unit plot size was 4m x 1m and seeds were sown in rows with spacing 40 cm. The seeds were sown on 30 October, 2010. The experiment involves two-hundred and twenty lentil germplasm and two varieties BARI Masur 4 and BARI Masur 6 were used as test genotypes. Fungicide, insecticide and other intercultural operations were done as and when necessary. Ten plants were sampled at random to study inter and intra-accession variation. Quantitative traits were recorded on 10 plants following IBPGR Descriptors (Anonymous, 1985). Morphological parameters of quantitative data were recorded on days to flowering, days to 50% flowering, plant height , number of pods per peduncle, number of pods per plant, number of seeds per pod, number of seeds per plant, 100 seed weight, yield per plant and different qualitative characteristics were recorded. Data were analyzed statistically by non-replication with computer and the means were separated. Genetic diversity was studied following Mahalanobis's generalized distance analysis (D2). Statistical analysis was done using the GENSTST 5 programme.

Results and Discussion

Qualitative characteristics

Qualitative characteristics of 110 genotypes are presented in Table 1. Variation between and within populations of crop species is useful for analyzing and monitoring germplasm during the maintenance phase and predicting potential genetic gain in a breeding programme (Hayward and Breese, 1993; Moore and Collins, 1983). For qualitative characteristics, a considerable level of variability was observed for growth habit, seedling stem pigmentation leaf pubescence, leaflet size, flower ground colour, seed coat colour, seed coat pattern and seed coat pattern colour, cotyledon colour and lodging susceptibility that could be exploited for developing future breeding material in lentil breeding programme (Table 1). Two classes were observed for stem colour, hairiness, tendril, pod pigmentation (anthocianin), pod indehiscence, beak on the pod and cotyledon colour, whereas for other traits more than one class were observed, especially for seed traits that represents the world classification as reported by Erskine and Witcombe (1984). Muehlbauer and Slinkard (1981) reviewed the genetics of *Lens* and listed 12 genes, which account for morphological and seed variation in lentil. It was observed critically that seed traits in lentil are difficult to record that need standardization.

Table 1. Method for measuring and classifying of plant descriptors in 110 accessions of *lens culinaris* qualitative characteristics of lentil germplasm

Sl. No.	Plant characteristics	When and where measured	Classification
01	Seedling stem pigmentation	Oberved at Seedling stage when plants were 2-3" high	Present (88), absent (22)
02	Leaf pubescence	Before maturity when plants were full grown	Absent (3), slight (88), dense (19)
03	Leaflet size	Observed on fully expanded leaves on the lower flowering nodes	Small (30), medium (28), large (52)
04	Tendril length	At the time of pod formation when plants were full grown	Rudimentary (35), prominent (75)
05	Flower ground colour	At 50% flowering. Ground colour of standard petal	Pink (12), white (98)
06	Pod pigmentation	Before maturity when pods were filled but not turned brown	Absent (54), present (56)
07	Pod shedding	Scored after or during harvesting a week after maturity	None (o), low (o), medium (o),High(o)
08	Pod dehiscence	At the time of maturity observed carefully to scored this trait up to one week after maturity	None (102), low (8), medium (o), high (o)
09	Ground colour of testa	To be observed on seed less than 3 months old	Green (2), grey (64),brown (18), black (14), pink (12)
10	Pattern of testa	To be observed on seed less than 3 months old	Dotted (77), spotted (17),marbled (16)
11	Colour Pattern on testa	To be observed on seed less than 3 months old	Olive (12), grey (67), brown (18), black (13)
12	Cotyledon colour	After harvesting but less than three months old	Yellow (22),orange/red (88)
13	Lodging susceptibility	Scored at maturity	Low (32), medium (60), high (14), none (04)
14	Pest and disease susceptibility	Before flowering when plants were full grown	Low (70), medium (32), high (4),none (4)

Quantitative characteristics

Yield and yield contributing characteristics

Yield and yield contributing characteristics of different lentil germplasm are presented in Table 2. Days to 1st flowering, days to 50% flowering, days to maturity, 100 seed weight and seed yield differed significantly among the entries. The first female flower initiation was noticed in BD-3812 (49 days). Maximum numbers of germplasm 104 were matured within 114-116 days where 6 germplasm matured within 117-118 days. Plant height varied from 35.80 cm to 48.60 cm, 64 germplasm were long (40.00-48.60 cm), 35 germplasm were medium (38.0-39.8 cm) and rest lines were dwarf (35.80-37.60 cm). Number of pods per plant varied from 96.0 to 325.0, 2 germplasm were highest number pods per plant (306.0-325.0), 52 germplasm were moderate number pods per plant (200.00-

298.40) and rest lines were lowest (96.0-196.0). Yield varied from1.52 g per plant to 12.24 g per plant, where BD-3894 and BD-3902 were high yielding (11.76 -12.24 g/plant), 14 germplasm were moderate yielding (7.03- 9.58 g/plant) and rest germplasm were low yielding (1.52-6.91 g/plant). The highest seed yield (12.24 g/plant) was recorded from BD-3894 lentil germplasm and the lowest yield (1.52 g/plant) from BD-3835 lentil germplasm. Sultana et al. (2010) also reported that eight lentil accessions gave seed weight more than 3.1 g and hence could be utilized for the manipulation of this trait as high seed weight in any grain crop. Variability for these traits in lentil germplasm was also reported by Agrawal et al. (1976), Tiwari and Singh (1980), Malik et al. (1984) and Toklu et al. (2009). Singh and Singh (1993) confirmed the wide range of variation in agronomic characteristics of lentil germplam, except for seeds per pod.

Table 2. Range, mean, SD and CV% of yield and yield contributing characteristics of 110 lentil germplasm

Characters	Range	Mean	SD	CV (%)
Days to 1st flowering	49.0-68.0	59.65	4.98	8.36
Days to 50% flowering	59.0-78.0	69.88	4.46	6.38
Days to maturity	114.0-118.0	115.24	0.98	0.85
Plant height(cm)	35.80-48.60	40.64	2.45	6.02
No. of pods/peduncle	1.0-3.0	1.87	0.38	20.60
No. of pods/plant	96.0-325.0	196.80	43.08	21.90
No. of seeds/plant	192.0-612.0	382.43	87.47	22.87
100 seed wt(g)	0.40-2.60	1.37	0.36	26.26
Yield per plant(g)	1.52-12.24	5.21	1.90	36.33

Genetic diversity in lentil germplasm

A considerable amount of genetic variability was observed and therefore diversity analysis was carried out through multivariate analysis.

The clustering pattern and distribution

Table 3. Distribution of 110 genotypes of lentil in 10 clusters

Clusters	No of genotypes	Genotypes
I	15	BD-3843, BD-3846, BD-3849, BD-3812, BD-3809, BD-3827, BD-3824, BD-3909, BD-3879, BD-3880, BD-3884, BD-3890, BD-3873, BD-3964, BD-3970
II	5	BD-3835, BD-3826, BD-3883, BD-3858, BD-3863,
III	11	BD-3856, BD-3853, BD-3852, BD-3839, BD-3840, BD-3805, BD-3829, BD-3874, BD-3892, BD-3867, BD-3966
IV	5	BD-3807, BD-3821, BD-3902, BD-3894, BD-3886
V	9	BD-3848, BD-3836, BD-3810, BD-3897 , BD-3887, BD-3908, BD-3859, BD-3986 , BD-3863
VI	10	BD-3844, BD-3834 , BD-3804 , BD-3820 , BD-3901 , BD-3877 , BD-3907 , BD-3869,BD-3866 , BD-3871
VII	16	BD-3847, BD-3837, BD-3838, BD-3808, BD-3817, BD-3810, BD-3876, BD-3881, BD-3882, BD-3898, BD-3900, BD-3857, BD-3861, BD-3865, BD-3968
VIII	9	BD-3841, BD-3842, BD-3833, BD-3815, BD-3823, BD-3819, BD-3885, BD-3891, BD-3987
IX	13	BD-3845, BD-3850, BD-3854,BD-3851, BD-3831, BD-3822, BD-3818, BD-3830, BD-3906, BD-3895, BD-3878, BD-3899, BD-3889
X	17	BD-3832, BD-3811, BD-3806, BD-3855, BD-3825, BD-3828, BD-3905, BD-3896, BD-3875, BD-3888 BD-3893 BD-3872 BD-3860, BD-3868, BD-3870, BD-3985,BD-3988

Table 4. Means for quantitative characteristics for 10 clusters in lentil germplasm

No. of clusters	Days to 1st flowering	Days to 50% flowering	Days to maturity	Plant height (cm)	No. of pods/ peduncle	No. of pods/ plant	No. of seeds /pod	No. of seeds /plant	100 seed weight (g)	Yield/ plant (g)
Cluster -1	60.60	70.53	115.07	39.64	1.93	135.83	1.95	258.08	1.36	3.48
Cluster -2	62.60	70.60	115.60	40.08	1.60	192.60	2.00	385.20	1.06	4.09
Cluster -3	59.91	71.18	115.73	41.78	2.18	152.47	1.98	290.31	1.41	4.09
Cluster -4	58.80	69.20	115.20	40.08	1.80	297.08	1.92	594.16	1.44	8.53
Cluster -5	59.67	69.33	115.11	41.36	1.89	213.44	1.87	421.78	1.40	5.91
Cluster -6	59.30	69.40	115.20	39.88	1.80	252.14	1.96	499.00	1.37	6.87
Cluster -7	59.38	69.87	115.37	39.38	1.94	174.51	1.94	335.16	1.37	4.59
Cluster -8	57.44	68.89	115.11	41.11	1.67	203.93	1.93	403.04	1.32	5.32
Cluster -9	59.92	69.54	115.08	40.34	1.85	189.89	1.94	364.48	1.30	4.74
Cluster -10	59.47	69.82	115.18	42.35	1.82	232.13	1.94	446.56	1.44	6.41

Non-hierarchical clustering

The covariance matrix gave non-hierarchical clustering among 110 genotypes and grouped them into ten clusters (Table 3). They coincided with the apparent grouping patterns performed by PCA. Cluster VII and X both contained the largest number of genotypes (sixteen), followed by clusters I, IX, III,V,VI, VIII, IV, and II. The genotypes of different geographic origin are accumulated in the same cluster indicating that the genotypes are not sharply diversified. Similar results were obtained by Alam *et al.* (2006) in Barley, Alam *et al.* (2011) in Lentil. These clusters lead to the highest cluster mean for maximum characters (Table 4). Among the ten characters, the highest mean values for four characters,viz. number of pods per plant (297.08), number of seeds per plant (594.16), 100 seed weight (1.41 g), and yield per plant (8.53 g) were found in cluster IV. Cluster IV had only five genotypes,viz. BD-3807, BD-3821, BD-3902, BD-3894 and BD-3886. Cluster X with sixteen genotypes was able to lead only for two traits in respect of cluster means of six characters. The highest cluster mean was recorded for 100-seed weight (1.41 g). Cluster VI, X and V was moderate yielding associated with desired characteristics like size and early maturity.

Canonical Variate Analysis

Average inter-cluster D2 values among the ten clusters are presented in Table 5. The highest inter-cluster distance was obtained between the cluster IV and I (24.61) followed by IV and III (22.33), while the lowest was between IX and II (1.63). The maximum value of inter-cluster distance indicated that genotypes belonging to cluster IV were far diverged from those of cluster I. Similarly, the highest inter-cluster values between cluster IV and III indicated that the genotypes between each pair of clusters were more diverged. Sultana *et al.* (2010) also reported that the inter-cluster distance among the accessions revealed that the cluster V consisting of five accessions was obviously very much different from all the other clusters with a genetic distance ranging from 1.99 (Cluster VIII) to 2.85 (Cluster X).

Table 5. Average inter and intra (bold) cluster distance (D2) for the 110 lentil germplasm obtained on the basis of the morphological characteristics

Clusters	I	II	III	IV	V	VI	VII	VIII	IX	X
I	**0.416**									
II	9.30	**0.717**								
III	2.68	7.25	**0.456**							
IV	24.61	15.35	22.33	**0.646**						
V	12.04	2.94	9.78	12.570	**0.438**					
VI	17.65	8.40	15.40	6.956	5.620	**0.485**				
VII	5.73	3.60	3.67	18.876	6.318	11.920	**0.415**			
VIII	10.71	1.74	8.45	13.905	1.336	6.954	4.983	**0.425**		
IX	7.87	1.63	5.68	16.740	4.176	9.784	2.142	2.841	**0.442**	
X	13.96	4.97	11.60	10.744	2.033	3.899	8.260	3.311	6.124	**0.351**

Principal Coordinate Analysis (PCO)

The results obtained from principal coordinate analysis showed that the highest inter-genotypic distance was between genotypes BD-3894 and BD-3970 (1.964), followed by BD-3849 and BD-3894 (1.868), and the lowest (0.083) between genotypes BD-3840 and BD-3805 as well as between BD-3909 and BD-3964 (0.102) (Table 5). The difference between the highest and lowest inter-genotypic distance indicated that moderate variability among 110 germplasm of grass pea. The highest intra-cluster distance was recorded in cluster II (0.717) containing five genotypes viz. BD-3835, BD-3826, BD-3883, BD-3858 and BD-

3863 (Table 6). The lowest intra-cluster distance was in cluster X (0.351) and containing also seventeen genotypes genotypes, viz. BD-3832, BD-3811, BD-3806, BD-3855, BD-3825, BD-3828, BD-3905, BD-3896, BD-3875, BD-3888 BD-3893 BD-3872 BD-3860, BD-3868, BD-

3870, BD-3985 and BD-3988. Similar types of research were conducted by Malik *et al.* (1984) who reported high genetic variance in cultivated lentil.

Table 6. Five highest and lowest inter-genotypic distances among one hundred seven genotypes of grasspea

| Genotypic combination | | Genotypic combination | |
A. Five lowest inter-genotypic distances	Distance	B. Five highest inter-genotypic distances	Distance
BD-3851 – BD-3863	0.130	BD-3894–BD-3970	1.964
BD-3893– BD-3988	0.123	BD-3849–BD-3894	1.868
BD-3875– BD-3888	0.118	BD-3894– BD-3966	1.861
BD-3909– BD-3964	0.102	BD-3821– BD-3970	1.844
BD-3840–BD-3805	0.083	BD-3849– BD-3902	1.810

Principal Component Analysis (PCA)

Principal Component Analysis (PCA) helps to assessment of diversity in multivariate scales. Principal Component Analysis was carried out with 110 genotypes of lentil. The first five Eigen values for five principal coordination axes of genotypes accounted for 83.48 % variation while only first two principal coordination axes of genotypes accounted for 46.77 % of total variation among the ten characteristics (Table 7).

Bozokalfa *et al.* (2009) also reported that the first six axes accounted for 54.29% of the variability among the 48 accessions and their lines. Alam *et al.* (2011) also reported that the first three Eigen values for three principal coordination axes of lentil genotypes accounted for 78.1% variation. The first two principal axes accounted for 61.2% of total variation among six characters.

Table 7. Eigen values and percentage of variation for corresponding 10 component characteristics of 110 lentil germplasm

Principal component axis	Eigen values	% of total variation accounted for	Cumulative percent
Days to 1st flowering	2.829	28.29	28.29
Days to 50% flowering	1.848	18.48	46.77
Days to maturity	1.389	13.89	60.66
Plant height(cm)	1.272	12.72	73.38
No. of pods/peduncle	1.010	10.10	83.48
No. of pods/plant	0.791	7.91	91.39
No. of seeds/pod	0.685	6.85	98.24
No. of seeds/plant	0.134	1.34	99.58
100 seed weight(g)	0.029	0.29	99.87
Yield per plant(g)	0.014	0.14	100.0

Contribution of characteristics towards the divergence of genotypes

The values of vector 1 and vector 2 are presented in Table 8. The value of vector 1 obtained from PCA for days to first flowering, days to 50% flowering, days to maturity, number of pods per peduncle and number of seeds per pod suggests it was the major characteristics that contributed to the genetic divergence. In vector 2 also obtained from PCA for days to 1st flowering (0.631), days to 50% flowering (0.622), days to maturity (0.115), Plant height (0.122), number of pods per peduncle(0.038), number of pods per plant (0.270), 100-seed weight (0.269) and yield per plant (0.080) showed their important role toward

genetic divergence. Both vector 1 and vector 2 revealed that vector 1 had positive values for days to first flowering, days to 50% flowering, days to maturity, number of pods per peduncle and number of seeds per pod and vector 2 had positive values for days to 1st flowering , days to 50% flowering, days to maturity, Plant height (cm), number of pods per plant, 100 seed weight (g) and yield per plant (g) are indicating highest contribution of these traits towards the divergence among 110 genotypes of lentil. Negative values in both vectors had lower contribution towards the divergence. Similar results were obtained by Alam *et al.* (2011) in lentil.

Table 8. Relative contributions of the 10 characteristics to the total divergence in Lentil germplasm

Characteristics	Vector 1	Vector 2
Days to 1st flowering	0.230	0.631
Days to 50% flowering	0.213	0.622
Days to maturity	0.070	0.115
Plant height(cm)	-0.081	0.122
No. of pods/peduncle	0.115	0.038
No. of pods/plant	-0.509	0.270
No. of seeds/pod	0.050	-0.098
No. of seeds/plant	-0.272	-0.150
100 seed weight(g)	-0.509	0.269
Yield per plant(g)	-0.529	0.080

Conclusion

Days to 1st flowering , days to 50% flowering, days to maturity, number of pods per peduncle, number of pods per plant, number of seeds per pod, 100 seed weight (g) and yield per plant (g) had highest contribution towards divergence among 10 characteristics for 110 lentil germplasm. Based on analysis the germplasm grouped into ten clusters. From morphological study, the highest inter-cluster distance was obtained between the cluster IV and I (24.61) followed by IV and III (22.33). The intra-cluster distance was highest (0.717) in cluster II and lowest (0.315) in cluster X. Considering yield performance, cluster distance and cluster mean, the genotypes BD-3894, BD-3821 and BD-3902 from cluster IV and BD-3804, BD-3902 and BD-3869 from cluster III may be considered better genotypes for recombination breeding due to their larger divergence.

References

Agrawal, S.C., Khare, M.N. and Agrawal, P.S. 1976. Field screening of lentil lines for resistance to rust. *Indian Phytopath.* 29 (2): 208.

Ahmad, M., McNeil, D.L. and Fautrier, A.G. 1997. Phylogenetic relationships in *Lens* species and parentage determination of their interspecific hybrids using RAPD markers. *Euphytica* 94: 101-110.

Alam, A.K.M.M., Naher, N. and Begum, M. 2006. Genetic divergence for some quantitative characters in hull-less barley. *Bangladesh J. Agril. Res.* 31 (3): 347-351.

Alam, A.K.M.M., Podder, R. and Sarker, A. 2011. Estimation of genetic diversity in lentil germplasm. *J. Agril. Sci. Agrivita* 33 (2): 103-110.

Amurrio, J.M., de Ron, A.M. and Zeven, A.C. 1995. Numerical taxonomy of Iberian pea landraces based on quantitative and qualitative characters. *Euphytica* 82: 195-205.

Anonymous. 1985. *Lentil Descriptors.* International Board for Plant Genetic Resources (IBPGR) and International Center for Agricultural Research in the Dry Areas (ICARDA). p. 13.

Anonymous. 2003. CGIAR Research: Areas of research; Lentil (*Lens culinaris* Medik) ttp://www.icarda.cgiar.org

Bozokalfa, M.K., Eşiyok, D. and Turhan, K. 2009. Patterns of phenotypic variation in a germplasm collection of pepper (*Capsicum annuum* L.) from Turkey. *Spanish J. Agril. Res.* 7 (1): 83-95.

Erskine, W. and Witcombe, J.R. 1984. 100 seed weight. *Lentil germplasm catalog*, ICARDA, Syria. p.45.

Fratini, R., Durán, Y., García, P. and Pérez de la Vega, M. 2007. Identification of quantitative trait loci (QTL) for plant structure, growth habit and yield in lentil. *Spanish J. Agric. Res.* 5 (3): 348-356.

Frederick, M., Cho, S., Sarker, A., McPhee, K., Coyne, C., Rajesh, P. and Ford, P. 2006. Application of biotechnology in breeding lentil for resistance to biotic and abiotic stress. *Euphytica* 147: 149-165.

Hayward, M.D. and Breese, E.L. 1993. Population structure and variability. pp. 17-29. *In:* Plant Breeding: Principles and Prospects, (Eds.): M.D. Hayward, N.O. Bosemark and I. Ramayosa, Chapman and Hall, London.

Malik, B.A., Tahir, M., Haqani, A.M. and Anwar, R. 1984. Documentation, characterization, and preliminary evaluation of lentil (*Lens culinaris*) germplasm in Pakistan. *LENS Newsl.* 11 (2): 8-11.

Moore, G.A. and Collins, G.B. 1983. New challenges confronting plant breeders. *In:* Isozyme in plant genetics and breeding, part A. (Ed.) S.D. Tanksley and T.J. Orton. Elsevier Science Publishers B.U., Amsterdam.

Muehlbauer, F.J. and Slinkard, A.E. 1981. *Genetics and breeding methodology.* pp. 69-90. *In:* Lentils. (Eds.): C. Webb & G. Hawtin, Commonwealth Agricultural Bureaux.

Nazir, S., Bashir, E., Bantel, R. and Habib-ul-Rehman Mian. 1994. *Crop Production.* National Book Foundation, Islamabad. pp. 294-300.

Pecetti, L., Annicchiario, P. and Damania, A.B. 1996. Geographic variation in tetraploid wheat (*Triticum turgidum* spp. *turgidum* convar. Durum) landraces from two provinces in Ethiopia. *Euphytica* 43: 395-407.

Singh, B.B. and Singh, D.P. 1993. Evaluation of lentil germplasm in Uttar Pradesh. *LENS Newsl.* 20 (2): 11-12.

Smith, S.E., Al Doss, A. and Warburton, M. 1991. Morphological and agronomic variation in North African and Arabian alfalfas. *Crop Sci.* 31: 1159-1163.

Smith, S.E., Guarino, L., Al Doss, A. and Conta, D.M. 1995. Morphological and agronomic affinities among Middle Eastern alfalfas accessions from Oman and Yemen. *Crop Sci.* 35: 1118-1194.

Sultana,T., Nadeem, S., Fatima, Z. and Ghafoor, A. 2010. Identification of elite pure-lines from locallentil germplasm using diversity index basedon quantitative traits. *Pak. J. Bot.* 42 (4): 2249-2256.

Tiwari, A.S. and Singh, B.R. 1980. Evaluation of lentil germplasm. *LENS Newsl.* 7: 20-22.

Toklu, F., Biçer, B.T. and Karaköy, T. 2009. Agro-morphological characterization of the Turkish lentil landraces. *African J. Biotech.* 8 (17): 4121-4127.

Tullu, A., Tar'an, B., Warkentin, T. and Vandenberg, A. 2008. Construction of an intraspecific linkage map and QTL analysis for earliness and plant height in lentil. *Crop Sci.* 48 (6): 2254-2264.

Warburton, M.L. and Smith, S.E. 1993. Regional diversity in nondormant alfalfas from India and the Middle East. *Crop Sci.* 33: 852-85.

24

NUTRIENT DISTRIBUTION IN FLOWERING STEMS, NUTRIENT REMOVAL AND FERTILIZATION OF THREE PROTEACEAE CULTIVARS

M.M. Hernández, M. Fernández-Falcón and C.E. Álvarez*

Abstract

Several parameters of flowering stems of two cultivars of protea *Leucospermum cordifolium* (Knight) Fourc ('Succession II' and 'Tango') and one of protea *Protea* L. (*Protea susannae x magnifica* 'Susara') were studied in different commercial plantations. These included length and weight measures, nutrient concentrations of detached parts (flower heads, leaves and stems), and removal of nutrients of harvested flowers. A base for fertilization was also calculated. Harvested flowers of 'Susara' removed more P, K and Cu than 'Succession II' and 'Tango', and more Ca and Fe than 'Tango'. 'Succession II' showed the highest Zn removal. The P and Cu removal by the crop of 'Susara' were significantly higher than those of both *Leucospermum* cultivars, while K, Ca, and Fe outputs of 'Susara' exceeded only those of 'Tango'. The nutrient top removals amounted to 4.55 g m^{-2} of N, 0.48 g m^{-2} of P, and 5.26 g m^{-2} of K. Data as a base to supply fertilizer to each cultivar are given, with N:P:K ratios of 1:0.08:0.87 for 'Succession II', 1:0.08:0.83 for 'Tango' and 1:0.10:1.27 for 'Susara'.

Keywords: *Leucospermum*, Succession II, Susara, Tango

Instituto de Productos Naturales y Agrobiología, Consejo Superior de Investigaciones Científicas, Avenida Astrofísico Francisco Sánchez 3, 38206 La Laguna, Tenerife, Canary Islands, Spain.

*Corresponding author's email: carlose@ipna.csic.es (C.E. Álvarez)

Introduction

Leucospermum and *Protea* are genera of the Proteaceae family that have obtained an increasing importance in the market of cut flowers worldwide, where they are known as proteas. Most of the species are native to Africa and Australia, which are also the main producers together with USA (Malan, 2012).

It is difficult to get a clear idea of the distribution of nutrients within the flowering stems in many proteas (genera *Leucadendron*, *Leucospermum* and *Protea*), because the chosen tissues are different depending on the authors. Cresswell (1991) took stems, young leaves and old leaves of *Leucadendron* cv. Harvest, and *Protea neriifolia* cv's Satin mink and Pink Ice, where he distinguished among optimum, high and toxic levels of phosphorous. Maier *et al.* (1995) analysed the nutrient contents of stems and leaves, axillary buds and flowers separately. The nitrogen and potassium contents of stems and leaves significantly exceeded those of the other fractions of the flowering stem, and the flowers showed the lowest calcium and magnesium concentrations. González *et al.* (2008) divided commercial flowers of five *Leucospermum* cultivars into flower head, half-distal and half-proximal stems, and half-proximal and half-distal leaves. Nutrient levels within every organ

presented significant variations among cultivars. The nutrients related with the length of the stems also varied. Fernández-Falcón *et al.* (2008) observed that nutrient percentages in flower heads, leaves and stem of the flowering stem varied within the cultivar of *Leucospermum*.

These studies may help to determine nutrient removal of the crop by taking into account the nutrient content of the plant part that is generally harvested (Mengel and Kirkby, 2001). In the case of protea for cut flower there is very limited information in the literature that includes wastage by pruning and fallen foliage, which could represent up to 54 % of the total nutrients removed by a protea plant (León, 2011).

Claassens (1986) reported removal of 5.3 kg ha^{-1} N by flowering stems of *Protea neriifolia*, and 7.5 kg ha^{-1} N by *Leucospermum cordifolium*. He also did assays with *L. cordifolium* in sand hydroponic cultures, and obtained higher nutrient removals 30.0 kg ha^{-1} N, as well as increased values of P, K, Ca, and Mg. He concluded that protea crop removes low quantities of nutrients, and that fertilization with N could increase the productions in very acid and sandy soils. However, in a controlled fertilization experiment, Hawkins *et al.* (2007) reported that *Leucospermun* 'Succession' and *Leucadendron* 'Safari Sunset' acquired more N amounts than were supplied, but increasing N fertilization decreased their growth. They also found than 'Safari Sunset' removed 18 g N per plant, and 'Succession' 5.,5 g N per plant. Maier *et al.* (1995) detected N removal by *Protea* cv. Pink Ice of 26-43 kg ha⁻¹. Fernández-Falcón *et al.* (2008) reported higher amounts of N removed by the crops of *L. patersonii* and four *L. cordifolium* cultivars that reached nearly 150 kg ha⁻¹ in some instances. In a compilation of works made in South Australia, Reid (2003) showed annual N exportations of *Protea* 'Pink Ice' and *Leucadendron* 'Silvan Red' which were 27.2 and 15.3 g plant⁻¹, respectively.

The study of nutrient removal of leaves and flowering stem of *Leucadendron* 'Silvan Red' and 'Safari Sunset' lead Cecil *et al.* (1995) to recommend annual applications of N and Ca at 20-30 g/plant, and Mg and K at 10-15 g/plant, that would meet the nutrient needs of both cultivars. Reid (2003) suggested that, to calculate the fertilizer needs based on the nutrient extractions of the crop, inefficiency of fertilizer by leaching should be taken into account, and proposed to apply at least double quantities of nutrients than the removed ones.

All these data show that nutrient removal and fertilization needs are different depending on genus, species and cultivars. The present paper aims to broaden the knowledge of nutrient distribution within the flowering stem of protea genera *Leucospermum* and *Protea*, and their nutrient removal. Besides, a base for fertilization was also calculated.

Materials and Methods

Two cultivars of *Leucospermum cordifolium* (Knight) Fourc ('Succession II' and 'Tango'), and one of *Protea* L. (*Protea susannae x magnifica* 'Susara') were studied in commercial plantations located in eight municipalities of La Palma (Canary Islands, Spain), distributed around the island, to obtain the general averages of the parameters listed below. The cultivars were selected by their very high commercial importance. Soils were Inceptisols Andepts and Ultisol Udults.

Flowering stems of commercial quality (30 cm or more of length, straight and without defects in the cases of 'Succession' and 'Tango' and equal to or longer than 45 cm for 'Susara') were taken in the field from each of the mentioned cultivars, during the harvest season (January to April) of the years 2009 and 2010. Samples of each farm consisted of three replications per cultivar, with three flowering stems per replication.

Lengths of the flowering stems were measured before cutting them into different parts: flower head, leaves, and stem. Fresh weight and dry weight of each plant part were determined, as well as their nutrient concentrations. Dry weights were measured after drying each plant part in an oven at 80ºC, and weighting them till no difference in weight was detected between two weightings.

For chemical analysis purposes, the samples were washed in distilled water and dried in an oven at 80ºC, after which they were ground to powder. One g of the powder was ashed in an oven at 480ºC and then mineralized by dry ashing with 6 M hydrochloric acid (Chapman and Pratt, 1961). The P, K, Ca, Mg, Cu, Fe, Mn and Zn concentrations were determined by ICP Perkin-Elmer. Nitrogen was determined by the Kjeldahl method (Cottenie, 1980).

The nutrient content of the flowering stem for each cultivar was obtained from the concentrations of nutrients of the flowering stems and dry mass weight. The nutrient removal of each cultivar was calculated from the nutrient content of the whole flower and the flower yields by square meter that amounted to 30 flowers in 'Succession', 25 in 'Tango' and 10 in 'Susara'.

Production data were collected from the Association of Protea Growers of La Palma Island. Data were subjected to one-way variance analysis, using Tuckey b test at p = 0.05, by SPSS 15.0 statistical software. Correlation and regression analysis were also performed, but no important relationship among nutrients and/or other studied parameters was found. Soil chemical analysis has been reported elsewhere (Álvarez *et al.* 2012).

Results and Discussion

Nutrient concentration in the different plant parts of the flowering stems of the different cultivar

Flower: Succession II cultivar showed the highest concentrations of N, Mg, Na, and Zn (Table 1), though not significantly higher than 'Tango' for Mg. The Na concentration of 'Susara' flowers was significantly lower than 'Succession' and 'Tango'. The other nutrients presented similar concentrations among the cultivars.

Table 1. Nutrient concentration of different plant parts for a protea cultivar on a dry mass base for 2009 and 2010

Cultivar	Plant part	g kg⁻¹						mg kg⁻¹			
		N	P	K	Ca	Mg	Na	Fe	Mn	Cu	Zn
Succession	Head	8.16a	0.64	7.04	2.44b	1.58	6.17	35	83b	5	19
	Leaf	10.56a	0.56	4.52	9.16a	2.01	6.26	60	315a	4	21
	Stem	5.31b	0.59	8.69	3.72b	1.45	3.27	46	104b	5	14
Tango	Head	6.08b	0.59	5.87a	1.94b	1.42b	5.22a	32b	86b	8	12
	Leaf	10.31a	0.52	4.28b	9.12a	2.11a	5.68a	61a	553a	7	15
	Stem	5.72b	0.39	6.19a	3.15b	1.51b	3.38b	37b	112b	7	11
Susara	Head	5.35	0.50	6.36	2.40	0.91b	1.23a	34	83b	11	10b
	Leaf	6.08	0.43	3.75	8.03	1.43a	1.21a	38	199a	7	11b
	Stem	4.92	0.63	7.34	2.70	0.62c	0.66b	36	200a	11	20a

Different letters following the data within each column of each cultivar denote significant differences at p = 0.05 level.

Maier *et al.* (1995) found in *Protea* 'Pink Ice' lower concentrations of N, K, Ca, Mg, Na, Cu, Mn and Zn than those detected in this study, although the level of Na equalled that of 'Susara', and the concentration of Zn was higher than those of 'Susara' and 'Tango'. Phosphorus presented similar levels than the cultivars of this assay, while Fe exceeded them.

González *et al.* (2008) determined in five *Leucospermum* cultivars N and Zn concentrations close to those of this study. The same happened with Fe but only in two of the five cultivars. On the other hand, they observed P, Ca, Mg, Na, and Mn higher levels, where Mg became two to three times higher. Potassium and Cu concentrations in this study exceeded the ones detected by these authors, though K levels were lower than in some cultivars.

There were differences of some nutrient concentrations between cultivars of *Leucospermum*, as well as between those of *Leucospermum* and *Protea*, confirming what the literature mentioned on this subject (Parvin 1986; Cecil *et al.*, 1995; Montarone *et al.*, 2003). Na presented very high concentrations, specially in the cultivar 'Succession II', a fact that Alvarez *et al.* (2012) had pointed out for proteas.

Leaf: The leaves of 'Susara' cultivar exhibited significantly lower N, Mg, Na, Fe, Mn and Zn compared to the other cultivars (Table 1), though Mn concentration resembled that of 'Succession II', and Zn that of 'Tango'.

The concentrations of N, P, K, Ca, Mg and Na obtained by González *et al.* (2008) in several *Leucospermum* cultivars exceeded those detected in this study. Iron followed the same trend, but its values in one of the cultivars matched the ones of 'Succession II' and 'Tango'. On the contrary, Mn levels of this study surpassed the concentrations found by these authors in three cultivars, while the levels of Zn and Cu were similar.

The great accumulation of Na observed in the flower head of 'Succession II' appears also in the leaves, though in the case of the leaves 'Tango' presents similar Na levels. They remained significantly lower in 'Susara'. Moreover, the two genera presented notable differences in the concentration of some nutrients.

Stems: The stems of both 'Succession II' and 'Tango' had higher levels of Mg and Na compared to 'Susara' (Table 1). On the contrary, 'Succession II' stems showed the lowest concentration of Cu, while 'Susara' presented the highest one, and that of 'Tango' was intermediate.

Nitrogen, K, Mg, Na, Cu and Mn concentrations fell below those reported by Maier *et al.* (1995) in *Protea* 'Pink Ice', though Mg level in 'Susara' was similar, and greater than that of Na. Phosphorus and Zn values resembled those of this study, while Ca and Fe exceeded them.

On the other hand, González *et al.* (2008), in several *Leucospermum* cultivars, obtained N, P, K, Ca, Mg, Na, Fe and Zn concentrations higher than those of this study, with the exception of K in 'Succession II', that was alike. Copper levels did not differ, while Mn behaved without a clear trend.

As in the previous two organs, the concentration of Na in the stems of both cultivars of *Leucospermum* exceeded by far that of the cultivar of *Protea*, while differences of the concentrations of some nutrients between the genera turned up once more.

Fernández-Falcón *et al.* (2008) found the highest N, Ca, Mg, Na and Mn contents in the leaves of five *Leucospermum* cultivars, and the lowest P, K and Cu in the stems, and Fe in the flower. Zinc was similar in the three organs. Nevertheless, they found different behaviours of the nutrient contents in some of the cultivars, as has happened in this study. For example, N was similar in the three studied flower organs of 'Susara', the flower head of 'Succession' had more than the stem, and

'Tango' showed similar values in the stem and the flower head.

Total nutrient concentration of flowering stem on a fresh mass base

Both Leucospermum cultivars presented similar concentration of N, Mg and Na in the entire flowering stems (Table 2), though they significantly exceeded those of the Protea cultivar Susara. No differences were detected in the concentrations of the rest of the nutrients. Their behaviour was closely related to those observed in the detached organs.

Flowering stem length, fresh and dry mass of the different cultivars

Table 3 shows the lengths of flowering stems, as well as the fresh and dry weights of the entire flowering stems and the different studied parts (heads, stems and leaves).

Table 2. Total nutrient concentrations of the flowering stem of different protea cultivars on dry mass base for 2009 and 2010

Cultivars	g kg^{-1}						mg kg^{-1}			
	N	P	K	Ca	Mg	Na	Fe	Mn	Cu	Zn
Succession	8.30 a	0.59	6.75	5.10	1.68 a	5.23 a	47	168	5	18
Tango	7.46 a	0.67	5.44	4.74	1.68 a	4.76 a	44	264	7	12
Susara	4.51 b	0.52	5.82	4.38	0.99 b	1.03 b	36	161	10	14

Different letters following the data within each column denote significant differences at p = 0.05 level.

Table 3. Flowering stem length, fresh and dry mass of different parts of different protea cultivars for 2009 and 2010

Cultivars	Length (cm)	g							
		FFM*	LFM	SFM	FSFM	FDM	LDM	SDM	FSDM
Succession	56.8 b	30.4 b	22.8 b	23.8 b	76.8 b	5.4 b	7.3 b	7.0 b	19.8 b
Tango	57.0 b	32.2 b	24.5 b	22.3 b	79.7 b	6.3 b	7.7 b	6.8 b	20.5 b
Susara	72.6 a	95.3 a	80.8 a	62.4 a	238.5 a	32.2 a	33.1 a	24.4 a	89.6 a

* FFM = flower fresh mass; LFM = leaf fresh mass; SFM = stem fresh mass; FSFM = flowering stem fresh mass; FDM = flower dry mass; LDM = leaf dry mass; SDM = stem dry mass; FSDM = flowering stem dry mass
Different letters following the data within each column denote significant differences at p = 0.05 level.

Cultivar 'Susara' presented significantly higher data of all these parameters than the other two cultivars, while 'Succession II' and 'Tango' did not exhibited appreciable differences between them. The observed findings could be expected because 'Succession II' and 'Tango' come from the same genus (*Leucospermum*), and 'Susara' from *Protea* genus that produces longer and heavier flowers.

Total nutrient content of flowering stem on a dry mass base: *Leucospermum* cultivars 'Succession II' and 'Tango' had similar nutrient contents of the entire flowering stems (Table 4), but 'Susara' *Protea* cultivar significantly exceeded them. This could be due to the greater size of the *Protea* 'Susara' cut flowers, as remarked previously.

Table 4. Total nutrient content of the whole flowering stem of different protea cultivars on a dry mass base for 2009 and 2010.

Cultivars	g flowering stem^{-1}						mg flowering stem^{-1}			
	N	P	K	Ca	Mg	Na	Fe	Mn	Cu	Zn
Succession	0.15 b	0.012 b	0.13 b	0.10 b	0.033 b	0.11	0.9 b	3.6 b	0.10 b	0.34 b
Tango	0.14 b	0.011 b	0.12 b	0.10 b	0.033 b	0.10	0.9 b	4.8 b	0.15 b	0.24 b
Susara	0.41 a	0.048 a	0.53 a	0.39 a	0.088 a	0.09	3.2 a	14.6 a	0.80a	1.28 a

Different letters following the data within each column denote significant differences at p = 0.05 level.

Fernández-Falcón *et al.* (2008) observed higher N, P, K, Ca, Mg, Fe and Mn contents compared to this study. Nevertheless, two of the five cultivars investigated by these authors presented lower Mn contents than those of 'Susara'. The levels of Cu and Zn detected by these researchers rose above the values found in 'Succession II' and 'Tango', but they were lower than in 'Susara'.

Nutrient removal of the flowering stems per m². Table 5 shows the nutrient removal of each cultivar. Nitrogen, Mg and Mn showed similar outputs in the three cultivars. 'Tango' presented the lowest outputs of K, Ca and Zn, while the other two cultivars had analogous values. On the other hand, 'Susara' flowers removed more P and Cu than the other two cultivars, and more Fe than 'Tango'.

Table 5. Nutrient removal per protea cultivar according to number of flowering stems m⁻² for 2009 and 2010

Cultivars	g m⁻²					mg m⁻²			
	N	P	K	Ca	Mg	Fe	Mn	Cu	Zn
Succession	4.55	0.35 b	3.96 b	3.12 a	1.00	27.8 ab	106.2	3.33 b	10.08 a
Tango	3.51	0.27 b	2.90 b	2.55 b	0.91	22.7 b	120.6	3.71 b	6.09 b
Susara	4.13	0.48 a	5.26 a	3.90 a	0.88	31.7 a	146.0	8.23 a	12.77 b

Different letters following the data within each column denote significant differences at p = 0.05 level.

The nutrient removal far exceeded the N, P, K, Ca and Mg outputs reported by Claassens (1986) for *P. neriifolia* and *L. cordifolium* grown in the field, though they approached those of N and P when these plants were grown in hydroponics. The outputs of N and P by *Protea* 'Pink Ice' observed by Maier *et al.* (1995) and Reid (2003) were more similar, and even some values of P and Ca surpassed the removals detected in this study. The removal of Zn and Fe exceeded that found by Reid (2003) for 'Pink Ice', while the opposite happened with Mg. Fernández-Falcón *et al.* (2008) reported different amounts of nutrients removed by crops of *Leucospermum*. They found nutrient removals higher, lower or similar to those observed in this study, depending upon the cultivar.

Though the nutrient content per cut flower of 'Susara' exceeded those of the other cultivars, as stated before, its yields were clearly lower. As a consequence, similar nutrient outputs among the cultivars resulted in some cases and 'Susara' even presented a lower Zn value than 'Succession II'.

Protea nutrient removal reported in the literature is based on the removal by the harvested flowers, but they do not include the nutrients removed by the crop wastes, so that they do not illustrate completely the real removals. León (2011) had observed that the plant mass of 'Succession II' wastes (pruning and pinching out) represented an average of 54 % of the total plant mass-produced throughout the crop cycle. In the present study, total nutrient removal could be calculated taking into account this percentage and extrapolating it to 'Tango' and 'Susara'. Such nutrient removal (flowering stems + crop wastes) would be 9.9 g m⁻² N, 0.76 g m⁻² P and 8.6 g m⁻² K by 'Succession II', 7.6 g m⁻² N, 0.59 g m⁻² P and 6.3 g m⁻² K by 'Tango', and 9.0 g m⁻² N, 1.01 g m⁻² P and 11.4 g m⁻² K by 'Susara'. These data are appropriate as a base to supply NPK fertilizers to avoid depletion of these nutrients in the soil. The N:P:K ratios should be 1:0.08:0.87 for 'Succession II', 1:0.08:0.83 for 'Tango' and 1:0.10:1,27 for 'Susara'. The remarks of Reid (2003) on the loss of effectiveness of fertilizers by leaching may be also taken into account.

Acknowledgements

This study was funded by the Agreement of the High Council of Scientific Research and the Cabildo Insular de La Palma. The authors are also grateful for the collaboration of the Proteas de La Palma Sociedad Cooperativa.

References

Álvarez, C.E., Fernández-Falcón, M. and Hernández, M.M. 2012. Plant nutrition, foliar standards, chlorophyll activity and soil nutrient status of two cultivars of *Leucospermum cordifolium* (Proteaceae). *Cien. Invest. Agrar.* 39: 105-116.

Cecil, J.S., Barth, G.E., Maier, N.A., Chviyl, W.L. and Bartetzko, M.N. 1995. Leaf chemical composition and nutrient removal by stems of *Leucadendron* cv. 'Silvan Red' and 'Safari Sunset'. *Aus. J. Exp. Agric.* 35: 547-55.

Chapman H.D. and Pratt, P.F. 1961. Methods of analysis for soils, plants and waters. California: University of California, Division of Agricultural Science. 309 p.

Claassens, A.S. 1986. Some aspects of the nutrition of proteas. *Acta Hort.* 185: 171-179.

Cottenie, A. 1980. Soil and plant testing as a basis of fertilizer recommendations. Rome: FAO Soil Bulletin. 118 p.

Cresswell, C. 1991. Assessing the phosphorus status of Proteas using plant analysis. *Sixth Biennial Conference*. Perth: International Protea Association. pp. 303-310.

Fernández-Falcón, M., Alvarez, C.E. and Hernández M. 2008. Nutrient removal, fertilization, needs and yields of Protea plants cultivated in subtropical conditions. *J. Plant Nutr.* 31: 1018-1032.

González, A.C.E., Hernández, M. and Fernández-Falcón, M. 2008. Nutrient distribution and stem length in flowering stems of protea plants. *J. Plant Nutr.* 31: 1624-1641.

Hawkins, H.J., Hettasch, H. and Cramer, M.D. 2007. Putting back what we take out, but how much? Phosphorus and nitrogen additions to farmed *Leucadendron* 'Safari Sunset' and *Leucospermum* 'Succession' (Proteaceae). *Sci. Hort.* 111: 378-388.

León, A.M.J. 2011. Study of growth, yield, photosynthetic capacity and fertilization of two cultivars of *Leucospermum* planted in clay soil directly or grafted on rootstock tolerant to clay soils. PhD thesis, University of La Laguna, Spain. 240 p.

Maier, N.A., Barth, G.E., Cecil, J.S., Chvyl, W.L. and Bartetzko, M.N. 1995. Effect of sampling time and leaf position on leaf nutrient composition of *Protea* 'Pink Ice'. *Aust. J. Exp. Agric.* 35: 275-283.

Malan, G. 2012. Protea cultivation. From concept to carton. Stellenbosch: SUN MeDIA. 297 p.

Mengel, K. and Kirkby, E.A. 2001. Principles of Plant Nutrition. Boston: Kluwer Academic Publishers. 849 p.

Montarone, M., Ziegler, M., Dridi, N. and Voisin, S. 2003. Comparison of mineral requirements of some cultivars in two proteaceae genera. *Int. Protea News* 44: 103-111.

Parvin, P.E. 1986. Use of tissue and soil samples to establish nutritional standards in protea. *Acta Hort.* 185: 145-153.

Reid, A. 2003. Nutrition and irrigation of proteas. *Hort. News* August, 2013. pp. 9-11.

25

STATUS OF COMPOST USAGE AND ITS PERFORMANCE ON VEGETABLE PRODUCTION IN MONGA AREAS OF BANGLADESH

G.K.M.M. Rahman[1], M.S.I. Afrad[2*] and M.M. Rahman[3]

Abstract

The present study was carried out to assess the existing status of compost usage on vegetable production and determine the overall effect of household waste compost (HWC) on growth and yield of vegetables and enhancement of soil fertility in the monga areas of Bangladesh. A field survey was conducted on 152 sampled farmers during 2010 to 2011. Questionnaire containing both closed and open-ended questions were used to assess existing production practices of vegetables using compost in both homestead and field conditions. Three field trials at Badargonj and Kawnia upazilas of Rangpur district were conducted taking four treatments i.e. control, recommended doses (RD) of fertilizers, HWC at the rate of 10 tha^{-1}, and HWC 10 t ha^{-1} plus RD as IPNS based with Lal shak, Palong shak, Pui shak and Tomato. Base line survey results indicated inadequate knowledge of the farmers on use and preparation of the household waste compost. Yield data of all vegetables i.e. Tomato, Lal shak, Palong shak and Pui shak indicated that the combined application of nutrients using organic and inorganic sources were significantly better than that of solitary application of inorganic fertilizers. The potential of household waste compost applied @ 10 t ha^{-1} along with inorganic fertilizers applied was found highly satisfactory in producing Tomato, where yield was recorded 75 t ha^{-1} in the study area. The fresh yield of Palong shak was found 16 t ha^{-1} when recommended doses of inorganic fertilizers were applied, but it was about 19 t ha^{-1} under combined application of HWC @ 10 t ha^{-1} and inorganic fertilizers following IPNS concept. The fresh yield of Pui shak was found about 49 t ha^{-1} under combined application of organic and inorganic nutrients. Considering the availability and costs of different composts, it is evinced that HWC contained good amount of NPK which indicates its potentiality to be used as a soil amendment, improving soil fertility and crop productivity. It can be an alternative to chemical fertilizer to increase soil microbial populations and enzyme activities and to promote the soil nutrient for horticultural crops in the unfertile areas especially in the monga areas of Bangladesh.

Keywords: Compost, Monga, Vegetable, Soil Fertility

[1&3]Department of Soil Science, Bangabandhu Sheikh Mujibur Rahman Agricultural University, Salna, Gazipur, Bangladesh
[2]Department of Agricultural Extension and Rural Development, Bangabandhu Sheikh Mujibur Rahman Agricultural University, Salna, Gazipur, Bangladesh

*Corresponding author's email: afrad69@gmail.com (M.S.I. Afrad)

Introduction

It is a great challenge to keep the agricultural production system sustainable under changing climate. Agriculture in Bangladesh faces the challenge of producing crops from its limited land resource to meet up the huge demand of foods for its ever-growing population. The scope is limited to bring new land under cultivation. Therefore, intensification of land with modern crop varieties for increased production is a necessary (Rahman and Azam, 2005). Heavy pressure on agricultural land through intensive cultivation of high yielding crop varieties using inorganic fertilizers solely and almost no recycling from organic residues has led to decline organic matter and mining out the inherent plant nutrients from soils (Karim *et al.*, 1994). Soil organic matter is an indicator of its fertility and thereby crop productivity. According to BARC (2005), a good soil should have at least 2.5% organic matter, but most of the soils in Bangladesh have less than 1.5%, and some soils even less than 1% organic matter. The soils of the Monga area of Bangladesh are seriously depleted and organic matter content is very low. Monga is a local term, which means seasonal food scarcity occurring in September – November and March – April in the northern part of Bangladesh. During the period people become jobless leading to poverty. Production of short duration vegetables during this crisis period using household waste compost can be a better option for increasing income, available of domestic foodstuff towards reduction of monga. The use of household waste, poultry manure and cow dung contributes to the requirement to maintain the long-term fertility of soils by improving physical and chemical properties of soils (Rahman, 2010). Cow dung and crop residues are used as fuel for cooking in many rural areas of Bangladesh and thereby are not available for using in the crop fields. However, household waste is available in everywhere and its composting is very easy. Household waste contains valuable plant nutrients especially N, P, S, K and many other macro and micronutrients, which may pollute the environment if not properly managed and utilized for crop production. Agricultural use of household waste may substantially cut down the cost of production reducing application of inorganic fertilizers especially for short duration vegetable production in the homestead and thus ensure an eco-friendly environment. Therefore, household waste compost could be a good alternative for replenishment of organic matter in soils. Therefore, the project was designed to (a) assess the existing status of using organic manures in vegetable production, (b) motivate the farmers for preparing household wastes compost for vegetable production, and (c) improve soil health in the project area.

Methodologies

Social survey

Interview schedule containing both closed and open-ended questions were prepared as per project objectives. A total of 152-farmers (68 from Kawnia and 84 from Badarganj upazilas of Rangpur district of Bangladesh) were selected based on some criteria viz., home near to well-communicated road; possess at least two decimal spare and sunny homestead area and willing to cultivate vegetables following guidelines provided by the project authority. Trained enumerators were engaged to collect primary data from the selected respondents. The field survey was conducted to delve into their livelihood activities including vegetables production practices, income and expenditure, education and health care, and other related aspects of food quality and production. Selected respondents were trained on preparation of household waste compost and its application for vegetable production in homestead areas. Of them twenty were chosen for conducting trial. As per project objectives, selected farmers prepared household wastes

compost in their own homestead and used for vegetable production in the same areas.

Field trials

A series of trials were laid out in a factorial (nutrients x locations) randomized complete block design with four replications during 2010-2011. The four levels of nutrient treatments were: (1) control, (2) recommended doses (RD) of fertilizers, (3) household waste compost (HWC) at the rate of 10 t ha^{-1}, and (4) HWC 10 t ha^{-1} plus RD as IPNS based. During the 2010-11 two trials (first one as per nutrient treatments mentioned and the second one using residual nutrients i.e. without applying either compost or inorganic fertilizers) with Palong shak (Spinach) were conducted in 17 locations/villages (Kawnia 7 and Badarganj 10). The dates of sowing of Palong shak seeds were November 3 to December 14, 2010 for the first trial, while for the second one were December 11, 2010 to January 20, 2011. The crop was harvested after 30-35 days of sowing. During the 2010-11, among 17 villages ten were selected (Kawnia 5 and Badarganj 5) to grow Lal shak (Red Amaranth) as first crop using nutrients as per treatments, while second crop Pui shak (Indian spinach) without applying nutrients. The seeds of Lal shak was sown on February 10, 2011 and harvested on March 18, 2011. The Pui shak was sown on April 10, 2011 and harvested on June 8, 2011. During 2010-11, one farmer from Kawnia grew Lal shak, Pui shak and Tomato as first, second and third crops, respectively in the same piece of land using above mentioned four treatments. The Lal shak was sown on July 1, 2010 and harvested on August 4, 2010, while Pui shak was sown on August 19, 2010 and harvested on September 25, 2010. Tomato seedlings were transplanted on November 23, 2010 and final harvest was done on March 25, 2011. In the treatment 4, where household waste compost was applied at the rate of 10 t ha^{-1} plus recommended doses of N, P and K as IPNS based before sowing Lal shak and after that Pui shak was grown without applying nutrients and Tomato was grown using only nitrogen (N) at the rate of 117 kg ha^{-1}. Before applying to soils house hold waste compost, and initial soil were analyzed for quantifying the nutrients (Table 1) and then rates of chemical fertilizers were calculated following IPNS concept.

Table 1. N, P and K content of household waste compost and initial soil in the study area

Waste material	Total N (%)	Available N (%)	Total P (mg kg^{-1})	Available P (mg kg^{-1} soil)	Exchangeable K (mg kg^{-1} soil)
Household waste compost	1.36	0.089	1650	450	1410
Plot soil	0.090	0.030	-	12.84	0.14 [*]

unit of K for plot soil was meq 100 g^{-1} soil that was for fertilizer recommendation of crops

After harvesting of the Lal shak and Tomato soil samples were collected from each plot at 0-15 cm depth and were analyzed for pH, OM and residual nutrients. Soil pH was determined by Glass Electrode pH Meter method with soil water ratio 1:2.5 (McLean, 1982), OM by Walkley-Black method (Nelson and Sommers, 1982), total N was determined by Kjeldhal systems (Bremner and Mulvaney, 1982), total P was determined by the Acid Digestion method (Jones and Case, 1990; Watson and Issac, 1990), available P by Olsen's method (Olsen and Sommers, 1982), and K by Ammonium acetate extraction method (Barker and Surh, 1982). Available N (NO_3^--N plus NH_4^+-N) was determined using Steam Distillation method (Keeney and Nelson, 1982). The analyses of variance for different parameters were done following the principle of F-statistics and the mean results was compared using Duncan's Multiple Range Test (DMRT).

Results and Discussion

Social survey

Base line survey was conducted on selected 152 farmers (68 from Kawnia and 84 from Badargonj, Rangpur). The main emphasis of the base line survey was on the knowledge of the respondents on preparation and use of household wastes compost and benefits of cultivating vegetables. The findings obtained have been presented in the following headings:

Preparation of household wastes compost with cow dung

Village people usually do not prepare and use scientifically household wastes compost. Rural women throw these here and there around their homestead. Household wastes composts can be prepared in many ways. Information displayed in Table 2 indicate that about four-fifths of the respondents (79%) are using cow dung compost including household wastes and interestingly about one-fifth (21%) used it in other ways.

Table   2.   Distribution   of   the   respondents according to types of preparation of household wastes compost with cow dung

Preparation of household wastes compost with cow dung	Respondents	
	Number	Per cent
Direct application in the field	0	0
Composting in separate pit	0	0
Composting with cow dung	120	79
Other ways (if any)	32	21

Time of preparation of cow dung compost

Findings available in Table 3 show that almost all of the respondents (91%) are found using cow dung compost during final land preparation. Though the next negligible portion of them (9%) are found using cow dung compost in other ways i.e. throw in the field whenever available, use during intercultural operations.

Table   3.   Distribution   of   the   respondents according to the time of preparation of cow dung compost

Time of using cow dung compost	Respondents	
	Number	Per cent
During final land preparation	138	91
After 15 days of planting	0	0
After 30 days of planting	0	0
Other ways (if any)	14	20

Doses of cow dung compost

Fertilizers and manures are applied in the field based on the demand of the crops grown. The respondents do not care for required doses of the cow dung compost. Information displayed in the Table 4 point out that more than half of the respondents (51%) use cow dung compost @ 10 t ha⁻¹ followed by near about half (45%) use @10 t ha⁻¹.

Table   4.   Distribution   of   the   respondents according to the using doses of cow dung compost

Doses of cow dung compost	Respondents	
	Number	Per cent
5 t ha⁻¹	6	4
10 t ha⁻¹	77	51
15 t ha⁻¹	69	45
20 t ha⁻¹	2	1

Benefits of using household wastes compost

There are lot of benefits can be harvested by using cow dung compost like increased soil fertility, sound environment, low production cost, higher yield of vegetables, better quality of products etc. Respondents were asked for demonstrating their opinions regarding the benefits of using cow dung compost. Results demonstrated in Table 5 specify that more than four-fifths of the respondents (84%) indicate low production cost, followed by 18 per cent higher yield of vegetables and 13 percent increased soil fertility as the benefits of using cow dung compost.

Table   5.   Distribution   of   the   respondents according to their responses on the benefits of using cow dung compost

Benefits of using cow dung compost	Respondents	
	Number	Per cent
Increase soil fertility	19	13
Low production cost	128	84
Higher yield of vegetables	28	18
Sound environment	2	1
Others	0	0

Usefulness of vegetables cultivation

Vegetable cultivation may harvest varieties of advantages like high nutritional value, short durational, prevent disease, higher income etc. However, almost all of the respondents (91%) under study point out higher income as one of the most benefit of vegetable cultivation. A tiny portion of them (8%) indicates prevention of diseases, followed by high nutritional value (5%) and short duration (3%) as the advantages of vegetable cultivation (Table 6).

Table 6. Distribution of the respondents according to their knowledge on the usefulness of vegetables cultivation

Benefits of cultivating vegetables	Respondents	
	Number	Percent
High nutritional value	8	5
Short durational	5	3
Prevent disease	12	8
Higher income	139	91
Others	0	0

Results of field trials in the study area

Fresh yields of Palong shak

Fresh yields of Palong shak were significantly affected by different nutrient levels over the control (Table 7), but location did not show any significant effect on its yield (Table 8).

Table 7. Fresh yield of Palong shak under different nutrient levels in Kawnia and Badarganj, Rangpur

Nutrient levels	Fresh yield of Palong shak (t ha^{-1})	
	First harvest	Second harvest
Control	3.62c	4.91b
Recommended doses of NPK	15.79b	16.09a
Household waste 10 t ha^{-1}	15.67b	17.99a
Household waste 10 t ha^{-1} + RD* of N (IPNS concept)	18.55a	18.86a
S.E. (±)	0.36	0.27
%CV	22.16	22.49

*RD = Recommended dose

Table 8. Fresh yield of Palong shak under different locations in Kawnia and Badarganj, Rangpur

Locations		Fresh yield of palong shak (t ha^{-1})	
Upazila	Village	First harvest	Second harvest
Kawnia	Kachu	10.86	11.17
	Shingerkura	13.21	10.53
	Nazirdah	13.14	14.46
	Shabdi	13.04	15.36
	Madhupur	13.99	14.30
	Shahabaz	13.24	16.56
	Haragach	14.57	14.89
Badarganj	Shonkorpur	13.31	17.63
	Amrulbari	12.84	13.11
	Kachabari	12.99	13.31
	Nataram	13.67	14.98
	Rajarampur	14.36	14.68
	Gopalpur	14.14	13.41
	Khiarpara	14.27	16.51
	Radanagar	14.11	12.42
	Osmanpur	13.62	15.94
	Ramnathpur	12.56	16.87
S.E. (±)		0.74	0.77
%CV		22.16	22.49

It should be noted that crops grown using combined nutrients i.e. household waste and recommended N following IPNS concept were found somewhat better over solitary application of either organic or inorganic nutrients. The fresh yield of Palong shak was found 16 t ha^{-1} when recommended doses of inorganic fertilizers were applied, but it was about 19 t ha^{-1} under combined application of organic and inorganic nutrients in both years.

Fresh yields of Lal shak and Pui shak were significantly influenced by different nutrient levels over the control, but location did not show any significant effect on its yield in Kawnia and Badarganj (Table 9 and 10). Yields of Lal shak grown using combined nutrients i.e. household waste and recommended N following IPNS concept were found significantly higher over recommended doses of fertilizers and only household waste. In case of Pui shak, application of recommended fertilizers and only household waste provided same amount of yield i.e. 45 t ha^{-1}, however combined applications of inorganic fertilizers and household waste gave significantly higher yield over solitary application of either organic or inorganic fertilizer. The fresh yield of Pui shak was found about 49 t ha^{-1} under combined application of organic and inorganic nutriments.

Table 9. Fresh yield of Lal shak and Pui shak under different nutrient levels in Kawnia and Badarganj, Rangpur

Nutrient levels	Fresh yield of Lal shak and Pui shak (t ha^{-1})	
	Lal shak	Pui shak
Control	9.91d	16.72c
Recommended doses of NPK	20.61c	45.19b
Household waste 10 t ha^{-1}	29.07b	45.36b
Household waste 10 t ha^{-1} + RD of N (IPNS concept)	34.79a	48.66a
S.E. (±)	0.65	1.13
%CV	17.51	18.41

Table 10. Fresh yield of Lal shak and Pui shak under different locations in Kawnia and Badarganj, Rangpur

Locations		Fresh yield of Lal shak and Pui shak (t ha^{-1})	
Upazila	Village	Lal shak	Pui shak
Kawnia	Kachu	22.75	39.34
	Shingerkura	23.22	37.42
	Nazirdah	25.11	39.80
	Shabdi	24.34	42.52
	Haragach	22.69	37.41
	Amrulbari	21.08	40.40
Badarganj	Nataram	23.59	39.31
	Rajarampur	23.42	35.56
	Gopalpur	24.84	38.78
	Ramnathpur	24.93	39.29
S.E. (±)		1.03	1.79
%CV		17.51	18.41

Treatments containing different levels of nutrients significantly improved soil pH and OM in soils after harvesting of Lal shak in Kawnia and Badargonj, Rangpur (Table 11). Though the treatments did not show any significant effects on N, P and K contents in soils after harvesting of the crop, however these parameters were improved in soils with the application of household waste. The increment of nutrients in soils with the application of household waste compost indicated the improvement of soil fertility. Different locations did not show significant effect on residual nutrients in soils after harvesting of Lal shak (Table 12).

Table 11. Soil properties after harvesting of Lal shak under different nutrient levels in Kawnia and Badarganj, Rangpur

Nutrient levels	Soil properties at crop harvest				
	pH	OM (%)	N (%)	P (mg kg^{-1})	K (meq 100 g^{-1})
Control	5.15b	1.13b	0.10	14.12	0.17
Recommended doses of NPK	5.18b	1.14a	0.13	15.23	0.23
Household waste 10 t ha^{-1}	5.85a	1.30a	0.12	18.54	0.23
Household waste 10 t ha^{-1} + RD of N (IPNS concept)	5.71a	1.35a	0.14	17.09	0.25
S.E. (±)	0.11	0.04	0.004	0.44	0.01
%CV	12.61	19.20	30.40	17.23	14.37

Table 12. Soil properties after harvesting of Lal shak under different locations in Kawnia and Badarganj, Rangpur

Locations		Soil properties at crop harvest				
Upazila	Village	pH	OM (%)	N (%)	P (mg kg^{-1})	K (meq 100 g^{-1})
Kawnia	Kachu	5.51	1.19	0.09	15.93	0.25
	Shingerkura	5.75	1.20	0.10	17.07	0.22
	Nazirdah	5.29	1.66	0.08	15.85	0.23
	Shabdi	5.61	1.28	0.11	16.20	0.20
Badarganj	Haragach	5.51	1.37	0.10	14.43	0.21
	Amrulbari	5.52	1.50	0.12	16.38	0.23
	Nataram	5.70	1.23	0.11	18.04	0.19
	Rajarampur	5.41	1.19	0.13	14.32	0.22
	Gopalpur	5.67	1.24	0.11	19.93	0.25
	Ramnathpur	5.53	1.18	0.15	16.85	0.20
S.E. (±)		0.17	0.06	0.01	0.69	0.01
%CV		12.61	19.20	30.40	17.23	14.37

Yields of lal shak, pui shak and tomato

Yields of Lal shak, Pui shak and Tomato were significantly affected by different nutrient levels over the control (Table 13). However, yield of Lal shak and Pui shak among recommended doses of chemical fertilizers, household waste compost alone and combined application of household waste and chemical fertilizer was statistically similar. Tomato yield was higher in household waste compost alone and combined application of household waste and chemical fertilizer over recommended doses of chemical fertilizers. From the results, it was observed that combined application of nutrients using organic and inorganic sources performed better.

Table 13. Fresh yield of Lal shak, Pui shak and Tomato under different nutrient levels in Kawnia, Rangpur

Nutrient levels	Yield of lal shak (t ha^{-1})	Yield of pui shak (t ha^{-1})	Tomato fruit yield (t ha^{-1})
Control	7.15b	14.33b	24.10c
Recommended doses of NPK	17.35a	30.80a	63.70b
Household waste 10 t ha^{-1}	20.75a	33.48a	78.67a
Household waste 10 t ha^{-1} + RD of N (IPNS concept)	20.85a	32.98a	75.35a
S.E. (±)	1.90	2.08	3.65
%CV	23.02	14.98	12.09

Treatments containing different levels of nutrients did not show significant effects on soil pH and nutrients except potassium in soils after harvesting of tomato in Kawnia, Rangpur (Table 14). However, soil pH, OM, nitrogen and phosphorous were improved in soils with the application of household waste compost. Application of household waste compost alone and in combination with inorganic fertilizers increased soil K level over the control and recommended doses of chemical fertilizers. The increment of nutrients in soils with the application of household waste compost indicated the improvement of soil fertility.

Table 14. Soil properties after harvesting of Tomato under different nutrient levels in Kawnia, Rangpur

Nutrient levels	Soil properties at crop harvest				
	pH	OM (%)	N (%)	P (mg kg^{-1})	K (meq 100 g^{-1})
Control	5.56	1.22	0.08	10.01	0.12b
Recommended doses of NPK	5.58	1.21	0.09	10.22	0.13b
Household waste 10 t ha^{-1}	5.71	1.29	0.11	11.31	0.20a
Household waste 10 t ha^{-1} + RD of N (IPNS concept)	5.63	1.27	0.11	12.11	0.22a
S.E. (±)	0.15	0.08	0.006	0.52	0.02
%CV	11.55	15.32	27.38	14.89	19.56

Case study

Punil Chandra Roy, a medium farmer of the Rajarampur village of Badarganj upazila was selected as a participatory farmer of the project. Mr. Roy actively participated in the project training, positively motivated and gave consent to work to achieve the project mission. He dug a pit in front of his house and started collecting and piling all his decomposable household wastes therein. As per instruction, he covered the pit with a thin layer of soil when sufficient household wastes were collected in the pit. Mr. Roy observed the decomposition of the household wastes into compost. After three months, the household wastes decomposed into household wastes compost which was 26 kg in weight. Mr. Roy prepared a land of 4 decimal subdivided into 8 plots using household wastes compost at different doses. He grew Palong shak in the plots and got the differences of using household wastes compost within a very short time. He found the plots with only household wastes composts (@ 10 t ha^{-1}) showed better result. Second time he continued it with Pui shak without any chemical fertilizers. Third time he produced Lal shak using only household wastes compost and got the similar result. Neighbors visited his household wastes pit and vegetables plots. They became interested to hear from him and motivated to do the same in their homestead. Mr. Khokon Chandra Roy, one of the sons of Mr. Punil Chandra Roy, took the leadership of convincing people in preparation and use of household wastes compost for vegetable production. Like Mr. Roy, other participating farmers also got benefit from this project and committed to continue it with increasing rate along with their neighboring farmers. The main outcome of the project is successful preparation of household wastes compost and better performance of household wastes compost in vegetable production. Rural household dwellers can have vegetable production without using chemical fertilizers, which is economic for the rural poor. It will also increase the soil health with added organic matter in the soil and will maintain sound environmental condition. The policy makers may include this idea during formulation of environmentally sound and economically viable vegetable production in the homestead.

Conclusion

Most of the farmers under study showed low-level knowledge on household compost preparation as well as benefits of vegetable cultivation. Household waste compost has all the potentials to be used as a soil amendment, improving soil fertility and crop productivity and farmers have been motivated to prepare and use household waste compost for vegetable production in their homestead. Application of household waste compost @ 10t ha^{-1} along with inorganic fertilizers as IPNS based was found the best for short duration vegetables like Lal shak, Palong shak, Pui shak and also Tomato production over the sole application of either household waste compost or recommended doses of inorganic fertilizers. Household waste is available in the rural areas. Its compost is well proved as economic and valued source of plant nutrients to be applied to soils for crop production. Once household waste compost is applied @ 10 t ha^{-1} along with inorganic fertilizers as IPNS based, cultivation of three successive short duration vegetables is recommended for farmers' practice without applying any organic or inorganic fertilizers. Policy makers may take into consideration to circulate the technology to the rural household dwellers through extension organization.

References

BARC. 2005. Fertilizer recommendation guide, Bangladesh Agricultural Research Council (BARC). Soil publication No. 45. Farm Gate, Dhaka 1215, Bangladesh. 260p.

Barker, D.E. and Surh, N.H. 1982. Atomic absorption and flame emission spectroscopy. *In:* Page AL, Miller RH, Keeny DR (ed.) Methods of soil analysis, Part 2, Chemical and microbiological properties. American Society of Agronomy and Soil Science Society of America, Inc., Madison, Wisconsin, USA. pp. 13-26.

Bremner, J.M. and Mulvaney, C.S. 1982. Total nitrogen. *In:* Page, A.L., R.H. Miller and D.R. Keeny (ed.). Methods of soil analysis, Part 2, Chemical and microbiological properties, 2nd edition. American Society of Agronomy and Soil Science Society of America, Inc., Madison, Wisconsin, USA. pp. 595-624.

Jones Jr, J.B. and Case, V.W. 1990. Sampling, handling and analyzing plant tissue samples. *In:* Westermaan WS (ed.). Soil testing and plant analysis, 3rd edition, SSSA Book series 3, Soil Science Society of America, Madison, WI, USA. pp. 389-427.

Karim, Z., Miha, M.M.U. and Razia, S. 1994. Fertilizer in the national economy and sustainable environmental development. *Asia Pacific J. Envir. Div.* 1: 48-67.

Keeney, D.R. and Nelson, D.W. 1982. Nitrogen-inorganic forms. *In:* Page, A.L., R.H. Miller and D.R. Keeny (ed.). Methods of soil analysis, Part 2, Chemical and microbiological properties. American Society of Agronomy and Soil Science Society of America, Inc., Madison, Wisconsin, USA. pp. 643-698.

McLean, E.O. 1982. Soil pH and lime requirement. *In:* Page, A.L., R.H. Miller and D.R. Keeny (ed.). Methods of soil analysis, Part 2, Chemical and microbiological properties. American Society of Agronomy and Soil Science Society of America, Inc., Madison, Wisconsin, USA. pp. 199-224.

Nelson, D.W. and Sommers, L.E. 1982. Total carbon, organic carbon, and organic matter. *In:* Page, A.L., R.H. Miller and D.R. Keeny (ed.). Methods of soil analysis, Part 2, Chemical and microbiological properties. American Society of Agronomy and Soil Science Society of America, Inc., Madison, Wisconsin, USA. pp. 539-579.

Olsen, S.R. and Sommers, L.E. 1982. Phosphorus. *In:* Page, A.L., R.H. Miller and D.R. Keeny (ed.). Methods of soil analysis, Part 2, Chemical and microbiological properties.

American Society of Agronomy and Soil Science Society of America, Inc., Madison, Wisconsin, USA. pp. 403-430.

Rahman, M.M. 2010. Effect of different organic wastes in tomato (*Lycopersicon esculentum* Mill.) cultivation. *J. Env. Sci & Nat. Res.* 3 (1): 247-251.

Rahman, M.M. and Azam, G.M. 2005. Fertility evaluation of Old Meghna River Floodplain soils in Bangladesh for sustainable agriculture. *Chiang Mai J. Sci.* 32 (2): 127-137.

Watson, M.E. and Isaac, R.A. 1990. Analytical instruments for soil and plant analysis. *In:* Westermaan WL (ed). Soil testing and plant analysis, 3rd edition. SSSA Book series 3, Soil Science Society of America, Madison, WI, USA. pp. 691-740.

26

EFFECTS OF ARTIFICIAL FEEDS ON GROWTH AND PRODUCTION OF FISHES IN POLYCULTURE

M.A. Hosen[1,2], M. Shahjahan[1*], M.S. Rahman[1] and M.J. Alam[1,2]

Abstract

A study on the effects of artificial feeds on growth and production of fishes along with some limnological conditions were conducted in polyculture system. Species of Indian major carp (*Cirrhinus mrigala*) and exotic fishes (*Hypophthalmicthys molitrix* and *Oreochromis niloticus*) were stocked in six ponds under two treatments, each with three replications. Stocking rate in both treatments was 100 fish per decimal at the ratio of silver carp: tilapia: mrigal = 2: 2: 1. Fertilization and artificial feeds were given in Ttreatment 1 (T1) and only fertilization was done in Treatment 2 (T2). Wheat bran, rice bran and soybean meal were given daily as artificial feed in T1 in the ratio of wheat bran: rice bran: soybean meal = 2: 2: 1 (by wt). Urea, T.S.P and cow dung were applied fortnightly at the rate of 60 g deci^{-1}, 90 g deci^{-1} and 2 kg deci^{-1} respectively. Water temperature, transparency, pH, dissolved oxygen, free CO_2, total alkalinity, PO_4-P and NO_3-N were determined fortnightly and phytoplankton and zooplankton were studied fortnightly. These limnological conditions were more or less similar in the ponds under two treatments and were within suitable ranges. Calculated gross and net yields of fish were 16.56 and 12.48 ton ha^{-1} respectively in case of fertilization and artificial feeding application (T1) and 9.99 and 5.91 ton ha^{-1} respectively in case of only fertilization (T2). Application of artificial feed in T1 significantly increased the growth and production of fish more than two times which indicates that artificial feeding in polyculture is very useful for increasing fish production.

Keywords: Artificial Feed, Aquaculture, Polyculture, Fertilizer, Water Quality

[1]Department of Fisheries Management, Bangladesh Agricultural University, Mymensingh-2202, Bangladesh
[2]Department of Fisheries, Ministry of Fisheries and Livestock, Dhaka, Bangladesh

*Corresponding author's email: mdshahjahan@bau.edu.bd (M. Shahjahan)

Introduction

Aquaculture in Bangladesh has rapidly progressed in recent years with a contribution of 44% to the annual fish production (DoF, 2012). Among different techniques of aquaculture, polyculture is one of the most important techniques. The basic principle of fish polyculture systems rests on the idea that when compatible species of different feeding habits are cultured together in the same pond, the maximum utilization of all natural food sources takes place without harmful effects. Polyculture or mixed culture of carps has been found as an economically viable and technically sustainable in perennial water bodies (Alikunhi, 1957; Chen, 1976). The selection of fish species is very important for polyculture systems. In the present study, tilapia (*Oreochromis niloticus*), silver carp (*Hypophthalmicthys molitrix*), and mrigal (*Cirrhinus mrigala*) were selected for polyculture. These species are suitable for low inputs culture system in small ponds and ditches for their quick growth and maximum production within short period (4-6 months). Bangladesh has numerous seasonal water bodies in the form of shallow ponds, ditches, roadside canals, pits in rice fields, which retain water for 4-6 months. The natural environment of Bangladesh is also suitable for growing these fish species, which can be cultured in both shallow seasonal ponds and deeper perennial ponds.

To increase the fish production, improved techniques should be applied and management practices should be developed. Fertilization is one of the most important techniques to increase the fish production. Through fertilization, natural food of fish i.e. plankton is increased. On the other hand, artificial feed application is the most important technique to increase the production. Application of fertilizers or use of supplementary feed can play a vital role to increase the fish production in polyculture system. Faluroti (1987) stated that fertilization increased the fish production in polyculture system. Gupta *et al.* (1990) found about double production (4917 kg ha^{-1} per annum) with supplementary feed than that of without supplementary feed (2583 kg ha^{-1} per annum) in the polyculture of Indian and exotic carp. The ingredients that are easily available and are of comparatively low price are used as supplementary feed. The most easily available ingredients are rice bran, wheat bran, soybean meal that is appropriate for Indian major carps and exotic fishes of the South Asian region. Developed and scientific fish culture and successful fisheries management depends on the various information of limnological factors of water body. These limnological factors are water quality, microorganisms, plankton, benthos etc. According to Hickling (1968), fish farming is a practical application of limnology and freshwater biology. Reid (1971) reported that fish culture could be enhanced by the improvement of substratum by the use of fertilizers along with other pond management measures. In view of the above facts, in the present study, we examined the effects of artificial feeding on the growth and production of fishes along with limnological conditions in polyculture system.

Materials and Methods

The experiment was conducted for a period of 105 days in the earthen ponds each measuring 1 decimal (40 m^2) area at Bangladesh Agricultural University, Mymensingh, Bangladesh.

Pond preparation

Before fish stocking water of the experimental ponds were drained out to eradicate all the undesirable fishes, renovated and liming was done in all the ponds at the rate of 1 kg 40 m^{-2}. Ponds were filled up with deep tube-well water and fertilized with cow dung 10 kg, urea 100 g and TSP 100 g per decimal as initial doses.

Stocking of fish

Ponds were stocked at a stocking density of 100 fingerlings per decimal (40 m^2) at the ratio of tilapia: silver carp: mrigal = 2: 2: 1.

Fertilization and supply of artificial feed

The ponds were fertilized fortnightly with cow dung 2 kg, urea 60 g and TSP 90 g per 40 m^2 to grow natural food, phytoplankton and zooplankton. Three ingredients such as wheat bran, rice bran and soybean meal were applied as supplementary feed once daily in the morning between 8.00 and 9.00 a.m. The required amount of feed was mixed with a little amount of water to make it into thick 'dough' rolled into balls. The balls were then thrown into the ponds. Supplementary feed was supplied in T1 and T2 was control (no supplementary feed was supplied). Feed was supplied every day at the rate of 5% of the total fish biomass. Fish sampling was carried out at an interval of 15 days in order to calculate the increase in total wt. and to adjust the amount of feed. Ratio of artificial feed ingredients was, wheat bran: rice bran: soybean meal = 4: 4: 2 (by wt.).

Physico-chemical parameters

Various physical and chemical parameters of the ponds such as water temperature (°C), transparency (cm), dissolved oxygen (mg L^{-1}), pH, free CO_2 (mg L^{-1}), total alkalinity (mg L^{-1}), PO_4-P (mg L^{-1}) and NO_3-N (mg L^{-1}) were estimated fortnightly. Water temperature was recorded with a Celsius thermometer and transparency was measured with a Secchi disc of 30 cm diameter. Dissolved oxygen was measured directly with a DO meter (Lutron, DO-5509) and a portable digital pH meter was used to measure pH. Free CO_2 and total alkalinity were determined by titrimetric method (APHA, 1992). PO_4-P (mg L^{-1}) and NO_3-N (mg L^{-1}) were determined by a Hach Kit (DR/2010, a direct reading Spectrophotometer).

Study of planktons

Biological parameters of the ponds such as phytoplankton density (cells L^{-1}) and zooplankton density (cells L^{-1}) were estimated fortnightly. The counting of plankton (both phytoplankton and zooplankton) was done with the help of Sedgwick-Rafter Counting Cell (S-R cell) under a compound binocular microscope. The plankton population was determined by using the formula of Rahman (1992). Identification of plankton (phytoplankton and zooplankton) up to generic level were made according to Prescott (1964), Needham and Needham (1963) and Belcher and Swale (1978).

Study of growth and production of fish

At the end of the experiment, all fish were harvested through repeated netting by a seine net to calculate gross and net production of fish.

Specific growth rate (SGR, % per day) was estimated by the following formula:

$$\text{SGR (\% per day)} = \frac{\text{Loge } W_2 - \text{Loge } W_1}{T_2 - T_1} \times 100$$

Where, W_1= Initial live body weight (g) at time T_1 (day)
W_2= Final live body weight (g) at time T_2 (day)

Statistical analysis

A computer using SPSS package programme did T-test of net fish production of the ponds under T1 and T2.

Results

Water quality parameters

The results of physico-chemical parameters are shown in Table 1. All physical and chemical parameters of the ponds were found to be within the acceptable ranges for fish culture in both treatments.

Table 1. Water quality parameters (Means ± SD; n = 3) of the ponds during the experimental period

Parameters	Treatment 1	Treatment 2
Water temperature (^{o}C)	30.80±1.52	31.28±1.47
Transparency (cm)	24.54±6.94	33.25±5.98
pH	7.21±0.16	7.11±0.30
Dissolved oxygen (mgL^{-1})	5.31±0.64	5.21±0.54
Free CO_2 (mgL^{-1})	4.65±0.17	4.58±0.27
Total alkalinity (mgL^{-1})	99.15±9.11	97.71±8.26
Phosphate-phosphorous (mgL^{-1})	0.29±0.12	0.35±0.07
Nitrate-nitrogen (mgL^{-1})	1.09±0.48	1.11±0.43

Plankton

Fortnightly fluctuation of phytoplankton density (cells L^{-1}) and zooplankton density (cells L^{-1}) are shown in Table 2. The average density of phytoplankton of the ponds under T1 was 60.95 ± 4.48 (x10^3) cells L^{-1} and that of the ponds under T2 was 58.87 ± 3.52 (x10^3) cells L^{-1}. The average density of zooplankton of the ponds under T1 was 8.23 ± 0.79 (x10^3) cells L^{-1} and that of the ponds under T2 was 7.92 ± 0.60 (x10^3) cells L^{-1}. The generic status of phytoplankton and zooplankton found during the tenure of experiment are shown in Table 3. During the study period, 32 genera of phytoplankton belonging to four groups and 11 genera of zooplankton belonging to 4 groups were found in all the experimental ponds.

Table 2. Fortnightly fluctuation of phytoplankton and zooplankton densities in the ponds during the experimental period

Parameters	Treat-ments	Sampling days								Mean ± SD
		1	2	3	4	5	6	7	8	
Phytoplankton	T1	67.78	59.75	60.50	61.15	59.95	54.68	56.86	66.90	60.95 ± 4.48
(x10^3 cells L^{-1})	T2	62.38	63.58	53.57	58.27	59.66	57.90	54.66	60.94	58.87 ± 3.52
Zooplankton	T1	9.90	8.37	7.51	6.98	8.67	8.17	7.97	8.58	8.23 ± 0.79
(x10^3 cells L^{-1})	T2	8.70	8.60	7.4	7.50	7.83	7.33	7.43	8.53	7.92 ± 0.60

Table 3. Generic status of phytoplankton and zooplankton found in the experimental ponds

Phytoplankton		Zooplankton
Bacillariophyceae	**Chlorophyceae**	**Cladocera**
Actinella	Chlorella	Daphnia
Asterionella	Closteriurn	Diaphanosoma
Coscinodiscus	Gonatozygon	**Copepoda**
Cyclotella	Oedogonium	Cyclops
Diatoma	Oocystis	Diaptomus
Fragilaria	Pediastrum	**Rotifera**
Frustularia	Scenedesmus	Asplanchna
Navicula	Sphaerocystis	Brachionus
Nitzchia	Spirogyra	Filinia
Tabellaria	Tetraedron	Keratella
Cyanophyceae	Ulothrix	Polyarthra
Anabaena	Volvox	Trichocerca
Aphanocapsa	Zygnema	**Crustacean larva**
Gleocapsa	**Euglenophyceae**	Nauplius
Gomphospaeria	Euglena	
Microcystis	Phacus	
Oscillatoria		
Pleurosigma		

Growth and production of fish

The details of mortality rate, growth and production of fish are presented in Tables 4 and 5. The calculated gross and net productions of fish of the ponds under T1 were 16.56 and 12.48 ton ha^{-1} year^{-1} and those of the ponds under T2 were 9.99 and 5.91 ton ha^{-1} year^{-1}, respectively.

Table 4. Stocking density, mortality (%), initial and final weight and net and gross yield of fishes in ponds under two treatments

Treatments	Species stocked	Stocking density (per decimal)	Mortality rate (%)	Average initial weight (g)	Average final weight (g)	Species wise yield (kg/decimal/3.5 month)	
						Gross	Net
T1	Silver carp	40	7.5	52.33	197.54	7.31	5.22
	Tilapia	40	2.5	49.68	142.25	5.55	3.56
	Mrigal	20	5.0	33.75	169.95	6.46	5.78
T2	Silver carp	40	12.5	52.33	115.88	4.06	1.96
	Tilapia	40	5.0	49.68	98.25	3.73	1.75
	Mrigal	20	5.0	33.75	101.48	3.86	3.18

Table 5. Calculated gross and net productions of fish

Treatments	Production (kg/deci/3.5 month)		Production (ton/ ha/yr)		% increase of net production of T1 over T2*
	Gross	Net	Gross	Net	
T1	19.32	14.56	16.56	12.48	211.17%
T2	11.65	6.89	9.99	5.91	

*Net production (ton ha^{-1} yr^{-1}) under T2 (control) was taken for 100%.

Discussion

The present study was conducted to evaluate the effects of artificial feed on growth and production of fishes in polyculture of tilapia, silver carp and mrigal. The supply of artificial feed showed better growth and production performance.

The water quality parameters of the experimental ponds were within the productive ranges for the growth of plankton and benthos during the tenure of experiment (Table 1). Within limit productive ranges of such water quality parameters have also been observed by a number of authors (Uddin et al., 2007; Chowdhury et al., 2008; Uddin et al., 2012; Rahman et al., 2012; Talukdar et al., 2012; Siddika et al., 2012; Nupur et al., 2013) in the aquaculture ponds of BAU area which are in conformity with those of the present study.

During the study period, the gross and net production of fish of T1 (with supply of artificial feeding) were 19.32 and 14.56 kg deci^{-1} per 3.5 months respectively and those of the ponds under T2 (without supply of artificial feeding) were 11.65 and 6.89 kg deci^{-1} per 3.5 months respectively. The calculated gross and net productions of fish of T1 and T2 were 16.56 and 12.48 ton ha^{-1} yr^{-1} and 9.99 and 5.91 ton ha^{-1} yr^{-1} respectively. Murty et al. (1978) obtained gross and net fish productions of 4096.09 and 3858 kg ha^{-1} yr^{-1} and 2512.67 and 2275.37 kg ha^{-1} yr^{-1} with and without supplementary feed respectively. Gupta et al. (1990) reported 4917 and 2583 kg ha^{-1} yr^{-1} productions with and without supplementary feed respectively. Rahman (1997) also obtained 7903 and 3374 kg ha^{-1} yr^{-1} productions with and without supplementary feed respectively. The results of fish productions with and without supplementary feeding obtained by Murty et al. (1978), Gupta et al. (1990) and Rahman (1997) are similar to those of the present research work. Higher production of fishes was found in T1 where fishes were fed with

supplementary feed. Highly significant difference in the production of fishes was found between T1 (with supply of supplementary feed) and T2 (without supply of supplementary feed). Higher fish productions from the ponds with supply of supplementary feed than those of the ponds without supply of supplementary feed were also observed by Chaudhuri et al. (1975), Singh and Singh (1975), Chakrabarty et al. (1976) and Gupta et al. (1990). Khan and Jhingran (1975), Gupta et al. (1990), Rahman (1997) reported double yield of fish of the ponds with supply of supplementary feed than that of the ponds having no supplementary feed. In the present study, more than two folds of yield of fish were obtained in the ponds of having supplementary feeding than that of the ponds without supplementary feed. Singh and Singh (1975) obtained 3 to 4 times yield from supplementary feeding ponds than that of the ponds without supplementary feeding. About four folds of yield from supplementary feeding ponds than that of the ponds without supplementary feeding were also found by Hepher et al. (1971). In another study, Chaudhuri et al. (1978) reported seven folds of yield from supplementary feeding ponds than that of the ponds without supplementary feeding.

In conclusion, effects of artificial feeds on the growth and production of fishes along with some limnological conditions were conducted in polyculture system under two treatments. In T1, the fishes were reared with artificial feeding and fertilization and in T2, fishes were reared with only fertilization. Physico-chemical and biological parameters were more or less similar in the ponds under two treatments and were within suitable ranges. The net fish production was more than two times higher in T1 compared to T2 indicated that artificial feeding might play a vital role in pond fish culture to increase production of fishes in polyculture system.

References

Alikunhi, K.H. 1957. Fish culture in India. *F.M. Bull. Indian Coun. Agric. Res.* 20: 144.

APHA. 1992. Standard Methods for the Examination of Water and Wastewater. American Public Health Association, Washington DC. p. 874.

Belcher, H. and Swale, E. 1978. A Beginner's Guide to Freshwater Algae. HMSO, London. 47p.

Chakrabarty, R.D., Sen, P.R., Rao, N.G.S. and Ghosh, S.R. 1976. Intensive culture of Indian major carps. *Aqu. Sci. & Fish. Abs.* 6: 12.

Chaudhuri, H., Chakrabarty, R.D., Sen, P.R., Rao, N.G.S. and Jena, S. 1975. A new thought in fish production in India with record yields by composite culture in freshwater ponds. *Aquaculture,* 5: 343-355.

Chaudhuri, H., Rao, N.G.S., Saha, G.N., Ront, M. and Ranavsia, D.R. 1978. Record of fish production through intensive fish culture in farmer ponds. *J. Inland Fish. Soc. India,* 10: 19-27.

Chen, T.P. 1976. Aquaculture practices in Taiwan. Fishing News (Books) Ltd., West Byfleet, Surrey, England. 236p.

Chowdhury, M.M.R., Shahjahan, M., Rahman, M.S. and Sadiqul Islam, M. 2008. Duckweed (*Lemna minor*) as supplementary feed in monoculture of nile tilapia, *Oreochromis niloticus. J. Fish. Aqu. Sci.* 3: 54–59.

DOF. 2012. Fish week 2012. Department of Fisheries, Ministiy of Fisheries and Livestock. Government of the People's Republic of Bangladesh, Ramna, Dhaka. 123p.

Faluroti, E.O. 1987. Performance of fertilization and supplementary feed on fish production under polyculture system in warm water fish ponds. *J. West Afr. Fish.* 3 (2): 162-170.

Gupta, V.K., Sharma, J.P. and Srivastava, J.B. 1990. Polyculture of Indian and exotic carps using cattle manure with and without supplementary feed. *Recent-Trends in Limnol.* pp. 439-446.

Hepher, B., Chervinski, T. and Tagari, H. 1971. Studies on carp and silver carp nutrition III. Experiments on the effect on fish of dietary protein source and concentration. *Bamidgeh,* 23 (1): 11-37.

Hickling, C.F. 1968. The Farming of Fish, 1st ed., Pergamon Press Ltd., Oxford, 88p.

Khan, H.A. and Jhingran, V.G. 1975. Synopsis of biological data on rohu, *Labeo rohita* (Hamilton, 1822). FAO, Fish. Synop., (III): 100p.

Murty, D.S., Salia, G.N., Selveraj, C., Reddy, P.V.G.K. and Dey, R.K. 1978. Studies on increased fish production in composite fish culture through nitrogenous fertilization, with and without supplementary feeding. *J. Inland Fish. Soc. India,* 10: 39-45.

Needham, J.G. and Needham, P.R. 1963. A Guide to Study of Freshwater Biology. 5th Edn., Holden-Day, Inc., San Francisco. 106p.

Nupur, N., Shahjahan, M., Rahman, M.S. and Fatema, M.K. 2013. Abundance of macrozoobenthos in relation to bottom soil textural types and water depth in aquaculture ponds. *Int. J. Agril. Res. Innov. & Tech.* 3 (2): 1-6.

Prescott, G.W. 1964. Algae of Western Great Lakes area. Wm. C. Brown Co. Dubuque, IOWA. 946p.

Rahman, A.M. 1997. Effects of artificial feeding on the growth and production of fishes in polyculture. M.S. Thesis. Dept. of Fish. Biol. and Genetics, BAU, Mymensingh. 34p.

Rahman, M.S. 1992. Water Quality Management in Aquaculture. BRAC Prokashana, Mohakhali, Dhaka, Bangladesh. 84p.

Rahman, M.S., Shahjahan, M., Haque, M.M. and Khan, S. 2012. Control of euglenophyte bloom and fish production enhancement using duckweed and lime. *Iranian J. Fish. Sci.* 11: 358-371.

Reid, G.K. 1971. Ecology of Inland Waters and Estuaries. Reinhold Publishing Corporation, New York, Amsterdam. 767p.

Siddika, F., Shahjahan, M. and Rahman, M.S. 2012. Abundance of plankton population densities in relation to bottom soil textural types in aquaculture ponds. *Int. J. Agril. Res. Innov. & Tech.* 2 (1): 56-61.

Singh, G.S. and Singh, K.P. 1975. Feeding experiments on Indian major carps in Taraj ponds. *J. Inland Fish. Soc. India,* 7: 212-21.

Talukdar, M.Z.H., Shahjahan, M. and Rahman, M.S. 2012. Suitability of duckweed (*Lemna minor*) as feed for fish in polyculture system. *Int. J. Agril. Res. Innov. & Tech.* 2 (1): 42-46.

Uddin, M.N., Rahman, M.S. and Shahjahan, M. 2007. Effects of duckweed (*Lemna minor*) as supplementary feed on monoculture of GIFT strain of tilapia (*Oreochromis niloticus*). *Prog. Agric.* 18: 183-188.

Uddin, M.N., Shahjahan, M. and Haque, M.M. 2012. Manipulation of species composition in small scale carp polyculture to enhance fish production. *Bangladesh J. Prog. Sci. & Tech.* 10: 9-12.

27

PESTICIDE USAGE PATTERN FOR VEGETABLE CULTIVATION IN MANMUNAI SOUTH & ERUVILPATTU DIVISIONAL SECRETARIAT DIVISION OF BATTICALOA DISTRICT, SRI LANKA

S. Sutharsan*, K. Sivakumar and S. Srikrishnah

Abstract

Batticaloa, is a coastal district in Sri Lanka. Vegetables except up-country vegetables sold in Batticaloa District are mainly grown in villages. Manmunai South and Eruvilpattu divisional secretariat (DS) division is a predominantly vegetable cultivating area in the Batticaloa district. Farmers in this region use variety of synthetic pesticides to protect vegetables. Recently public concern related to health risks associated with pesticide residues has been increased, substantially. Therefore, a study was conducted to find out pesticide usage practices of farmers on vegetable cultivation in Manmunai South and Eruvilpattu DS division. Stratified random sampling method was used to select respondents' for the survey and the collected data were analyzed statistically. It was observed that, the usage of pesticides was higher in the study area. Vegetable farmers use more than 14 Insecticides to control pest infestation. Farmers in the study area apply pesticides more frequently. Highly pesticide sprayed crop is Brinjal. About 66% of the Chilli producing farmers and 84% of the Brinjal producing farmers apply pesticide more than 22 times per cropping season. Around 90% of the farmers apply more than the recommended dosage and frequency of the pesticides. It was noticed that more than 89% of the farmers harvest the produce before the recommended pre harvest interval. It was found out that farmers in the study area are not following recommended pesticide usage practices. Hence, it is essential to educate the farmers on recommended pesticide usage practices, reduced usage of synthetic pesticides and use of organic farming practices to reduce the ill effects of synthetic pesticides.

Keywords: Pesticide Residues, Questionnaire Survey, Insecticides, Fungicides

Department of Crop Science, Faculty of Agriculture, Eastern University, Sri Lanka, Chenkalady- 30350, Sri Lanka

*Corresponding author's email: sutharsans@esn.ac.lk (S. Sutharsan)

Introduction

Sri Lanka is an agricultural country as majority of the rural people is still engaged in agriculture for their main livelihood (Vidanapathirana, 2008). Agriculture is an important sector in the economy contributing to 11.1% as a share of the Gross Domestic Product for the year 2012 (Anonymous, 2013a). In the agricultural sector of Sri Lanka, vegetables are the second most important sub-sector after rice. Vegetables are produced on a year round basis and a large number of farmers are involved in the production process. Vegetable production is important in many aspects to Sri Lanka and recognized as priority crops in national production enhancement programs (Illankoon *et al.*, 2011).

Batticaloa, a coastal District and central part of Eastern Province of Sri Lanka, falls under the dry zone climatic conditions. Total population in the District is 586,400 (Anonymous, 2013b). Vegetables except up-country vegetables sold in

Batticaloa District are mainly grown in villages. and are harvested and sold fresh. Manmunai south and Eruvilpattu divisional secretariat (DS) division is a predominantly vegetable cultivating area among the 14 DS divisions in Batticaloa District. Farmers in this area use variety of synthetic pesticides to protect vegetable crops from pests and diseases. Recently public concern related to health risks associated with pesticide residues remain in food products has been increased, substantially. Hence, it is essential to find out the current pesticide usage practices of farmers in that area. It is important to carry out awareness programmes to minimize pesticides residues in vegetables. Therefore, this study was performed to find out the pesticide usage pattern for vegetable cultivation by the farmers in the Manmunai South & Eruvilpattu DS division, the major vegetable producing area, of the Batticaloa District.

Methodology

Manmunai South and Eruvil pattu DS division is located in the coastal side of Batticaloa district. There are 45 Grama Niladari (GN) Divisions included under this DS division. Total agricultural land extent in the DS division is 3,555 ha in which 330 ha is used for vegetable cultivation. Major agricultural villages in this DS division are Kaluthavalai, Mankadu, Thettativu and Cheddipalayam.

This survey was carried out from February to April 2013. A number of 387 respondents were selected among the vegetable farmers through stratified random sampling method from the study area. In each GN division, about 10% of the farm families were randomly selected as sample. Structured questionnaires were designed to gather required information and pretested to assess their suitability. Selected respondents were interviewed at their doorsteps and field observations were also made. The questionnaires were checked for completeness and the data were analyzed using SPSS 11.0.

Results and Discussion

Major vegetable crops cultivated in the study area

Many economically important vegetable crops are cultivated in the study area. Among them brinjal (*Solanum melongena* L.), chilli (*Capsicum annum* L.), long bean (*Vigna unguiculata* L.) and okra (*Abelmoschus esculentus* L.) are the major vegetable crops, which are cultivated by many farmers in larger extent.

Types of pesticides

It was observed that, the usage of pesticides was higher in the study area. The vegetable farmers use more than 14 type's pesticides to control pest infestation and minimize crop losses. It was also found that, pesticides were applied without adequate knowledge of pest ecology, economic injury levels and type of pesticides to control specific insect pest, their quantities and method of application, pre harvest interval and protective measures. Farmers assume that the only solution to pest problems is to spray more frequently and using different types of pesticides (Dunham, 1995). Several authors (Ngowi *et al.*, 2007; Legutowska *et al.*, 2002) reported that insecticides were the most used pesticides as insect pests are the serious problems in vegetable cultivation. Insecticides were followed by fungicides in usage, indicating that fungal attacks rank second to insect pests. The significant finding is that, vegetable farmers do not spray herbicides. They practice manual weeding with the help of family members.

Table 1. Pesticides used by farmers in the study area

Pesticides	Generic name	Crops
Insecticide	Abamectin	Chilli
	Thimethoxam 20% +Chlorantraniliprole 20%(w/w)	Brinjal
	Chlorphyrifos	Brinjal, Chilli, Long bean
	Imidachlorprid	Brinjal, Chilli, Long bean
	Acetamiprid	Brinjal
	Profenophos	Chilli, Long bean
Fungicide	Propineb	All vegetables
	Mancozeb	All vegetables
	Chlorothalonil	All vegetables
	Hexaconazole	All vegetables

It was common in this study that the labour division including women, children and all farming community exposed to pesticides. The high dependence on pesticides by vegetable farmers in the study area is an indication that they are not aware of other pest management strategies that are effective, inexpensive and environment friendly. There is an urgent need to bring the attention of these farmers to existing alternative pest management strategies that are cost effective and environment friendly (Hummel *et al.*, 2002).

Frequency of application

Farmers in the study area apply pesticides more frequently since pest infestation is relatively higher in vegetable crops, particularly in chilli, brinjal, long bean and okra. (Fig. 1). Highly pesticide sprayed crop is brinjal. About 84% of the brinjal farmers apply pesticides more than 22 times for one growing season. Varela *et al.* (1988) reported that, farmers tend to overuse chemicals by increasing pesticide quantity and spray frequency as well as applying pesticide cocktails. In general, the frequencies of pesticides application by farmers were higher in the study area. Such heavy use of pesticides may result in frequent contact with pesticides, which can lead to significant health problems. Excessive application of pesticides may lead to high levels of concentration on the plants, which may be dangerous to the farmers themselves and to consumers of the final product.

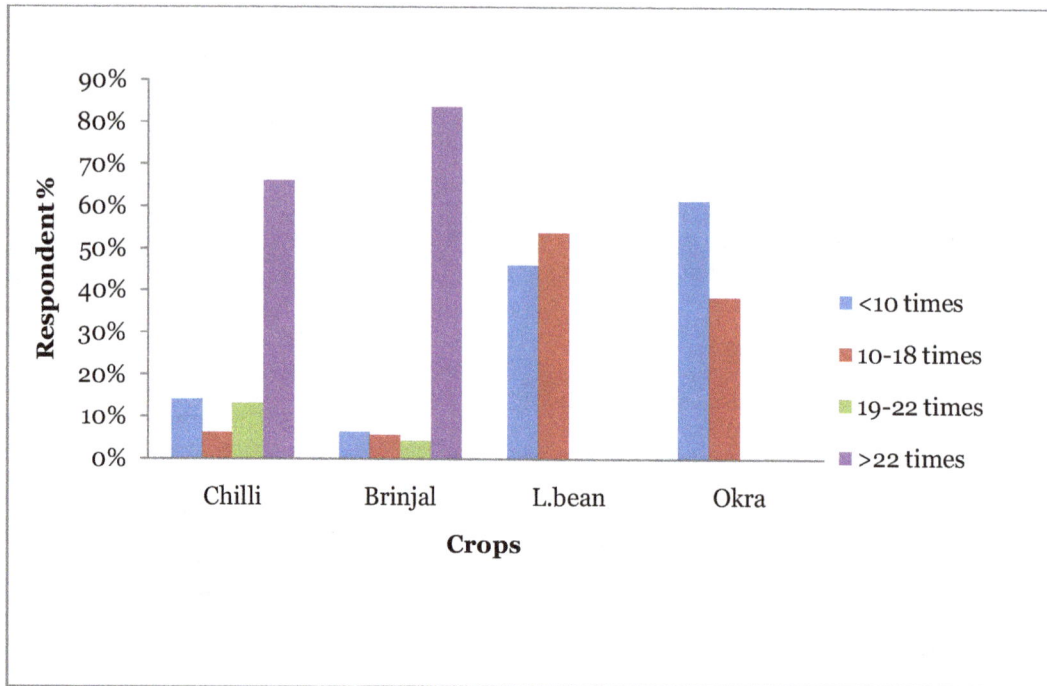

Fig.1. Frequency of pesticide application on the selected vegetables

Dosage of pesticide

About 90% of the farmers use more amount of pesticide than the recommendation to dilute with water in the study area. About 4% of the respondents follow the instruction given by the label, they dilute to the recommended level (Fig. 2). Especially for chilli and brinjal farmers use higher dosages of pesticides than recommendation. Farmers reluctant to follow the recommended dilution as prescribed on the label.

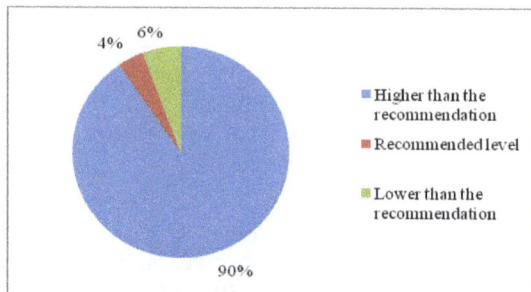

Fig. 2. Dosage of pesticides used by farmers

Some respondent said that they use excessive dosage of chemicals as their neighbour is using the same dosage. They also believe that excessive use of pesticides will give more yields. Resistance of pests to chemicals due to long-term application also made the farmers to use over dosages. Farmers use over dosages of pesticides in many developing countries. The major reasons stated by the farmers for less use of pesticides were lack of funds, low pest incidence and lack of knowledge (Ngowi et al., 2007).

Mixing of Pesticide

Farmers in the survey applied pesticides by both single and mixed method. Few farmers (about 9 %) apply one chemical at a time. However, majority of (91%) the farmers applied the pesticides in mixtures (Fig. 3). Farmers believe that a "cocktail" application is always more effective and reduce labour cost (Jipanin et al., 2001).

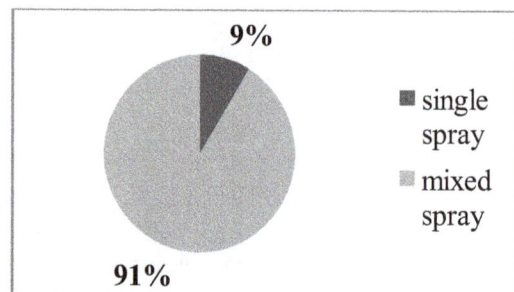

Fig. 3. Method of application practiced by the farmers

Pre-harvest intervals

It was noticed that more than 89% of the farmers in the selected GN divisions harvest the produce before the recommended pre-harvest interval (Fig. 4). Among the selected respondents, 100% of farmers from Thettativu-south, Kaluthavalai-4 and Shanthipuram GN divisions harvest the produce before the period of pre-harvest interval.

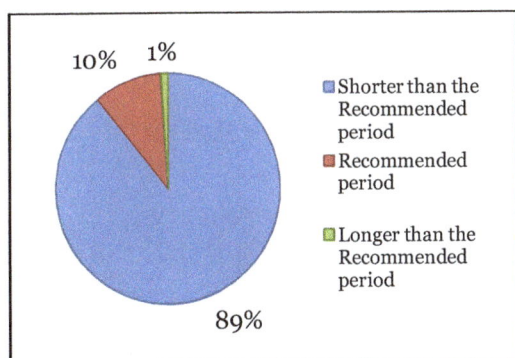

Fig. 4. Pre-harvest interval practiced by the farmers

In the study area, it is a regular practice to spray pesticides immediately before or after the harvest. When the pesticide spray was done immediately before harvesting, the danger of pesticide residue on produce was more (Jeyanthi and Kombairaju, 2005).

Conclusion and Recommendations

Findings of this survey highlighted that, the pesticide usage pattern in vegetable production in Manmunai South and Eruvilpattu Divisional Secretariat division of Batticaloa district, which is the major vegetable producing area of Batticaloa District. Farmers in the study area use heavy dosages of pesticides to protect their vegetable crops. The awareness of farmers related to pesticide usage is very less. Most of the farmers apply pesticide regularly to their crops irrespective of incidents of pests and crops diseases. Majority of the farmers use pesticides higher than the recommended level. Further, farmers mix more than one pesticide (Cocktails) to control many pest and diseases simultaneously and they do not know its effect on crop and environment. From this study, it could be concluded that, farmers in the Manmunai South and Eruvilpattu Divisional Secretariat division depend on chemical pest control method and use heavy dosages of pesticides. Therefore, awareness programmes are essential to change the attitudes of farmers to shift them to environment friendly pest control methods. It is essential for the sustainability of vegetable production in the study area and to safe guard the health of the people. The farmers need to be educated on the concepts of natural and organic farming for better yield and better health.

References

Anonymous. 2013a. Sri Lanka economy profile 2013. Index Mundi. Available at: http://www.indexmundi.com/sri_lanka/economy_profile.html (Accesses on 28.11.2013)

Anonymous. 2013b. Statistical handbook of Batticaloa district 2012/2013. Batticaloa. District Planning Secretariat. 18p.

Dunham, D. 1995. Contract Farming and Export Horticulture: Can Agribusiness Revitalise the Peasant Sector in Sri Lanka? IPS Agricultural Policy Series, No. 3. Colombo. Institute of Policy Studies. 34 p.

Hummel, R.L., Walgenbach, J.F., Hoyt, G.D. and Kennedy, G.G. 2002. Effects of production system on vegetable arthropods and their natural Enemies. *Agric. Ecosyst. Environ.* 93 (1–3): 165–176.

Illankoon, I.M.N., Abeynayake, N.R. and Kadupitiya, H.K. 2011. Forecasting vegetable extent and production in Sri Lanka: ARIMA model approach. Statistical Concepts and Methods for the Modern World. Available at: http://at.yorku.ca/c/b/d/h/29.htm (Accessed on 03.06.2013).

Jeyanthi, H. and Kombairaju, S. 2005. Pesticide use in vegetable crops: frequency, intensity and determinant factors. *Agril. Econ. Res. Rev.* 18: 209-221.

Jipanin, J., Rahman, A.A., Jaimi, J.R. and Phua, P.K. 2001. Management of pesticide use on vegetable production: Role of Department of Agriculture Sabah. 6th SITE Research Seminar. pp. 1-21.

Legutowska, H., Kucharczyk, H. and Surowiec, J. 2002. Control of thrips infestation on leek by intercropping with clover, carrot or bean. *Acta Hort. (ISHS).* 579: 571-574.

Ngowi, A.V.F., Mbise, T.J., Ijani, A.S.M., London, L. and Ajayi, O.C. 2007. Pesticides use by smallholder farmers in vegetable production in Northern Tanzania. *Crop Protection (Guildford, Surrey).* 26 (11): 1617.

Varela, G. and Navarro, M.P. 1988. Influence of pesticides on the utilization of food. *Bibl. Nutr. Dieta.* 4: 40-54.

Vidanapathirana, R.P. 2008. Marketing margins of the domestic vegetable trade in Sri Lanka. *Sri Lanka J. Agrarian Studies* 12 (2): 36-60.

28

FARMERS PERCEPTIONS ON DUAL-PURPOSE SORGHUM AND IT'S POTENTIAL IN ZAMBIA

S. Chikuta[1*], T. Odong[1], F. Kabi[1], M. Mwala[2] and P. Rubaihayo[1]

Abstract

Food feed crops play a cardinal role in mixed crop-livestock production systems yet views of farmers on their usage are limited. Farmers' perceptions in predominant sorghum growing areas of Zambia were solicited on socio-economic factors affecting sorghum production, awareness and willingness to adopt dual-purpose sorghum cultivars for food and feed. Preferred traits of a "model" dual-purpose cultivar were identified. The aim of the study was to generate information that would support the genetic improvement of dual-purpose sorghum. Questionnaires were used to generate this information. Results showed that less than 50% of sorghum growing SSFs had limited knowledge on the use of sorghum to produce feed silage; however, there was full awareness among the LSFs. Among other traits, farmers' "ideal" variety should combine high grain yield potential (100 %) with high biomass (100 % of LSFs and 80 % of SSFs) and high stem sugar content (100 % of LSF and 70 % of SSFs). All the SSFs and 20 % of the LSFs indicated that adequate production could be hampered by low grain yield, poor access to improved seed and unavailability of farmers'-preferred cultivars.

Keywords: Dual Purpose Sorghum, Farmer Perceptions, Feed, Grain, Silage, Zambia

[1]Department of Agricultural production, Makerere University, P .O. Box 7062, Uganda
[2]Department of Crop Science, University of Zambia, P. O. Box 32379, Zambia

*Corresponding author's email: sallychikuta@yahoo.com (S. Chikuta)

Introduction

Within the mixed crop-livestock systems of the tropics, the role of dual-purpose or food-feed crops is an area of substantial interest (Thornton *et al.*, 2003). With ever expanding croplands, small-scale farmers now rely on crop residues as a major source of feed (Sibanda *et al.*, 2011). In commercial farming systems, the cattle industry is centered on stocker cattle grazing systems and confined cattle feeding regimes, which utilize hay and silage (Rooney *et al.*, 2007).

Sorghum (*Sorghum bicolor* L. Moench) is an important cereal crop in the semi-arid tropics for human food, animal feed and raw material in commercial food industries. Improving the nutritive value of its stover is an important objective in the tropics where sorghum residues are extensively used for livestock feed (Rattunde *et al.*, 2001). However, the comparative advantage of sorghum over the competing crops and technologies has not been clearly identified and although there are no reports of farmer's acceptance and adoption on dual-purpose sorghum incidences have been reported with grain sorghums. Farmers' perceptions that may affect dual-purpose sorghum cultivar adoption and production have also not been reported yet this is important in any breeding program because farmers are the users of the varieties, regardless of the views of the researchers (Röling *et al.*, 2004). The purpose of this study was to generate information from farmers to support the genetic improvement of dualpurpose sorghum for grain and forage by identifying awareness levels, willingness of farmers to use dual purpose sorghum cultivars for food and feed silage and to identify the traits of a "model" dual-purpose cultivar that was desired by farmers in Zambia.

Methodology

Surveys were conducted in the sorghum growing areas of Siavonga, Chisamba, Mazabuka and Masaiti districts in Zambia between February and March 2013. The area represents low (less than 700 mm) to medium (800-1000 mm) rainfall and drought-prone environments in agro ecological zones I and II, respectively. The farmer group Participatory Rural Appraisal (PRA) approach was used in this study. Purposive selection of districts was done in collaboration with the sorghum-breeding programme at the Zambia Agricultural Research Institute (ZARI) while that of small-scale farmers was done in liaison with district extension officers. Fifty five (91.7%) Small Scale Farmers (SSF) (22 in Lusitu, 15 in Chisamba and 18 in Masaiti districts) and five

(8.3%) Large Scale farmers (LSF) (known to make silage from maize and/or sorghum to feed their dairy and beef cattle) were selected as respondents in Mazabuka and Chisamba. Sixty structured questionnaires were administered and data was collected on the farmers' socio-economic status, awareness and perceptions on use and potential of dual-purpose sorghum cultivars for food and feed, types of livestock kept and cropped area, type of feed fed to livestock, benefits and constraints of sorghum production, and the preferred traits for dual-purpose sorghum. Data analysis was based on descriptive and inferential analyses using SPSS 16.0 computer package (SPSS Inc., 2007).

Results and Discussion

Social economic characteristics of interviewed farmers

Factor levels of some of the socio-economic characteristics of farmers are presented in Table 1. More men (70.9%) than women among the SSFs were involved in farming activities due to their position as heads of households and higher access to farmland owing to previous land ownership systems, which discriminated against women. According to Opio (2003), in most parts of Africa, women have traditionally been responsible for producing food for the family on land to which they gain access upon marriage but do not necessarily control. This study observed that all the LSFs were men. Opio (2003) also observed that it was usually men who were responsible for large-scale cash cropping, especially when it was highly mechanized. The high percentage of the farmers that were married (93%) may be as a result of trying to raise families that would supply labour on the farm (Olweny et al., 2013). Approximately 29% of the SSF respondents had secondary level education but all the LSFs had attained some form of tertiary level training. The World Bank (2009) indicates that involving young women and men in training opportunities is a successful strategy in ensuring food security and sustainable livelihoods for households.

Table 1. Socio-economic characteristics of the farmers interviewed in the survey

Variables	factors	Counts (%)	
		SSF	LSF
Gender	Female	16 (29)	0
	Male	39 (71)	5 (100)
Age	≥ 35	6 (15) M / 2 (12.5) F	0
	36 to 49	18 (46) M / 12 (75) F	3 (60) M
	50 ≤	15(39) M / 2 F	2 (40) M
Marital status	Married	39 (100) M / 12 (75) F	5 (100) M
	Single	0 M / 2 (12.5) F	0
	Widowed	0 M / 2 (12.5) F	0
	Divorced	0 M / 0 F	0
Education	Primary	28 (72) M / 11 (69) F	0
	Secondary	11 (28) M / 5 (31) F	0
	Tertiary	0 M / 0 F	5 (100) M
Occupation	Farming only	39 (100) M / 16 (100) F	0
	Farming & trading	18 (46) M / 9 (56) F	0

SSF-Small scale farmers
LSF- Large scale farmers

Other than practicing agriculture alone as a source of livelihood, some of the SSFs were also involved in the business of trading (Table 1). Conroy and Sutherland (2004) reported that economic activities among small-scale farmers included many farming enterprises ranging from crop production to animal husbandry designed to minimize or spread the risk of crop failure due to drought and other constraints to production.

Livestock ownership, cropped area and feeding regimes followed by farmers

The results of interrelationships (integrated crop-livestock) as presented in table 2 showed significant differences (P < 0.05 and P < 0.001)) in means of area under crop and number of livestock kept by SSFs and LSFs. All the SSFs indicated that their prominent feed for livestock was pasture and crop residues (100%) followed by hay at 51% (Fig.1). According to Sibanda et al. (2011) and Mativavarira et al. (2011), the reliance of small-scale farmers on crop residues for animal feed is a serious constraint, which prevents them from adequately feeding their animals throughout the year. The informal group discussions revealed that SSFs ran out of feed for the animals three months after harvest, which was consistent with the findings of Mapiye et al. (2006). All the LSFs indicated that silage, hay and green chop were important feed sources (Fig. 1).

Table 2. Means of cropped area and number of livestock kept by farmers in the survey

Variables	Mean		T test$_{0.05}$
	SSF	LSF	2 tailed
Area under maize (ha)	2.1 ± 0.512	11 ± 2.64	0.002
Area under sorghum (ha)	0.8 ± 0.403	7.8 ± 1.92	0.001
Area under Millet (ha)	0.25 ± 0.00	4.5 ± 2.12	0.016
Area under legumes (ha)	0.48 ± 0.318	7.25 ± 1.26	0.002
No. of large ruminants	12.39 ± 7.93	1154 ± 599	0.005
No. of small ruminants	17.32 ± 9.99	1522 ± 1620	0.000
No. of non ruminants	25 ± 18	2700 ± 1131	0.000

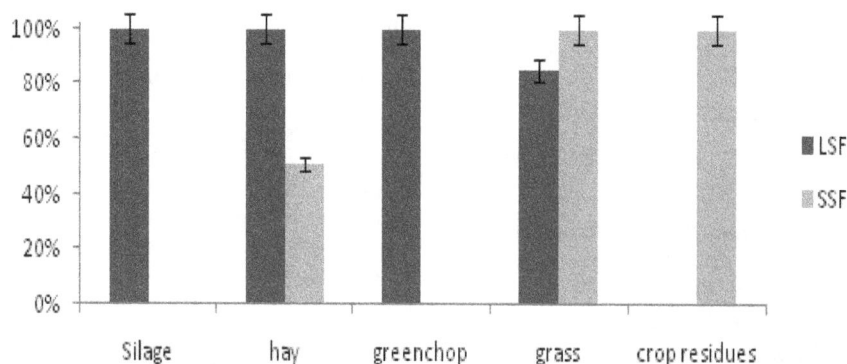

Fig.1. Feeding regimes followed by interviewed farmers' who keep large ruminants

Respondent's knowledge, attitude and practice on the use of sorghum to produce silage

The results in Table 3 showed the respondent's knowledge, attitude and perceptions on the use of sorghum as feed silage which suggested that a significant number of SSFs (54%) were not aware that sorghum could be used for production of silage or that such varieties of sorghum existed indicating that the technology requires a sustained promotion through demonstrations and trainings in the area. The SSFs' willingness to grow sorghum solely as a silage crop varied considerably with 76.4% saying they could do it while 12.7% were unsure but 10.9% of the farmers were sure that they could not (Table 3). The results indicated that farmers were willing to change their practices when exposed to appropriate technologies that met their needs.

Moreover, the literacy levels (Table 1) of people in the area gave a strong combination characteristic that when fully utilized would lead to high awareness and hence high adoption and productivity of technologies. Marsalis (2011) argued that improvements in varieties and a better understanding of proper management could lead to a greater acceptance and willingness to grow sorghum as an alternative silage crop; He, however, identified lack of water and desperation as likely the main drivers behind any major cropping changes. The high percentage (100%) of awareness among the LSFs was due to high access to information. The World Bank (2009) reported that the overwhelming majority of SSFs were not clients for private extension services but relied on public extension services and farmer-to-farmer information exchange.

Table 3. Awareness and willingness of farmers to use sorghum as a silage feed crop

Variables	Farmer Category	Responses (%)		
		Yes	No	Maybe
Awareness on use	SSF	25.5	74.5	0
	LSF	100	0	0
Awareness on existence of varieties	SSF	47.3	52.7	0
	LSF	100	0	0
Willingness to grow silage crop	SSF	76.4	10.9	12.7
	LSF	100	0	0
capacity to produce silage	SSF	61.8	38.2	0
	LSF	100	0	0
Willingness to promote use of silage	SSF	100	0	0
	LSF	100	0	0
Willingness to be contracted to produce silage	SSF	100	0	0
	LSF	100	0	0

Benefits and constraints of cultivating sorghum

The results of farmers perception on the benefits and challenges associated with use of sorghum as a food and feed crop are presented in Fig. 2 and 3 respectively. The farmers in this study indicated that drought tolerance was a major advantage of sorghum in comparison to other cereals such as maize (Fig.2). Marsalis *et al.* (2010) reported that the drought and heat tolerance of forage sorghum combined with the ability to resume growth after drought made the crop an ideal candidate for silage systems in dry climates. Reddy *et al.* (2011) also indicated that drought tolerance made sorghum especially important in dry regions and that it was among the climate resilient crops that could better adapt to climate change conditions. The farmers (50% SSFs and 20% LSFs) also reported that sorghum was a high energy feed crop owing to its nutritious stover and that livestock fed on sorghum had more energy than when fed on other cereal residues. In contrast, Marsalis *et al.* (2010) observed that the perception that all sorghums are low in nutritive value and that they are more difficult to manage than maize were the main arguments against sorghum given by producers and feeders. Results presented in figure 3 indicated that major constraints faced by both SSFs and LSFs in sorghum production were associated with low yield, limited availability of improved sorghum varieties, poor access to improved seed, inconsistent grain market for the crop and pests and diseases. Muui *et al.* (2013) also reported similar results including lack of inputs. The poor grain yields in sorghum were partly due to the SSFs consistent use of unimproved seed and cultivation on small parcels of land. Ochieng *et al.* (2011) reported low sorghum grain yields ranging from 0.5 to 2.5 t ha^{-1} for 92% of the SSFs interviewed compared to the research potential yield of $\geq$ 4 t ha^{-1}. Most SSFs in sub-Saharan Africa who plant unimproved varieties (landraces) used on-farm produced and saved seed whose quality was usually poor (Ashiono *et al.*, 2005). The challenge of inconsistent sorghum markets in Zambia was raised by 79% of the SSFs and 60% of the LSFs. This was one reason why all the LSF who produced sorghum in this study used it as animal feed while the SSFs used it both as food and feed. A study be Ochieng *et al.*, (2011) also reported that all the farmers interviewed acknowledged the adaptability of the crop in the region but its production was constrained by lack of its marketability.

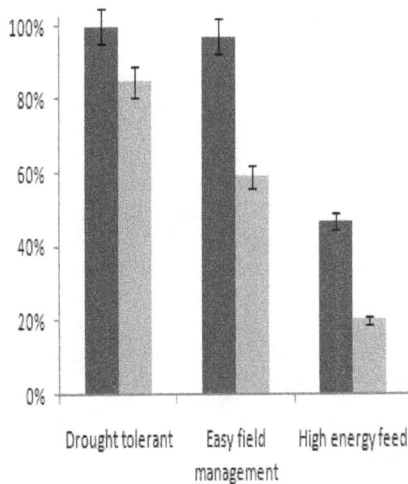

Fig. 2. Views of farmers on the benefits of sorghum production

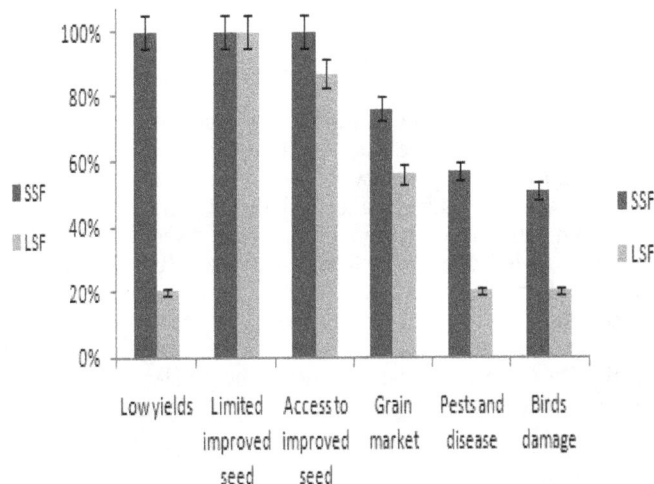

Fig.3. Views of farmers on the constraints of sorghum production

Farmer preferred quality attributes for dual purpose sorghums

The results of farmer perceptions on desired food and feed quality traits for dual-purpose sorghum cultivars are presented in Fig. 4. Both LSFs and SSFs regarded grain yield and biomass as top traits of importance in dual-purpose sorghum cultivars for grain and forage. Similar results were observed in a study done by Mativavarira *et al.* (2011) in Zimbabwe where results of farmer preference rankings pointed to grain yield being more important to 75% of the farmers' needs than

biomass production, although for a sub-set of farmers (25% of total) this preference was the reverse. Other observed traits of interest were the stay green trait which is essentially a trait associated with post rainy drought tolerance in sorghum. Delaying the onset of leaf senescence and reducing its rate were two elements of the stay-green trait, which offered an effective strategy for increasing grain production, fodder quality and crop residues particularly under water-limited conditions (Reddy *et al.*, 2007).

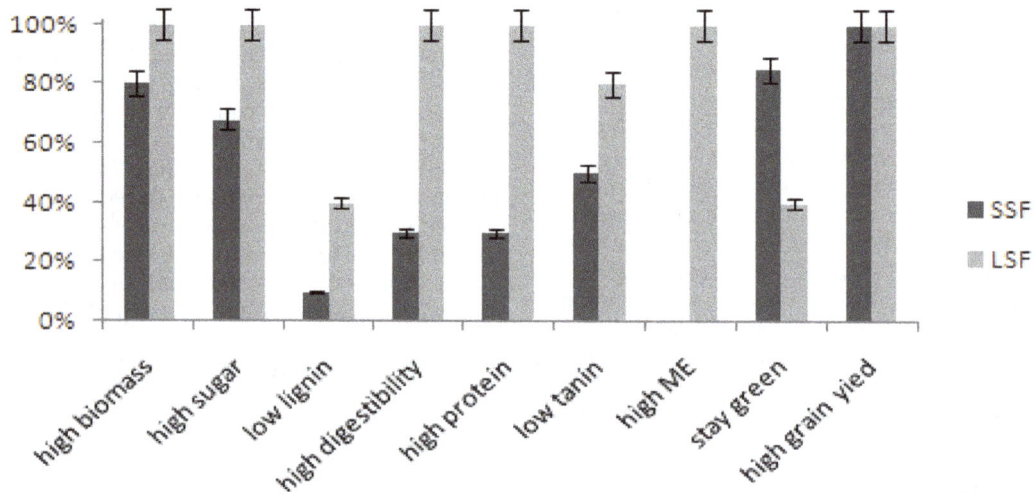

Fig. 1. Farmer perceptions of desired quality traits for dual purpose sorghum cultivars

High stem sugar was also indicated as a desirable trait in dual purpose sorghum as it makes feed more palatable as was observed by Kumar *et al.* (2010) where the daily intake and digestibility were high in large ruminants when sweet sorghum was fed directly as forage. All the LSFs and 35% of the SSFs indicated high digestibility of sorghum as desirable because livestock performance was improved by increasing digestibility of feeds as was observed by Casler and Vogel (1999).

It was observed in this study that the attributes that farmers chose were closely related to utilization. LSFs preferred to use sorghum as a bioenergy feed crop and they were able to meet the huge investment costs associated with silage production given the large number of animals that they owned. However, SSFs clearly used sorghum first as food then feed hence most of them indicated that they wanted high grain yield. Their interest in the stay green trait as well as high biomass showed that they also used sorghum stover and hay as animal feed. However, even though the SSF relied on stover and hay, they were still not able to feed their animals adequately throughout the year because the varieties that they planted did not have sufficient

biomass. Hence, the direction for technology delivery could be to enhance production, reduce postharvest losses, store feed in forms that it can stay for long periods while maintaining nutritional benefits, breed cultivars for such purposes as high grain yield, high biomass, sweet stems and stay green. The potential to use such cultivars was high as evidenced by the type of quality attributes desired justifying the need for the breeding programmes to address these demands.

Acknowledgemens

Gratitude is extended to the German Academic Exchange Programme (DAAD) and Regional Universities Forum (RUF) for financing this study and to all the survey respondents.

References

Ashiono, G.B., Gatuiku, S., Mwangi, P. and Akuja, T.E. 2005. Effect of nitrogen and phosphorus application on growth and yield of dual purpose sorghum in the dry highlands of Kenya. *Asian J. Plant Sci.* 4: 368-371.

Casler, M.D., and Vogel, K.P. 1999. Accomplishments and impact from breeding for increased forage nutritional value. *Crop Sci.* 39: 12–20.

Conroy, C. and Sutherland, A. 2004. Participatory technology development with resource-poor farmers: maximizing impact through the use of recommendation domains. Agricultural Research & Extension Network (AgREN), Network Paper No. 133.

Kumar, A.A., Reddy, B.V.S., Blümmel, M., Anandan, S., Reddy, Y.R., Reddy, C.R., Rao, P.S. and Reddy, P.S. 2010. On-farm evaluation of elite sweet sorghum genotypes for grain and stover yields and fodder quality. *Animal Nutr. & Feed Tech.* 10: 69-78.

Mapiye, C., Mwale, M., Chikumba, N., Poshiwa, X., Mupangwa, J.F. and Mugabe, P.H. 2006. A review of improved forage grasses in Zimbabwe. *Trop. and Subtrop. Agroeco.* 6: 125–131.

Marsalis, M.A. 2011. Advantages of forage sorghum for silage in limited input systems. *In*: Proceedings, Western Alfalfa & Forage Conference, Las Vegas. UC Cooperative Extension, Plant Sciences Department, University of California, Davis, CA 95616.

Marsalis, M.A., Angadi, S.V. and Contreras-Govea, F.E. 2010. Dry matter yield and nutritive value of corn, forage sorghum, and BMR forage sorghum at different plant populations and nitrogen rates. *Field Crops Res.* 116: 52-57.

Mativavarira, M., Dimes, J., Masikati, P., Van Rooyen, A., Mwenje, E., Sikosana, J.L.N. and Tui, S.H.K. 2011. Evaluation of water productivity, stover feed quality and farmers' preferences on sweet sorghum cultivar types in the semi-arid regions of Zimbabwe. *J. SAT Agril. Res.* 9: 1-9.

Muui, C.W., Muasya, R.M. and Kirubi, D.T. 2013. Baseline survey on factors affecting sorghum production and use in eastern Kenya. African *J.* Food Agric. & Nutr. Dev. 13 (1): 7339-7342.

Ochieng, L.A., Mathenge, P.W. and Muasya, R. 2011. A survey of on-farm seed production practices of sorghum (*Sorghum bicolor* L. Moench) in bomet district of Kenya. *African J. Food Agric. Nutr. Dev.* 11 (5): 5232-5253.

Olweny, C., Onga'la, J., Dida, M. and Okori, P. 2013. Farmers' perception on sweet sorghum (*Sorghum bicolo*r [L] Moench) and potential of its utilization in Kenya. *World J. Agril. Sci.* 1 (2): 65-75.

Opio, F. 2003. Gender mainstreaming in agriculture with special reference to Uganda: Challenges and prospects. *J. African Crop Sci. Soc.* 6: 699-703.

Rattunde, H.F.W., Zerbunu, E., Chandra, S. and Flower, D.J. 2001. Stover quality of dual-purpose sorghums: genetic and environmental sources of variation. *J. Field Crops Res.* 71: 1-8

Reddy, B.V.S., Kumar, A.A., Ramesh, S. and Reddy, P.S. 2011. Breeding sorghum for coping with climate change. pp. 326-339. *In*: S. S. Yadav, R. Redden, J. L. Hatfield and H. Lotze- Campen, Eds., Crop Adaptation to Climate Change, John Wiley & Sons Inc., Iowa.

Reddy, B.V.S., Ramaiah, B., Kumar, A.A. and Reddy, P.S. 2007. Evaluation of sorghum genotypes for the stay-green trait and grain yield. International Crops Research Institute for the Semi-Arid Tropics (ICRISAT), Patancheru 502 324, Andhra Pradesh, India.

Röling, N.G., Hounkonnou, D., Offei, S.K., Tossou, R. and Van Huis, A. 2004. Linking science and farmers' innovative capacity: diagnostic studies from Ghana and Benin. *Netherlands J. Agril. Sci.* 52 (3): 211-235.

Rooney, W.L., Blumenthal, J., Bean, B. and Mullet, J.E. 2007. Designing sorghum as a dedicated bioenergy feedstock. Biofuels, *Bioproducts and Biorefining*, 1: 147-157.

Sibanda, A., Homann-Kee Tui, S., Van Rooyen, A., Dimes, J., Nkomboni, D. and Sisito, G. 2011. Understanding user communities' perception of changes in rangeland use and productivity: Evidence from Nkayi district, Zimbabwe. *Exptl. Agric.* 47 (S1): 153-168.

SPSS Inc. 2007. SPSS for Windows, Version 16.0. Chicago, SPSS Inc.

Thornton, P.K., Kristjanson, P.M. and Thorne P.J. 2003. Measuring the potential impacts of improved food-feed crops: methods for ex ante assessment. *Field Crops Res.* 84: 199–212.

World Bank. 2009. Gender in agriculture, source book. Module 1-12. The International Bank for Reconstruction and Development. 1818 H Street, NW. Washington, DC 20433.

29

SCREENING OF TOMATO VARIETIES FOR FRUIT TREE BASED AGROFORESTRY SYSTEM

J. Hossain[1], T. Ahmed[2], M.Z. Hasnat[3]* and D. Karim[4]

Abstract

An experiment was conducted with four tomato varieties under a six year old orchard was accomplished at the Bangabandhu Sheikh Mujibur Rahman Agricultural University (BSMRAU) research farm during October 2011 to April 2012. The experiment was laid out in a Randomized Complete Block Design with three replications. Four tomato varieties (BARI Tomato 2, BARI Tomato 8, BARI Tomato 14 and BARI Tomato 15) were grown under guava, mango, olive and control. Results showed that light availability in control plot (999.75 μ mol $m^{-2}s^{-1}$) was remarkably higher over fruit tree based agroforestry systems and it was 58.8, 43.9 and 31.5% of the control for guava, mango and olive based systems, respectively. The shortest tomato plant was observed in olive based system (54.91 cm), while the tallest plant was observed in mango based system (60.09 cm). The highest SPAD value and number of primary branches per plant was recorded in control plot. Fruit length, fruit girth was found lowest in olive based system. The highest yield (34.06 t ha^{-1}) was recorded in control plot while the lowest yield (10.26 t ha^{-1}) was recorded in olive based system. The economic performance of fruit tree based tomato production system showed that both the net return and BCR of mango and guava based system was higher over control and olive based system. The contents of organic carbon, nitrogen, available phosphorus, potassium and sulfur of before experimentation soil were slightly higher in fruit tree based agroforestry systems than the control. After experimentation, nutrient elements in soil were found increased slightly than initial soils. Fruit tree based agroforestry systems could be ranked based on the economic performance as mango> guava> control> olive based system with BARI Tomato 15, BARI Tomato 2, BARI Tomato 14 and BARI Tomato 8, respectively.

Keywords: Fruit Tree Based Agroforestry System, tomato plant growth and yield.

[1]Lecturer, Dept. of Agroforestry and Environmental Science, Exim Bank Agricultural University, Chapainawabganj, Bangladesh
[2]Associate Professor, Dept. of Agroforestry and Environment, BSMRAU, Gazipur, Bangladesh
[3]Information Officer (Plant Protection), Agriculture Information Service, Ministry of Agriculture, Bangladesh
[4]Senior Scientific Officer, Soil Resource Development Institute, Ministry of Agriculture, Bangladesh

*Corresponding author's email: zhasnat@yahoo.com (M.Z. Hasnat)

Introduction

Fruit tree based vegetable production system would be very good options for maximizing and diversifying as well as sustaining the production system with high vitamin source for the country. However, before giving any policy options on the selection of vegetable varieties for the fruit tree based agroforestry production system, adequate information on various aspects of the system at farm level is required. Information on this regards is very limited due to lack of adequate research on this aspect. Some sporadic research has been done on fruit tree based vegetable production system but information about performance of different vegetable varieties is limited. To identify the compatible tree-vegetable combination, particular understory species i.e. different vegetable varieties should be screened out in terms of their adaptability and yield under different tree canopies is needed. In Bangladesh, a large number of vegetable are grown of which most of them are grown in winter season. Among them Tomato (*Lycopersicon esculentum* L.) is very popular vegetable grown successfully throughout the Bangladesh. Tomato has good production potential in our climate. Miah (2001) observed that tomato (single variety) could be grown successfully without yield loss up to 25% shade level, but a lot of varieties were not systematically tested in agroforestry system or in natural shade condition to see their production ability. Very little scientific research work has been done in this field. To screen out suitable tomato variety, the best way to grow different tomato varieties under different tree species. It would be very useful information in selecting the best fruit tree-tomato combination in order to promote agroforestry at field level. The experiment was therefore undertaken to find out the light availability for tomato varieties under different fruit tree based agroforestry system, quantify the growth and yield of four tomato varieties under fruit tree based agroforestry system and determine soil fertility changes in different fruit tree based agroforestry system.

Materials and Methods

The experiment was conducted in a six years old orchard of the Department of Agroforestry and Environment, Bangabandhu Sheikh Mujibar Rahman Agricultural University (BSMRAU), Gazipur during the period from October 2011 to April 2012. The experiment was laid out in a Randomized Complete Block Design (factorial) with three replications.

The treatments were as follows:

Factor A: Fruit tree species (4) - Guava (*Psidium guajava*), Mango (*Mangifera indica*), Olive (*Olea europaea*) and Control (No tree).

Factor B: Tomato varieties (4) - BARI Tomato 2, BARI Tomato 8, BARI Tomato 14, BARI Tomato 15. The variety of guava was Bari Peyara 2, mango was BARI Aam 3 and olive was local variety. Twelve pits were prepared in each block and spacing for all trees were 4m × 4m.

Seeds were sown in the seedbed on 9 October 2011. Thirty five days old seedlings of tomato were transplanted at 60 cm × 50 cm spacing on 14 November 2011. The experimental plots were fertilized with 12 ton cow dung and 600 kg urea, 500 kg TSP, 300 kg MP per hectare. All cow dung, TSP and one third of Urea and MP were applied during the final land preparation and the rest of the Urea and MP were applied in two equal installments at 20 and 40 days after transplanting (Hussain and Miah, 2004). Various intercultural operations (weeding, rouging, bamboo sticking and pesticides application) were done in appropriate time. The harvesting started at 115 days and ended at 155 days after transplanting. During experiment SPAD (Soil Plant Analysis Development) values,

the plant height and number of branches, number of fruits per plant, fruit length (mm), fruit girth (mm) and fruit weight (g) and yield (t ha⁻¹) was measured for tomato plant. For fruit tree component data on tree height (m), canopy spreading were recorded from every tree species. Chlorophyll content of the leaf was measured from selected plant by SPAD 502 plus Chlorophyll meter. Light was measured by Sunflect ceptometer (LP-80 Accu PAR ceptometer) from each plot. From collected soil sample organic carbon (%), total nitrogen (%), available phosphorus (ppm), exchangeable potassium (meq 100 g⁻¹) and sulphur (meq 100 g⁻¹) estimation were done. After 12 months of the experimentation, soil samples were again collected and same properties were analyzed.

Data recorded for different parameters of plant and soil were processed by Excel and statistically analyzed by "CROPstat" and MSTAT software and means were compared by DMRT at 5% level of significance.

Result and Discussion

Light availability over crop canopy

The light availability over four tomato varieties in fruit tree based agroforestry system were collected in three sampling dates at 9:00 AM, 12:00 PM and 3:00 PM. Results showed that the light availability over the tomato plants grown in control plots were higher (999.75 μ mol m⁻²s⁻¹) than the fruit tree based agroforestry system. Among the tree species, light availability over tomato plants grown in guava trees (588.72 μ mol m⁻²s⁻¹) were higher than mango (438.97 μ mol m⁻²s⁻¹) and olive (308.29 μ mol m⁻²s⁻¹) (Table1). The light availability over the tomato plants grown in guava, mango and olive based agroforestry system were 58.88, 43.90 and 31.51% of the control respectively. However, among the four tomato varieties, light availability did not vary much when they were grown within a tree.

Table 1. Light availability (PAR) over the tomato varieties grown in fruit tree based agroforestry system

Tree species	Average light on tomato plant grown under different fruit trees (μ mol m⁻²s⁻¹)				Mean
	BARI Tomato 15	BARI Tomato 14	BARI Tomato 8	BARI Tomato 2	
Guava	568.81	553.17	651.17	581.73	588.72
Mango	450.12	460.32	420.66	424.78	438.97
Olive	316.83	331.58	314.78	296.99	308.29
Control	981	1038	1030	950	999.75
Mean	579.19	595.76	604.15	563.37	

Performance of tomato grown in association with different tree species

Plant height

In the study, the shortest plant was observed under olive tree where light availability was only 31% compare to control. Plant height of BARI Tomato 15 was significantly influenced when they were grown under different tree species. However, the tallest tomato plant was recorded under mango tree (58.27 cm) (Table 2) but it did not vary significantly with guava and control. However, significantly the shortest tomato plant (54.50 cm) (Table 2) was recorded under olive tree. Plant height of BARI Tomato 14, BARI Tomato 8 and BARI Tomato 2 showed similar trend of variation where the tallest plant was recorded under mango tree insignificantly followed by guava and control. On the other hand, the shortest plant was found under olive tree but it did not vary with control.

Table 2. Effect of different tree species on the plant height of tomato varieties

Tree species	Height of tomato plant (cm)				Mean (cm)
	BARI Tomato 15	BARI Tomato14	BARI Tomato 8	BARI Tomato 2	
Guava	57.63 a A	58.13 a A	57.80 a A	57.87 a A	57.86
Mango	58.27 a A	60.60 a A	60.37 a A	61.13 a A	60.09
Olive	54.50 b A	54.73 b A	54.90 b A	55.50 b A	54.91
Control	55.83 a A	56.50 ab A	55.86 ab A	56.46 ab A	56.17
Mean±SE	56.56± 0.86	57.49±1.25	57.23±1.21	57.74±1.23	

In a column, means followed by a common small letter and in a row, means followed by a common capital latter are not significantly different at the 5% level by DMRT.

SPAD value

In the present study, the highest SPAD value was recorded in tomato plants grown in control plots, while, the lowest value was recorded under olive tree. The SPAD value of BARI Tomato 15, BARI Tomato 14 and BARI Tomato 2 showed the highest value in control plots (Table 3) which did not differ significantly with the SPAD value recorded under guava and mango trees. The SPAD value recorded in tomato plants grown under olive tree were the lowest which did not vary significantly with the SPAD value recorded in tomato plants grown under guava and mango trees. However, the SPAD value of BARI Tomato 8 did not vary significantly when they were grown under different tree species. Among the varieties, the SPAD values did not vary in each tree species, except under olive tree. The SPAD value of BARI Tomato 8 was found highest which was identical with the SPAD value found in BARI Tomato 15. The SPAD values of BARI Tomato 14 (36.80) and BARI Tomato 2 (37.58) were found lowest which were also identical with BARI Tomato 15.

Table 3. Effect of different fruit tree species on the SPAD value of tomato varieties

Tree species	SPAD value of tomato plant				Mean
	BARI Tomato 15	BARI Tomato 14	BARI Tomato 8	BARI Tomato 2	
Guava	43.93 ab A	45.00 ab A	40.83 a A	41.39 ab A	42.79
Mango	40.75 ab A	42.47 ab A	44.17 a A	40.91 ab A	42.08
Olive	39.76 b AB	36.80 b B	43.23 a A	37.58 b B	39.34
Control	47.18 a A	48.35 a A	45.14 a A	43.00 a A	45.92
Mean±SE	42.90±1.68	43.16±2.44	43.34±0.92	40.72±1.14	

In a column, means followed by a common small letter and in a row, means followed by a common capital latter are not significantly different at the 5% level by DMRT.

Number of primary and secondary branch

In general, the highest number of branches per plant was recorded in control plot while the lowest value was found under olive tree. The branch number of BARI Tomato 15 and BARI Tomato 14 (Table 4) showed similar trend of variation where the highest number were recorded in control plots and the number of branches per plant recorded in other treatments were identical. In case of BARI Tomato 8 the number of branches per plant recorded in control, mango and guava trees were similar. However, significantly the lowest number of branches per plant was recorded under olive tree but it was identical to the number of branches per plant recorded under guava. On the other hand, number of branches per plant recorded in BARI Tomato 2 was more or less similar as observed in BARI Tomato 8 (Table 4) with little exception. Among the varieties in each tree species, the number of branches per plant did not vary, except BARI Tomato 8 in control. BARI Tomato 8 gave lower number of branches per plant compare to other varieties. The lower number of primary branches under shaded conditions might be due to higher auxin production in plant grown under shaded condition, which ultimately suppressed the growth of lateral branches (Miah *et al.*, 1994).

Table 4. Effect of different fruit tree species on the number of primary branches of tomato varieties

Tree species	Number of primary branches of tomato				Mean
	BARI Tomato 15	BARI Tomato 14	BARI Tomato 8	BARI Tomato 2	
Guava	2.00 b A	2.66 b A	2.33 ab A	2.66 b A	2.42
Mango	2.66 b A	3.00 b A	3.33 a A	3.66 ab A	3.17
Olive	1.33 b A	2.00 b A	2.00 b A	1.66 b A	1.75
Control	4.33	4.66	3.33	4.00	4.08
Mean±SE	2.58±0.64	3.08±0.57	2.75±0.34	3.00±0.53	

In a column, means followed by a common small letter and in a row, means followed by a common capital latter are not significantly different at the 5% level by DMRT.

Secondary branches per plant

The number of secondary branches per plant of tomato was also influenced by different tree species (Table 5). BARI Tomato 15, BARI Tomato 14 and BARI Tomato 2 showed significantly lowest number of branches per plant when plant grown under olive tree. While the other treatment gave identical number of branches per plant. However, the number of branches per plant did not vary in case of BARI Tomato 8. Among the varieties the number of branches per plant did not vary.

Table 5. Effect of different fruit tree species on the secondary branches of tomato varieties

Tree species	Number of secondary branches of tomato				Mean
	BARI Tomato 15	BARI Tomato 14	BARI Tomato 8	BARI Tomato 2	
Guava	3.00 a A	3.88 a A	4.00 a A	4.00 a A	3.58
Mango	3.98 a A	3.67 a A	4.00 a A	3.89 a A	3.88
Olive	2.00 b A	2.33 b A	2.67 a A	2.00 b A	2.25
Control	3.67	4.00	4.66	3.77	3.92
Mean±SE	3.17±0.44	3.58±0.39	3.92±0.42	3.42±0.47	

In a column, means followed by a common small letter and in a row, means followed by a common capital latter are not significantly different at the 5% level by DMRT.

Fruit length

The influence of different tree species on the fruit length of tomato varieties was similar and the lowest fruit length was observed in the tomato varieties grown under olive tree (Table 6). The fruit length of tomato grown under mango, guava and control produced identical fruit length and significantly higher over olive. However, fruit length of tomato varieties did not vary when they were grown under each tree species. Different experiment showed similar effect on fruit length. Miah (2001) observed the longest length of carrot (17.59 mm) and radish (16.25 mm) under 75% PAR.

Table 6. Effect of different tree species on the fruit length of different tomato varieties

Tree species	Length of tomato (mm)				Mean
	BARI Tomato 15	BARI Tomato 14	BARI Tomato 8	BARI Tomato 2	
Guava	50.67 a A	50.16 a A	48.83 a A	52.14 a A	50.45
Mango	59.96 a A	59.05 a A	56.07 a A	50.34 a A	56.35
Olive	36.19 b A	38.15 b A	39.46 b A	38.85 b A	38.17
Control	53.33	54.67	56.00	52.00	54.00
Mean±SE	50.04±5.01	50.51±4.5	50.09±3.93	48.33±3.19	

In a column, means followed by a common small letter and in a row, means followed by a common capital latter are not significantly different at the 5% level by DMRT.

Fruit girth

The fruit girth of BARI Tomato 15 and BARI Tomato 14 showed that the highest fruit girth was recorded in control, which was identical with the fruit girth recorded under mango tree (Table 7). On the other hand, fruit girth of tomato recorded under mango tree was also similar to the fruit girth recorded under guava but these values were significantly higher over olive. Fruit girth recorded in case of BARI Tomato 8 did not vary among the tree species, except under olive. Fruit girth of tomato grown under olive tree was significantly the lowest compare to other treatments including control. In case of BARI Tomato 2, the highest fruit girth was recorded in control plot, which did not vary with the fruit girth recorded under guava tree. Among the tomato varieties, fruit girth did not vary when they were grown under guava and olive but BARI Tomato 2 produced the lowest fruit girth when they were grown under mango tree and control.

Table 7. Effect of different fruit tree species on the girth of tomato varieties

Tree species	Girth of tomato (mm)				Mean
	BARI Tomato 15	BARI Tomato 14	BARI Tomato 8	BARI Tomato 2	
Guava	55.20 b A	51.59 b A	54.67 a A	49.37 ab A	52.71
Mango	59.16 ab A	56.86 ab A	53.16 a A	46.71 b B	53.97
Olive	31.02 c A	34.41 c A	36.85 b A	37.55 c A	34.96
Control	66.67	62.00	60.33	57.67	61.67
Mean±SE	53.01±7.71	51.22±5.99	51.25±5.04	47.82±4.14	

In a column, means followed by a common small letter and in a row, means followed by a common capital latter are not significantly different at the 5% level by DMRT.

Tomato yield

Fruit yield of tomato was influenced when the tomato varieties were grown under different tree species and different varieties responded differently as well. In general, the highest tomato yield was recorded (34.06 t ha⁻¹) (Table 8) in control plot while the lowest yield (10.20 t ha⁻¹) was recorded under olive tree. The yield of tomato grown under guava (23.47 t ha⁻¹) and mango (19.94 t ha⁻¹) (Table 8) were higher over olive but lower over control. The yield performance of BARI Tomato 15 grown under different tree species showed that the highest yield was observed in control plot (26.94 t ha⁻¹) which was identical with the yield obtained from guava (22.68 t ha⁻¹) and mango (23.85 t ha⁻¹) tree. However, yield obtained from olive tree was the lowest (8.28 t ha⁻¹). In case of BARI Tomato 14 and BARI Tomato 8 the highest and the lowest yield was recorded in control and olive tree. Tomato yield recorded under guava and mango tree were significantly lower than control but higher than olive. In case of BARI Tomato 2, though the highest yield was recorded in control plot (31.43 t ha⁻¹) (Table 8) but the value was similar to the tomato yield obtained under guava tree. Tomato yield obtained under olive tree was the lowest and yield obtained from mango tree was higher over olive but lower than control and guava. The yield of tomato varieties did not vary significantly when they were grown under guava, mango and olive tree. In control, BARI Tomato 14 gave the highest yield (44.60 t ha⁻¹) compare to the other varieties. The average yield of tomato grown under different tree species showed that the highest yield was recorded in control plot. The yield of tomato grown under olive tree was found to suffer severely and it was 69.87% lower than control. The yield of tomato grown under mango and guava were also suffered and these values were 41.45 and 31.09% lower than control, respectively.

Table 8. Effect of different fruit tree species on the fruit yield of different tomato varieties

Tree species	Tomato variety yield				Mean
	BARI Tomato 15	BARI Tomato 14	BARI Tomato 8	BARI Tomato 2	
Guava	22.68 a A	21.56 b A	21.86 b A	27.79 a A	23.47
Mango	23.85 a A	19.40 b A	19.04 b A	17.46 b A	19.94
Olive	8.28 b A	10.17 c A	12.02 c A	10.54 c A	10.26
Control	26.94 a B	44.60 a A	33.27 a B	31.43 a B	34.06
Mean±SE	20.44±4.15	23.93±7.32	21.55±4.42	21.81±4.78	

In a column, means followed by a common small letter and in a row, means followed by a common capital latter are not significantly different at the 5% level by DMRT

Relationship graph between light and yield of tomato in this experiment

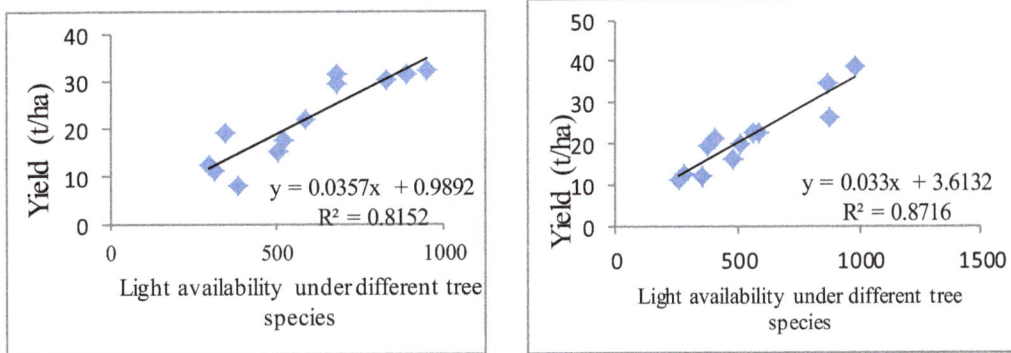

Fig. 1. Relationship between light availability under different tree species and yield of BARI Tomato 2 and BARI Tomato 8

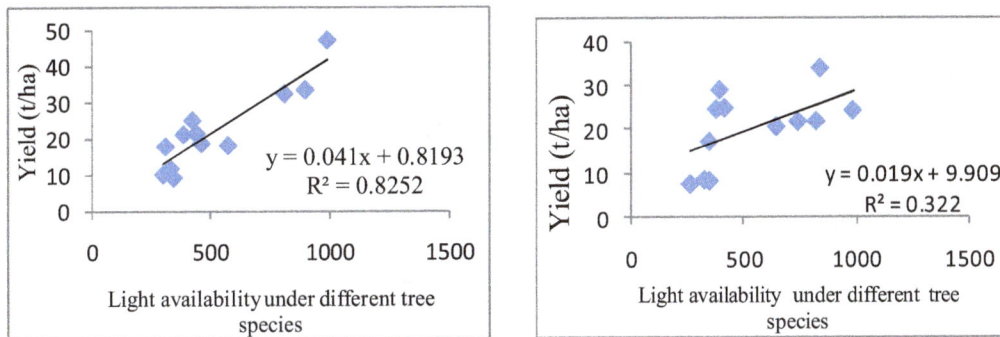

Fig. 3. Relationship between light availability under different tree species and yield of BARI Tomato 14 and BARI Tomato 15

Economic performance of agroforestry systems

The performance of tomato fruit tree based agroforestry systems in terms of economic performance was estimated and is presented in Tables 9-10. The overall economic performance of mango-tomato based system was found to outperform over other systems. The average net return of mango-tomato based system were (Tk. 515956.9) and BCR (4.9). Among the tested tomato varieties both the net return (TK. 555081.88) and BCR (5.3) of BARI Tomato 15 was found the highest compared to other varieties. The economic performance of guava-tomato based system was higher over control and olive based systems, but slightly lower than mango based system with the average net return (Tk. 442955.20) and BCR (4.3), respectively. Both the net return (Tk. 486130.19) and BCR (4.7) of BARI Tomato 2 was found highest in guava based system. The average net return control and olive based system were (Tk. 244019.50), (Tk. 62836.09) and BCR were (2.5), (0.61), respectively. In control and olive based systems BARI Tomato 14 and BARI Tomato 8 gave the highest net return (Tk. 349419.49), (Tk. 80511.09) and BCR (3.6), (0.79), respectively.

Table 9. Total cost and return of different agroforestry system and control (BSMRAU 2011-2012)

System	Tomato varieties (Tk)	Tomato fruit (Tk)	Return from tree (Tk)	Cost (Tk)	Total return (Tk)	Net return (Tk)	BCR
			System productivity				BCR
Guava	BARI Tomato 2	277900	310879	102649	588779	486130.19	4.73
	BARI Tomato 8	218600	310879	102649	529479	426830.19	4.15
	BARI Tomato 14	215600	310879	102649	526479	423830.19	4.12
	BARI Tomato 15	226800	310879	102649	537679	435030.19	4.23
Mango	BARI Tomato 2	174600	419940	103358	594540	491181.88	4.75
	BARI Tomato 8	190400	419940	103358	610340	506981.88	4.90
	BARI Tomato 14	194000	419940	103358	613940	510581.88	4.93
	BARI Tomato 15	238500	419940	103358	658440	555081.88	5.37
Olive	BARI Tomato 2	105400	61800	101489	167200	65711.09	0.64
	BARI Tomato 8	120200	61800	101489	182000	80511.09	0.79
	BARI Tomato 14	101700	61800	101489	163500	62011.09	0.61
	BARI Tomato 15	82800	61800	101489	144600	43111.09	0.42
Control	BARI Tomato 2	314300	-	96580.5	314300	217719.49	2.25
	BARI Tomato 8	332700	-	96580.5	332700	236119.49	2.44
	BARI Tomato 14	446000	-	96580.5	446000	349419.49	3.61
	BARI Tomato 15	269400	-	96580.5	269400	172819.49	1.78

Table 10. Total cost and return of four tomato varieties in agroforestry and non agroforestry system (BSMRAU 2011-2012)

Tree species	Item	Price (Tk) cost and income of four varieties (Average)	BCR
Guava	Cost	102648.91	4.3
	Gross return	545604.10	
	Net return	442955.20	
Mango	Cost	103358.12	4.9
	Gross return	619315.00	
	Net return	515956.90	
Olive	Cost	101488.91	0.61
	Gross return	164325.00	
	Net return	62836.09	
Control	Cost	96580.51	2.5
	Gross return	340600.00	
	Net return	244019.50	

Soil fertility changes

Soil organic carbon

The SOC content of the experimental field before experimentation varied between 0.54% to 0.71%. The highest SOC content was estimated from guava based system was 0.71%, followed by mango (0.66%) (Fig. 1) and olive (0.59%) based system, respectively. The lowest SOC content was estimated from control plot (0.54%). The SOC content of soil increased slightly after one season and it varied from 0.81% to 0.95%. The highest SOC content was estimated from guava (0.95%) based system followed by mango (0.91%) and olive (0.89%) (Fig. 1). However, the lowest value was estimated from control plot (0.81%). The changes in SOC content of the soil collected from guava, mango and olive based agroforestry systems were higher over control, but the changes were more or less similar among the tree species. Organic matter accumulation under trees was due to a better stability of litter from tree leaves (Bernhard, 1982).

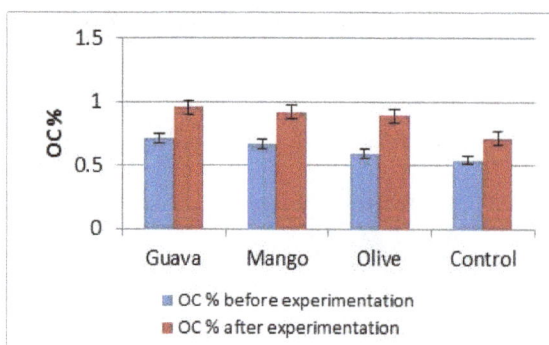

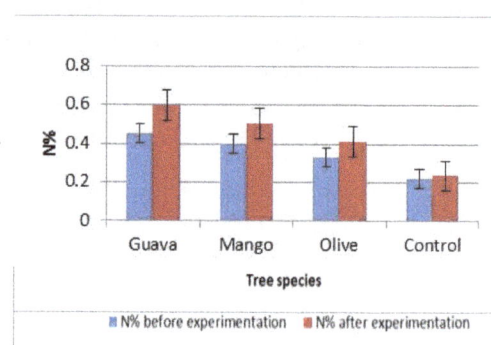

Fig. 1. Effect of fruit tree species on the organic carbon of soil Fig. 2. Effect of fruit tree species on the nitrogen content of soil

Soil nitrogen content

The Nitrogen content of the soil of different fruit tree based agroforestry systems before experimentation varied between 0.22 % to 0.45 %. The highest nitrogen content was estimated from the guava based agroforestry system (0.45%) (Fig. 2), followed by mango (0.40%) and olive (0.33%) based agroforestry system. However, the lowest nitrogen content was established from control (0.22%). The nitrogen content of soil of the same field increase slightly after experimentation and it varied from 0.23% to 0.60%. The highest N content was estimated from guava based agroforestry system (0.60%) followed by mango (0.50%), olive (0.41%) and control (0.23%).

Potassium

The K content of the soil of different fruit tree based agroforestry systems was higher over control plot, both before and after experimentation (Fig. 3). The highest total K content was recorded in mango based agroforestry system (0.22 meq 100g soil^{-1}) which was closely followed by olive (0.19 meq 100g soil^{-1}) and guava (0.19 meq 100g soil^{-1}) based agroforestry system. In control plot, K content was 0.18 meq 100g soil^{-1}, which was slightly lower than fruit based agroforestry systems. After experimentation, the total K content was found the highest in mango based system (0.27 meq 100g soil^{-1}), which was followed by olive (0.24 meq 100g soil^{-1}), guava (0.22 meq 100 g soil^{-1}) and control (0.20 meq 100 g soil^{-1}).

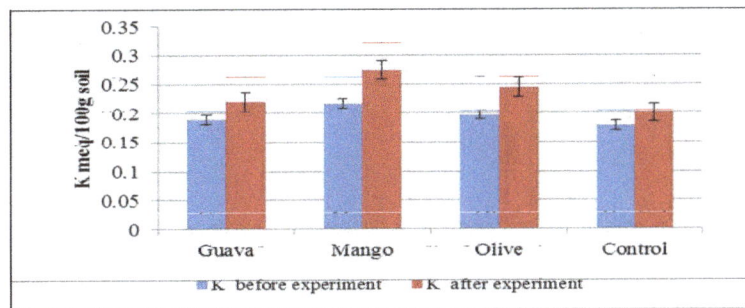

Fig. 3. Effect of fruit tree species on the potassium content of soil

Phosphorus

Available P content before experimentation of fruit tree based agroforestry plots were also slightly higher than control plot (11.87 ppm) (Fig. 4). Among the fruit tree based agroforestry systems, the available P content did not vary much where the highest P content was in guava based system (12.57 ppm) and the lowest P content was in olive based system (12.17 ppm). After harvesting of tomato, the available P content increased remarkably in fruit tree based agroforestry systems. However, it did not increase too much in control plot. The highest increase of available phosphorus was in olive based system (7.14 ppm) but it did not vary too much in guava (6.78 ppm) and mango (6.67 ppm) based systems.

Sulfur

The sulfur content (ppm) of the experimental soil before and after experimentation were very much distinct. Irrespective of fruit tree based agroforestry systems, the initial soil sulfur content was higher in the fruit tree based agroforestry systems compare to the control plot, (Fig. 5). Before experimentation, soil S content was the highest in mango based system (15.86 ppm) which was followed by guava (15.51 ppm) and olive (15.13 ppm) based systems, whereas in control plot it was (11.14 ppm). After harvesting of tomato, the highest soil S content was recorded in olive (28.16 ppm) followed by mango (26.89 ppm) and guava (26.79 ppm), whereas in control plot it was (17.12 ppm).

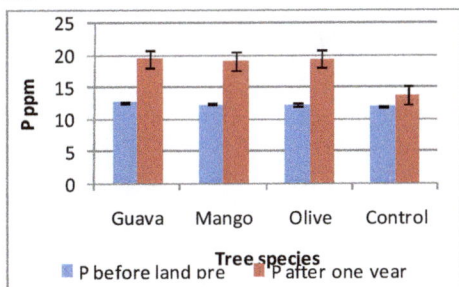

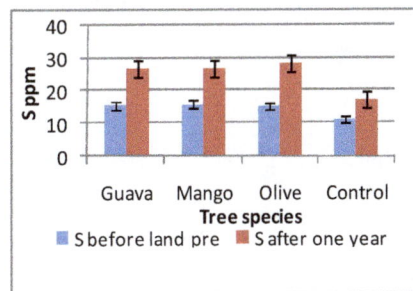

Fig. 4. Effect of fruit tree species on the P content of soil   Fig. 5. Effect of fruit tree species on the S content of soil

Conclusion

With the findings of the present study it may be concluded that different fruit tree based agroforestry systems affected the light availability for tomato cultivation. Light availability in olive, mango and guava based systems were only 31.5, 43.9 and 58.8% of control, respectively. The yield of tomato grown in olive, mango and guava based systems were 10.26 t ha[-1], 19.94 t ha[-1], 23.47 t ha[-1], when yield of tomato in open field was 34.06 t ha[-1]. The net return and BCR of mango and guava based system were higher over control and olive based system and BARI Tomato 15 and BARI Tomato 2 gave the highest net return (Tk.555081.88), (Tk.486130.19) and BCR (5.3), (4.7) in mango and guava based system respectively. The increment in organic carbon, nitrogen, available phosphorus, potassium and sulfur in soil indicated the fertility improvement of soil under tree based agroforestry system.

References

Bernhard, R.F. 1982. Biogeochemical cycle of nitrogen in a semi- arid savanna. *Oikos,* 3: 321-332.

Hussain, M.J. and Miah, M.G. 2004. Homestead agroforestry production and management manual. The small farmers and agroforestry development programme (SADP), GTZ and DAE, Rangpur, Bangladesh. pp. 437-463.

Miah, M.G., Garrity, D.P. and Argon, M.L. 1994. Light availability to the understory annual crops in an agroforestry system. pp. 99-107. *In:* Sinoquet, H. and Cruz P. (ed) Ecophysiology of tropical inter cropping. IRNA editions, Paris, France.

Miah, M.M. 2001. Performance of five winter vegetables under different light conditions for agroforestry systems. MS Thesis, BSMRAU, Gazipur, Bangladesh. p.82

30

IN VIVO EVALUATION OF WHEAT (*Triticum aestivum* L.) CULTIVARS FOR MOISTURE STRESS

S. Tamiru* and H. Ashagre

Abstract

In Ethiopia, wheat productivity is constrained by water stress especially at germination and seedling stage. The objective of this research was to determine the effect of moisture stress on bread wheat (*Triticum aestivum*) cultivars. Four wheat cultivars (Danda'a, Kubsa, Huluka and Local) were treated with six levels of water stress (0, 50, 100, 150, 200 and 250 g L^{-1}) using PEG-6000. The experiment was arranged factorially in completely randomized design with three replications. Data on germination and growth indices were collected and analyzed using one way ANOVA. The result revealed that highest germination percentage (99.7%), germination rate (14.6 plants day^{-1}) and seedling vigor index (9.67) were obtained from the local cultivar. Progressive increase in water stress had also significantly reduced germination indices. There was no significant difference among the cultivars in producing taller root and shoot length. The local cultivar had significantly produced highest root number (4.3 plant^{-1}), shoot fresh weight (0.214 g), seedling fresh (0.314 g) and dry weight (0.097 g). Even though, all the growth parameters of wheat cultivars showed a diminishing trend with increasing the level of PEG-6000 induced water stress, a highly significant decrease in the parameters were observed starting from 150 g L^{-1} PEG concentration. Higher shoot length inhibition was observed for Danda'a cultivar followed by Huluka. Huluka's root growth was more inhibited than Danda'a. According to the growth and germination indices results, the local cultivar is the drought tolerant wheat cultivar.

Keywords: Bread Wheat, Moisture Stress, Cultivars, PEG-6000

College of Agriculture and Veterinary Sciences, Department of Plant Sciences, Ambo University, Ambo, Ethiopia

*Corresponding author's email: solmill2000@gmail.com (S. Tamiru)

Introduction

Wheat (*Triticum aestivum* L.) is a staple food for more than 35% of the world population and is also the leading grain crops in most of the developing countries (Metwali *et al.*, 2011). In Ethiopia, the total area devoted for wheat production was estimated to be 1.68 million ha with an average yield of 1827 kg ha^{-1} (CSA, 2011). Wheat productivity in Ethiopia is influenced by biotic and abiotic factors. Moisture stress is among the major limiting abiotic factors hinder the productivity of *Triticum aestivum*. Abiotic stress, especially drought stress is a worldwide problem, seriously constraining global crop production (Pan *et al.*, 2002). It is one of the major causes of crop loss worldwide, which commonly reduces average yield for many crop plants by more than 50% (Bayoumi *et al.*, 2008). Plants may be affected by drought at any time of life, but certain stages such as germination and seedling growth are critical (Dhanda *et al.*, 2004). One of the most important abiotic factors limiting plant germination and early seedling stages is water stress brought about by drought and salinity (Almansouri *et al.*, 2001), which are widespread problems around the world (Soltani *et al.*, 2006)

Germination and seedling growth stages are the most critical periods in the life cycle of plants. Under water stress, low water potential is a determining factor inhibiting seed germination (Xu *et al.*, 2006). It is also important to determine the potential for seed germination in osmotic stress conditions, because, in this phase resistance against osmotic stress is a genetic quality and it is a good criterion for selection of stress resistant populations (Gharoobi *et al.*, 2012). According to Boureima *et al.* (2011), good germination capacity and seedling growth in water deficit conditions are drought tolerance indices, which allow better prediction on the crop establishment. In addition, field experiments related to water stress has been difficult to handle due to significant environmental or drought interactions with other abiotic stresses (Rauf *et al.*, 2008). An alternative approach is to induce water stress through polyethylene glycol 6000 (PEG 6000) solutions for screening of the germplasm (Khodarahmpour, 2011). Therefore, the objective of the current experiment was to determine moisture stress effect on bread wheat cultivars at germination and seedling growth stage.

Materials and Methods

In vivo experiment was conducted in November 2013 at the Department of Plant Sciences, Ambo University, to investigate the effect of moisture stress on germination and seedling growth of wheat. The experiment was arranged factorially in completely randomized design with three replications. Four wheat cultivars (Danda'a, Kubsa, Hulluka, and Local) were treated with six levels of water stress (0, 50, 100, 150, 200 and 250 g L^{-1} PEG) for the experiment; de-ionized water was used for the control treatment. Polyethylene Glycol Solution (PEG-6000) used as a source of moisture stress was prepared by dissolving the respective treatment amount at 25 °C with deionized water. Seeds were surface sterilized with 0.01 % HgCl$_2$ solution for 1 min, and rinsed with deionized water.

Twenty seeds were uniformly placed on Watman filter paper covered the Petri dish (9.5 cm diameter) using a forceps per treatment, and well soaked by adding 8 ml with the respective solutions. All the Petri dishes were covered with lids and kept at room temperature (22 ± 2°C). Germination continued for 10 days, and germinated seeds were counted daily. Germination was considered to have occurred when radicles attained a length of 2 mm. After 10 days, parameters such as per-cent germination and rate of germination were calculated according to ISTA (1999); and root and shoot lengths of seedling were measured using a scale. Root and shoot dry weights were recorded after oven drying for 72 h at 60°C. The seedling vigor index (SVI) was determined as Hosseein and Kasra (2011). Statistical analysis of the data was performed using one-way ANOVA using SAS statistical software (Version 9). Based on the ANOVA results, mean separations were performed by LSD test at 5% level.

Results and Discussion

Germination percentage

Among all test varieties, the local variety produced the highest germination percentage (99.7%). This indicates the superiority of the local variety over the improved varieties in terms of germination capacity under water stress condition (Table 1). Water stress level induced by 50 g L^{-1} PEG resulted the highest germination percentage (100%) and it showed a decreasing trend with increase in PEG concentration where 250 g L^{-1} PEG produced the lowest percentage (92.5%). Progressive increase in moisture stress had adversely affected germination capacity of wheat cultivars (Table 2). Similar results were found for linseed *(Linum usitatissimum* L) and cotton cultivars seeds as a consequence of increasing PEG concentration, because of diminished movement and accessibility of water

for seed imbibitions (Guo *et al.*, 2012; Meneses *et al.*, 2011).

Germination rate

The highest numbers of seeds germinated were recorded for local variety with 14.6 plants day^{-1}, which showed its high genetic capacity for fast velocity of germination. Variety Kubsa showed the lowest number of seeds germinated thus slow velocity of germination followed by Danda'a (Table 1). Kubsa and Danda'a were highly susceptible to moisture stress induced by PEG as compared to the local cultivar. Regarding different PEG induced water stress, the highest numbers of seeds were germinated for non-stress (control) treatment with 17.5 plants day^{-1}, and it followed diminishing trend when PEG concentration was increased from 50 to 250 g L^{-1} (Table 2). This delay in seed germination result obtained at more water stress levels were also found in cotton cultivars (Meneses *et al.*, 2011).

Seedlings vigor index

The analysis result revealed that the seedling vigor index of the local variety was 23.2, 14.6 and 6% higher than the vigor index of Huluka, Danda'a and Kubsa cultivars, respectively (Table 1). The vigor index of wheat cultivar seeds produced by inducing water stress at 50 and 100 g L^{-1} PEG were not significantly different from non-stress (0 g L^{-1} PEG) treatment. Seedling vigor index showed significant difference and decrease starting from seeds treated with water stress at 150 g L^{-1} to 250 g L^{-1} PEG where the lowest vigor index (2.793) was produced at the maximum water stress level (Table 2). Cokkizgin (2013) observed that seedling vigor index of pea (*Pisum sativum* L.) was decreased with increasing PEG concentration. Moraes *et al.* (2005) observed that the bean (*Phaseolus vulgaris* L.) seeds also presented a progressive decrease of vigor when PEG-6000 concentrations increase. Decreasing seed vigor index of wheat at high moisture levels was probably due to decreasing trend in shoot and root lengths (Table 2).

Shoot and root length

Both shoot and root lengths were affected by moisture stress induced by PEG but there was no significant difference among the wheat cultivars in producing taller shoot and root under moisture stress. Increasing PEG concentration has significantly reduced root and shoot lengths except at control and 50 g L^{-1} PEG where maximum shoot and root length were obtained. The reductions in shoot lengths were more drastic than root growth for each increase in PEG concentration. Drought stress suppressed shoot growth more than root growth and in certain cases root growth increased (Bibi *et al.*, 2010; Younis *et al.*, 2000).

Table 1.Mean comparison of wheat cultivars' response to PEG induced water stress

Varieties	Germination (%)	Germination rate (plants day[-1])	Seedling vigor index	Shoot length (cm)	Root length (cm)	Root number
Danda'a	97.2[a]	11.1(4.2)[b]	8.26[bc]	4.9(3.0)[a]	9.3(3.9)[a]	3.3(2.8)[c]
Kubsa	98.6[ab]	10.7(4.2)[b]	9.09[ab]	5.2(3.1)[a]	9.2(3.9)[a]	3.6(2.9)[b]
Hulluka	97.2[b]	12.6(4.5)[ab]	7.43[c]	5.7(3.1)[a]	8.9(3.9)[a]	3.6(2.9)[b]
Local	99.7[b]	14.6(4.8)a	9.67[a]	5.6(3.2)[a]	9.6(4.0)[a]	4.3(3.0)[a]

Means with the same letter are not significantly different; data in parenthesis are square root transformed

Table 2. Effect of PEG induced water stress on germination and growth indices of wheat

PEG (g L[-1])	Germination (%)	Germination rate (plants day[-1])	Seedling vigor index	Shoot length (cm)	Root length (cm)	Root number
0.0	99.6[a]	17.5 (5.1)[a]	11.389[a]	10.4(4.2)[a]	13.9(4.7)[a]	4.9(3.2)[a]
50	100[a]	16.7(5.0)[ab]	11.359[a]	9.5(4.1)[a]	13.8(4.7)[a]	4.8(3.2)[a]
100	99.6[a]	14.1(4.7)[b]	11.856[a]	6.8(3.6)[b]	11.4(4.3)[b]	4.6(3.1)[a]
150	98.3[a]	10.4(4.2)[c]	8.703[b]	3.3(2.8)[c]	8.8(3.9)[c]	3.5(2.9)[b]
200	99.2[a]	9.1(4.1)[c]	5.559[c]	1.6(2.2)[d]	5.8(3.4)[d]	2.8(2.7)[c]
250	92.5[b]	5.8(3.3)[d]	2.793[d]	0.6(1.7)[e]	1.8(2.3)[e]	1.6(2.3)[d]

Means with the same letter are not significantly different; data in parenthesis are square root transformed

Root number

Result in table 3 showed that the local cultivar had produced the highest root numbers (4.3) per plant, which indicated its high capacity to tolerate water stress conditions. In contrast, lower root numbers were produced by cultivar Danda'a. There was no significant difference in root number between Kubsa and Huluka cultivars. The reduction in root number became very low when PEG concentration was increased from 0 to 100 g L[-1]. Wheat cultivars root numbers were reduced by 28.4, 43 and 67.4 % when PEG concentration was increased from control (distilled water) to 150, 200 and 250 g L[-1], respectively. Root number become the parameter most affected by water stress i.e. under severe water stress, a preference should be given to wheat cultivars, which could produce more number of roots per plant.

Shoot fresh and dry weight

Shoot fresh and dry weight (Table 3) differed among the four wheat cultivars grown under water stress conditions. The local cultivar produced high shoot fresh weight, which was significantly different from the rest cultivars. Even though, the local cultivar produced high shoot fresh weight, its ability to convert into dry matter was lower than Huluka and Danda'a, which recorded low shoot fresh weight and relatively high dry weight. Regarding the water stress levels, shoot fresh and dry weight followed a decreasing trend when PEG was increased from 0 to 250 g L[-1]. Shoot fresh weight decreased significantly at 50 g L[-1] and by half when PEG concentrations were increased from 150 to 200 and 200 to 250 g L[-1]. The reduction in dry weight followed the same trend as fresh weight from 150 to 250 g L[-1] PEG concentration. But, the reduction followed slight decrease from 0 to 100 g L[-1] PEG concentration.

Root fresh and dry weight

Root fresh and dry weights were less influenced by osmotic concentration of PEG, but both parameters decreased with increase in PEG concentration. Significant difference in root fresh weight was observed between control (0 g L[-1] PEG) and water stress greater than 150 g L[-1] PEG. Sharp decrease in root fresh weight was noted than root dry weight with increasing the severity of moisture stress to wheat cultivars, suggesting that root dry weight is less affected by increasing moisture stress than fresh weight.

Table 3. Mean comparison of wheat cultivars' growth response to PEG induced water stress

Varieties	Shoot fresh weight (g)	Shoot dry weight (g)	Root fresh weight (g)	Root dry weight (g)	Seedling fresh weight (g)	Seedling dry weight (g)
Danda'a	0.144(1.339)[b]	0.0246(1.149)[c]	0.113(1.319)[ab]	0.0587(1.232)[a]	0.257(1.470)[bc]	0.083(1.276)[bc]
Kubsa	0.163(1.370)[b]	0.0279 (1.159)[bc]	0.123(1.339)[a]	0.0635(1.246)[a]	0.287(1.507)[ab]	0.092(1.293)[ab]
Hulluka	0.145(1.342)[b]	0.0286(1.161)[b]	0.085(1.280)[c]	0.0466(1.209)[b]	0.230(1.446)[c]	0.075(1.265)[c]
Local	0.214(1.419)[a]	0.0393(1.187)[a]	0.099(1.307)[bc]	0.0574(1.235)[a]	0.314(1.526)[a]	0.097(1.303)[a]

Means with the same letter are not significantly different; data in parenthesis are square root transformed

Table 4. Effect of PEG induced water stress on growth indexes of wheat

PEG (g L^{-1})	Shoot fresh weight (g)	Shoot dry weight (g)	Root fresh weight (g)	Root dry weight (g)	Seedling fresh weight (g)	Seedling dry weight (g)
0.0	0.308 (1.55)[a]	0.047(1.22)[a]	0.15(1.38)[a]	0.067(1.26)[ab]	0.454 (1.671)a	0.114 (1.338)[a]
50	0.281(1.52)[a]	0.043(1.21)[ab]	0.14(1.37)[a]	0.070(1.26)[a]	0.414 (1.638)[ab]	0.114 (1.336)[a]
100	0.222(1.47)[b]	0.042(1.20)[b]	0.13(1.36)[a]	0.078 (1.28)[a]	0.361 (1.597)[b]	0.119 (1.344)[a]
150	0.125(1.34)[c]	0.029(1.17)[c]	0.10(1.32)[b]	0.059 (1.24)[b]	0.229 (1.467)[c]	0.088 (1.294)[b]
200	0.048(1.21)[d]	0.013(1.11)[d]	0.08(1.28)[c]	0.042 (1.20)[c]	0.129 (1.349)[d]	0.056 (1.233)[c]
250	0.014(1.12)[d]	0.006(1.08)[e]	0.03(1.16)[d]	0.023 (1.14)[d]	0.042 (1.201)[e]	0.029 (1.163)[d]

Means with the same letter are not significantly different; data in parenthesis are square root transformed

Seedling fresh and dry weight

High seedling fresh and dry weight of 0.314 and 0.097 g, respectively were obtained from local cultivar. Low seedling fresh weight was recorded by Huluka. Inducing progressive moderate water stress (100 and 150 g L^{-1}) caused a significant decrease in seedling fresh weight whereas drastic decreases in seedling fresh and dry weights were observed at progressive severe stress (200 and 250 g L^{-1}) level.

Shoot and root growth inhibition

Higher shoot length inhibition were observed for Danda'a cultivar followed by Huluka. On the other hand, Huluka's root growth was more inhibited than Danda'a. Moderately, more root and shoot growth (less inhibition) was recorded by Kubsa. Higher difference inhibition percentage between shoot and root was observed for Danda'a and Kubsa cultivar thus shoot growth was more inhibited than root growth. Huluka and local cultivar shoot and root were less affected by water stress (Fig. 1). Wheat shoot growth was more inhibited than root length under PEG induced water stress. Shoot growth inhibition was more at 150 g L^{-1} PEG than shoot at 200 g L^{-1} at PEG. At maximum water stress, the inhibition power of PEG for shoot was slightly higher than root length indicating that PEG induced water stress had inhibited root elongation at lesser rate than shoot growth. Similar results were also reported in pearl millet and chickpea by Kalefetogllu et al. (2009) and Govindaraj et al. (2010), respectively. Shoot and root growth inhibition increased with increase in water stress levels (Fig. 2).

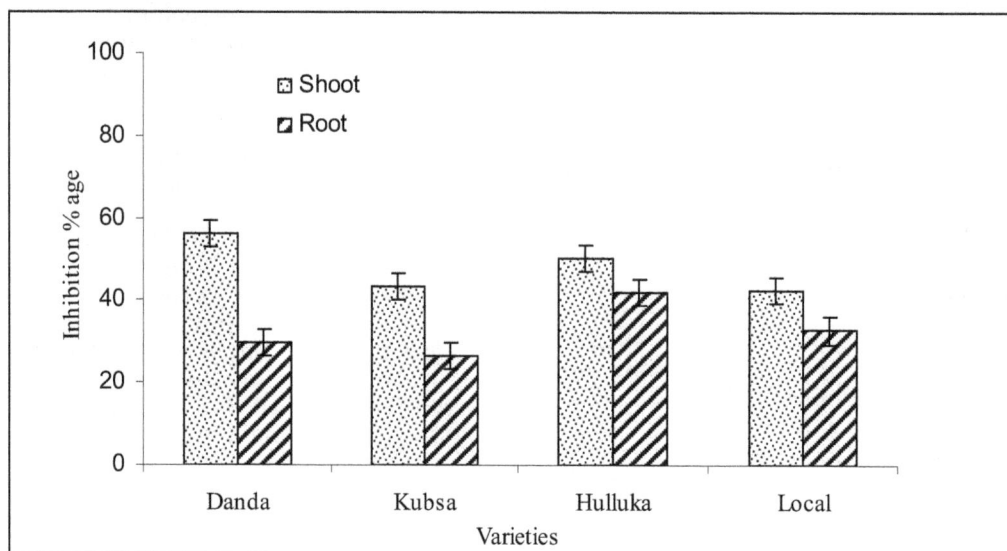

Fig. 1. Shoot and root length inhibition response of wheat cultivars under water stress

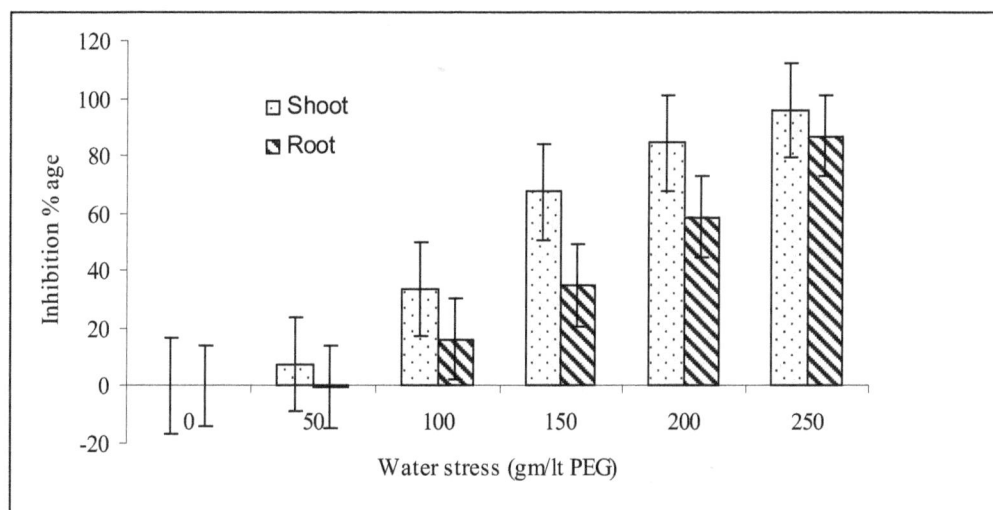

Fig. 2. Inhibition effects of different PEG concentration on wheat shoot and root length

Conclusion

All the wheat cultivars showed significant difference in germination percentage and rate. The local variety was superior over the released varieties in producing highest germination percentage and rapid rate of germination. In addition, the seedling vigor index of the local variety was superior to the rest cultivars. The germination indexes of wheat cultivars showed a diminishing trend with progressive increase in PEG concentration. The result showed that there was no significant difference among the wheat cultivars in producing taller shoot and root length. Increasing water stress levels has significantly reduced both root and shoot lengths. PEG induced water stress inhibited shoot growth more than root growth. The local cultivar was also able to produce more root numbers, shoot fresh and dry weight, seedling fresh and dry weight. The progressive moisture stress significantly reduced the wheat seedling growth parameters. Therefore, the current research suggests that the local cultivar is more tolerant to moisture stress and it could be used for breading tolerant cultivars. In addition, both germination indices and seedling growth parameters could be used for screening wheat cultivars for drought tolerance.

References

Almansouri, M., Kinet, J.M. and Lutts, S. 2001. Effect of salt and osmotic stresses on germination in durum wheat (*Triticum durum* Desf.). *Plant Soil*, 231: 243-254.

Bayoumi, T.Y., Manal, H. and Metwali, E.M. 2008. Application of physiological and biochemical indices as a screening technique for drought tolerance in wheat genotypes. *African J. Biotech.* 7 (14): 2341-2352.

Bibi, A., Sadaqat, H.A., Akram, H. M. and Mohammed, M.I. 2010. Physiological markers for screening sorghum (*Sorghum bicolor*) germplasm under water stress condition. *Int. J. Agric. Biol.* 12: 451–455.

Boureima, S., Eyletters, M., Diouf, M., Dio, T.A. and Damme, P.V. 2011. Sensitivity of seed germination and seedling radicle growth to drought stress in sesame (*Sesamum indicum* L). *Res. J. Envi. Sci.* 5 (6): 557-564.

Cokkizgin, A. 2013 Effects of hydro and osmo-priming on seed vigor of pea (*Pisum sativum* L). *Agric. Forestry & Fish.* 2 (6): 225-228.

CSA, 2011. Statistical abstract of 2009/10. Central Statistical Agency, Addis Ababa, Ethiopia. p. 106.

Dhanda, S.S., Sethi, G.S. and Behl, R.K. 2004. Indices of drought tolerance in wheat genotypes at early stages of plant growth. *J. Agron. Crop. Sci.* 190: 6-12.

Gharoobi, B., Ghorbani, M. and Ghasemi Nezhad, M. 2012. Effects of different levels of osmotic potential on germination percentage and germination rate of barley, corn and canola. *Iranian J. Plant Physiol.* 2 (2): 413-417.

Govindaraj, M., Shanmugasundaram, P., Sumathi, P. and Muthiah, A.R. 2010. Simple, rapid and cost effective screening method for drought resistant breeding in pearl millet. *Electronic J. Plant Breed.* 1 (4): 590- 599.

Guo, R., Hao, W. and Gong, D.Z. 2012. Effects of water stress on germination and growth of linseed seedlings (*Linum usitatissimum*), photosynthetic efficiency and accumulation of metabolites. *J. Agril. Sci.* 4 (10): 253-265.

Hosseein, A.F. and Kasra, M. 2011. Effect of hydropriming on seedling vigour in basil (*Ocimum basilicum* L.) under salinity conditions. *Adv. Env. Biol.* 5 (5): 828-833.

ISTA. 1999. International Rules for Seed Testing. Seed Science and Technology, 27,

International Seed Testing Association. p. 324.

Kalefetogllu, M.T., Turan, O. and Ekmekci, Y. 2009. Effect of water deficit induced by PEG and NaCl on Chickpea (*Cicer arieitinum* L.) cultivar and lines at early seedling stage. *G.U. Sci.* 22: 5-14.

Khodarahmpour, Z. 2011. Effect of drought stress induced by polyethylene glycol (PEG) on germination indices in corn (*Zea mays* L.) hybrids. *African J. Biotech.* 10 (79): 18222-18227

Meneses, C.H.S.G., Alcantara Bruno, R.D. Fernandes, D., Pereira, W.E. Morais Lima, L.H.G., Andrade Lima, M.M. and Vidal, M.S. 2011. Germination of cotton cultivar seeds under water stress induced by polyethyleneglycol-6000. *Sci. Agric. (Piracicaba, Braz.)* 68 (2): 131-138.

Metwali, M.R., Manal, H.E. and Tarek, Y.B. 2011. Agronomical traits and biochemical genetic markers associated with salt tolerance in wheat cultivars (*Triticum aestivum* L). *Australian J. Basic Appl. Sci.* 5 (5): 174-183.

Moraes, G.A.F., Menezes, N.L. and Pasqualli, L.L. 2005. Bean seed performance under different osmotic potentials. *Ciencia Rural,* 35: 776-780.

Pan, X.Y., Wang, Y.F., Wang, G.X., Cao, Q.D. and Wang, J. 2002. Relationship between growth redundancy and size inequality in spring wheat population mulched with clear plastic film. *Acta Phytoecol Sinica,* 26: 177-184.

Rauf, S., Sadaqat, H.A. and Khan, I.A. 2008. Effect of moisture regimes on combining ability variations of seedling traits in sunflower (*Helianthus annuus* L.). *Candian J. Plant Sci.* 88: 323-329.

Soltani, A., Gholipoor, M. and Zeinali, E. 2006. Seed reserve utilization and seedling growth of wheat as affected by drought and salinity. *Environ. Exp. Bot.* 55: 195-200.

Xu, S.G., Wang, J.H. and Bao, L.J. 2006. Effect of water stress on seed germination and seedling growth of wheat. *J. Anhui Agri. Sci.* 34: 5784-5787.

Younis, M.E., El-Shahaby, O.A., Abo-Hamed, S.A. and Ibrahim, A.H. 2000. Effects of water stress on growth, pigments and assimilation in three sorghum cultivars. *J. Agron. Crop Sci.* 185: 73–82.

31

CARBON DIOXIDE EMISSION FROM BRICKFIELDS AROUND BANGLADESH

M.A. Imran[1*], M.A. Baten[2], B.S. Nahar[3] and N. Morshed[4]

Abstract

The study was undertaken at six divisions of Bangladesh to investigate the CO_2 emission from brickfields. to explore the rate of carbon emission over the last 10 years, based on existing technology for brick production. The finding reveals that there were more than 45,000 Brick kilns in Bangladesh which together account for about 95% of operating kilns including Bull's Trench Kiln, Fixed Chimney Kiln, Zigzag Kiln and Hoffman Kiln. These kilns were the most carbon emitting source but it varies on fuel type, kiln type and also for location. It has been found that, maximum carbon emission area was Chittagong, which was 93.150 with percentage of last 10 years and 9.310 per cent per year. Whereas Sylhet was lower carbon emission area indicating percentage 17.172 of last 10 years and 4.218 percent per year. It has been found that total annual amount of CO_2 emission for 4 types brick kilns from Dhaka, Chittagong, Rajshahi, Khulana, Sylhet and Barisal were 8.862 Mt yr^{-1}, 10.048 Mt yr^{-1}, 12.783 Mt yr^{-1}, 15.250 Mt yr^{-1}, in the year of 2002, 2005, 2007 and 2010 respectively. In Mymensingh district, the maximum CO_2 emission and coal consumption was obtained in Chamak brick field, which was 1882 tons and 950 tons, respectively and minimum was obtained in Zhalak brick field, which was 1039.5 tons and 525.0 tons, respectively during the year of 2013. The percentage in last 10 years of CO_2 emission was 72.784 and per cent per year 7.970, which is very alarming for us. The estimates obtained from surveys and on-site investigations indicate that these kilns consume an average of 240 tons of coal to produce 1 million bricks. This type of coal has a measured calorific value of 6,400 KJ, heating value of coal is 20.93 GJ t^{-1} and it produces 94.61 TJ t^{-1} and 56.1 TJ t^{-1} CO_2 from coal and natural gas, respectively.

Keywords: CO_2 Emission, Carbon, Brick Kiln, Consumption

[1, 2,4]Department of Environmental Science, Bangladesh Agricultural University, Mymensingh, Bangladesh
[3]Principal Scientific Officer, Bangladesh Agricultural University, Mymensingh, Bangladesh

*Corresponding author's email: imran.pstu@gmail.com (M.A. Imran)

Introduction

Brick making is a significant sector in Bangladesh, contributing about one percent to the country's gross domestic product (GDP) (BUET, 2007) and generating employment for about one million people. Due to the unavailability of stone aggregate, brick is the main building material for the country's construction industry, which grew an average of about 5.6 per cent per year (Arifur, 2006).

Despite the importance of brick making, the vast majority of kilns use outdated, energy-intensive technologies that are highly polluting the environment. In the North Dhaka cluster, brick kilns are the city's main source of fine particulate pollution, accounting for nearly 40 per cent of total emissions (Biswas et al., 2009) during the 5-month operating period. It leads to harmful impacts on health, agricultural yields and global warming. The New technologies, such as the Vertical Shaft Brick Kiln (VSBK) and the Hybrid Hoffmann Kiln (HHK), are substantially cleaner than the Fixed Chimney Kiln (FCK) currently used. These improved technologies consume less energy and emit lower levels of pollutants and greenhouse gases (GHGs) (BUET, 2007; Heirli and Maithel, 2008). The existing brick kilns are the number one cause for fine particulate pollution in Bangladesh and its total greenhouse gas (GHG) emission is estimated to be 15.67 million tons of carbon dioxide (CO_2) equivalent (tCO$_2$e) per annum.

Global warming is an issue that calls for a global response. The rapid change in climate will be too great to allow many eco-systems to suitably adapt, since the change has direct impact on biodiversity, agriculture, forestry, dry land, water resources and human health. In addition, Bangladesh is one of the most climate change vulnerable countries. In Bangladesh, 92% of the 4,880 (Butler et al., 2004) brickfields are highly polluting Fixed Chimney Kilns (FCKs) because of a combination of low capital cost requirement and high investment return. However, these kinds of kiln use more coal/wooded fuel, which emits more carbon. Brick making significantly contributes to local air pollution including emission of various harmful gases such as Sulphur Oxides (SO_x), Nitrogen Oxides (NO_x), Carbon dioxide (CO_2) and Suspended Particulate Matter (SPM) and PM10 (Iqbal, 2007). About half of Bangladesh's bricks are baked with the use of coal, which is now considered the source of some 20 per cent of global greenhouse-gas emissions (Enters, 2000).

Realizing the importance of estimating the level of green house gas emission to combat against climate change and related ill consequences the present research was taken to fulfill the following objectives:

- To estimate the amount of green house gas mainly CO_2 emission from brick kilns in Bangladesh;
- To compare CO_2 emission among four types of brick kilns running in Bangladesh;
- To evaluate the future prediction of environmental condition and/or problems associated with the present level of GHG emission to help the policy makers to take necessary steps in time.

Equation (1) direct CO_2 = FC× CEF × f_o × 44/12
Equation (2) direct e_i = FC × CEF

Where,

FC = Total annual natural gas and/or coal consumption in energy conservation unit of brick kiln during a year (TJ)
CEF = Carbon emission factor of natural gas and coal (tC/TJ)
f_o = Carbon fraction of natural gas and coal that has been oxidized during combustion process
  44/12= Mass conversion factor of mass carbon to mass CO_2 generated during combustion processes
e_i = emissions level of non-CO_2 GHGs and other gas pollutant component (metric tons)

[BCAS, 2011]

Materials and Methods

To determine the Carbon emission from brick kiln, six divisions has been selected all over Bangladesh (24° 00' N and 90° 00' E) on the basis of BTK, FCK, Zigzag, Hoffman technology for brick production from January 2013 to June, 2013. Currently, there is no recommended estimation method to estimate CO_2, non-CO_2 GHGs and other gas pollutant emission from brick kilns. In this research study, CO_2, non-CO_2 GHGs and other gas pollutant emissions have been estimated based on natural gas and coal consumptions. CO_2, non-CO_2 GHGs and other gas pollutant emissions of brick kilns are divided into two categories, namely direct and indirect emissions. In this research study, direct gas emissions have been estimated independently. Direct an emission, which is due to natural gas and coal combustion from energy conservation units, were calculate based on Intergovernmental Panel on Climate Change (IPCC) guideline by using natural gas and other fossil fuel consumption. The CO_2, non-CO_2 GHGs and other gas pollutant emissions from brick kiln was calculated by the following equations: 44/12

Firstly, the quantities of natural gas and coal consumption are converted into energy units as tera joules, (TJ) using appropriate conversion factor, and then transformed into carbon emissions based on carbon emission factor (CEF). Though the IPCC has established CEF values, which can be used for general cases, data regarding fuel combustion in particular country has not been determined. As an approximation IPCC provides CEF value of natural gas, diesel, and coal that are 15.3 tC/TJ, 20.2 tC/TJ, and 26.4 tC/TJ (the average value of CEF for anthracite, coking coal, other bituminous coal, sub-bituminous coal and lignite). The fraction-oxidized value is used to account the carbon compound of natural gas and other fossil fuel that are not oxidized during combustion process. As an approximation, fraction-oxidized value of natural gas, diesel, and coal that are 0.995, 0.99, and 0.98, respectively (IPCC, 2000).

Results and Discussion

The experiments were conducted to study the rate of carbon emission from brick kilns around Bangladesh. Results of the experiments are presented and discussed as follows. The present situation of brick kilns in Bangladesh is as follows.

Table 1. A comparative study of the four kilns being used in Bangladesh

Parameter	Unit	Bull's Trench Kiln	Fixed Chimney Kiln	Zigzag Kiln	Hoffman Kiln
1.Initial Investment	Taka (Tk.) US\$=Tk. 77	2,500,000	4.000,000	4,000.000	32.000.000
2. Working Capital	Tk.	1.000,000	900,000	900.000	7.500,000
3. Land	acres	2.5	2.5	2.5	Min 10 year round
4. Raw Material	Clay ft3	100,000	95,000	95,000	425,000
	Labor	200 (5% skilled, 10% semiskilled, rest unskilled)	200 (15% skilled, 15% semi-skilled, rest unskilled)	200 (15% skilled. 15% semi-skilled, rest unskilled)	400 (25% skilled, 45% semi-skilled, rest unskilled)
	Electricity	Not essential	Not essential	Necessary in small scale	Necessary
	Fuel	Coal	Coal	Coal	Natural Gas
5. Fuel Consumption	Tones Per 100.000 Bricks	22-26	20-24	20-24	15000-17000 m3
6. Pollution		Severe pollution	Pollution	Pollution	Very little pollution
7. Production Period		Nov to mid-Apr	Nov to mid-April	Nov to mid-April	Round the year
8. Estimated Annual Production	Million bricks	2.0 to 2.5	2.0 to 2.5	2.0 to 2.5	7.5 to 9.0
9. Wastage	%	10 - 12	5 - 8	5 - 8	15 - 18
10. Quality of Bricks		Medium	Good	Good	Very good
11. Bricks Sale Price	Tk./1000 Bricks	3000-3500	3000-3500	3200-3800	3500-4000

Source: (DOE, 2010)

From the table 1 it is showed that, Initial Investment is varies on the basis of land type, location, kiln type, production period, etc. Eleven parameter has been identified on the table to understand the Study of the four kilns being used in Bangladesh.

Data table of CO_2 emission from four types of brick kilns

The major concerning greenhouse gas (CO_2) emission from four types of brick kilns, around Bangladesh, is presented in Table 2. The amount of CO_2 emission is calculated based on 100,000 bricks production.

Table 2. Amount of CO_2 emission from four types of brick kilns

Parameter	BTK	FCK	Zigzag	Hoffman
Fuel	coal	coal	coal	natural gas
Total fuel (tons)	28 t	20 t	18 t	16320 m3
Total energy consumption	586 GJ	4180 GJ	376 GJ	571 GJ
CO_2 emission(tons)	55.58 t	39.8 t	35.7 t	31.86 t

Basis : 100,000 bricks produced
Conversion factors: Heating value of coal = 20.93 GJ/t (Giga Joules/tons)
Calorific value of coal= 6,400 KJ
Heat value of natural gas = 35 MJ/m3 (Mega Joules/m3)
CO_2 emission of coal = 94.61 CO_2/TJ (Tera Joules/tons)
CO_2 emission of natural gas = 56.1tCO_2/TJ (Tera Joules/tons)

(Source: BCAS, 2005)

Fuel consumption and carbon emission rate of different types of kilns

Coal was used in the BTK, FCK and zigzag kilns. Natural gas was used only in Hoffman kilns. The most coal demanding kiln is BTK (Bull trench kiln) which consumes 28 tons coal per 100000 bricks production where as FCK and Zigzag kilns consume 20 tones and 18 tons respectively for same number of brick production (BCAS, 2011). The production period for BTK, FCK and Zigzag was November to mid-April. But Hoffman kiln can run all over the year. It concludes that to reduce CO_2 emission, at first Hoffman kiln should be chosen for brick manufacturing then gradually Zigzag kiln, FCK and then after BTK. (BCAS, 2005).

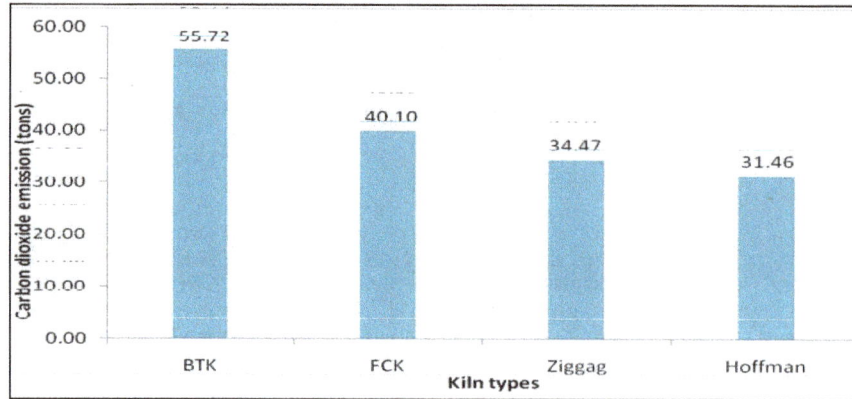

Fig.1. Coal consumption and CO_2 emission for individual brick kiln

CO$_2$ emission from different brick field of Mymensingh district

The statistical data of CO_2 emission from different brick field of Mymensingh and its surroundings are presented in Fig. 2. In those brick field they normally used different types of kilns such as Zigzag, Hoffman brick kiln etc. It has been found that highest amount of CO_2 is being emitted from Chamak brick field was 1882 t during the year of 2013 and the lowest amount of CO_2 is being emitted from Zhalak brick field was 1039.5t during the year of 2013.

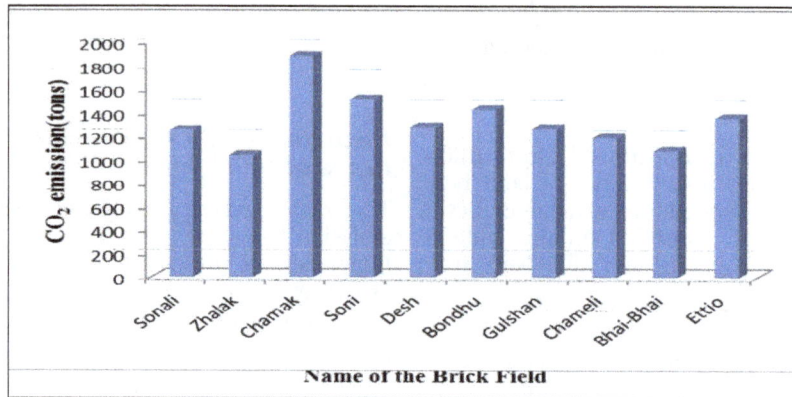

Fig. 2. CO_2 emission from different brick field of Mymensingh

Coal consumption from different brick field of Mymensingh district

The statistical data of coal consumption from different brick field of Mymensingh and its surroundings are presented in Fig. 3. The data shows that the maximum coal was consumed in Chamak brick field, which was 950 t and the minimum was in Zhalak brick field, which was 525 t during the year of 2013.

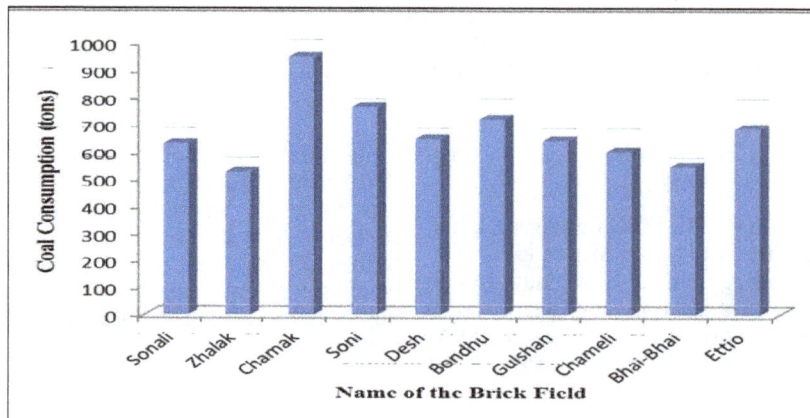

Fig. 3. Coal consumption from different brick field of Mymensingh

Total CO_2 emission from brick kilns

Total annual amount of CO_2 emission for six divisions are presented in Table 12. It has been found that total annual amount of CO_2 emission for 4 types brick kilns from Dhaka, Chittagong, Rajshahi, Khulana, Sylhet and Barisal are 8.862 Mt yr^{-1}, 10.048 Mt yr^{-1}, 12.783 Mt yr^{-1}, 15.250 Mt

yr^{-1}, in the year of 2002, 2005, 2007 and 2010, respectively. The percentage in last 10 years of CO_2 emission geographically increasing which is very alarming for us.

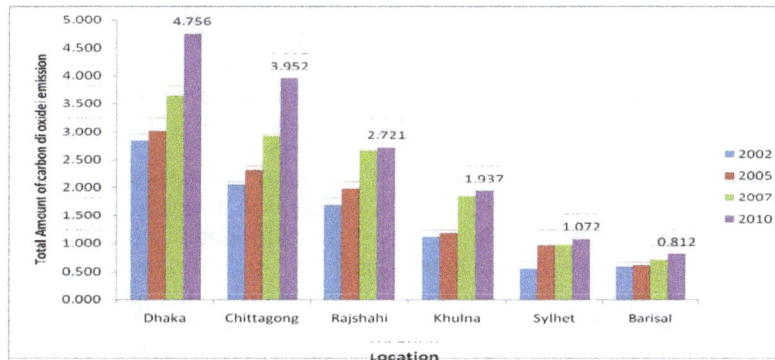

Fig. 4. Total CO_2 emission from brick kiln for six divisions for different years

All of the statistical graphs of total CO_2 emission in 2002, 2005, 2007 and 2010 of six divisions are presented in Fig. 4 shows the increasing rate of CO_2 emission in all of those years simultaneously around Bangladesh. The above result concludes that CO_2 emission rate are increasing day by day.

Conclusion

In the country such as Bangladesh with high level of poverty, malnutrition, and low human development index, it is unlikely that climate change will be a central focus for policies and measures. However, there are many interesting; conventional as well as innovative options, polices and measures those can benefit GHG (Green House Gas) reduction, better adaptation and increase sustainable livelihood potential. In spite of the issue of climate change, environment and resource management becomes important for sustainable development, which practically required integrated approach for accomplishment. Almost all the coal being used is imported from the Indian State of Meghalaya. This is because the brick industry has been growing at about 5.28% over the last decade with this trend leveling after 2014.

This baseline analysis indicates that GHG emissions from the brick industry are already at a high level and are expected to increase by at least 5.28% every year for the foreseeable future. This means that direct carbon emissions from kilns alone will rise to 8.7 million tons annually by 2014 or earlier depending on the growth rate of the industry. In addition, the brick industry is contributing in various ways to growing carbon emissions from other sources. Most notable, is the impact of brick making on land degradation

and deforestation. In a country where the pressure of population growth on a relatively small land mass is significant, farmland depletion can have alarming prospects for food security. Total farmland in Bangladesh is about 14 million hectares and this is depleting by about 80,000 hectares every year, a 0.05% depletion rate. Moreover, wood fuel is used as a secondary fuel for brick making accelerating the depletion of scarce carbon sinks in Bangladesh.

The results consist of different types of information regarding number of brick kiln, annual amount of fuel consumption (natural gas and coal) and CO_2 emission from brick kilns around Bangladesh and presented in the following sections. According to Kyoto protocol, being a developing country it is not obligatory for Bangladesh to reduce green house gas emission but clean development mechanism (CDM) should be promoted here. Finally, Government needs to push people by creating awareness against traditional kilns and make the technology simply available to the brick manufacturers. Electronic and print media should come forward to encourage people for using such kinds of bricks. More marketing is required to familiarize people with green bricks and to reduce carbon emission for better future and sustainable development.

References

Arifur, R. 2006. Introduction of brick kiln and their carbon emission in Bangladesh. CASE Project. Funded by the Energy Sector Management Assistance Program (ESMAP) of the World Bank. Final project report. vol.6. pp. 14-29.

BCAS. 2005. "Energy savings in brick units of Bangladesh", a report prepared by Bangladesh Centre for Advanced Studies, Dhaka, vol. 8. pp. 13-18.

BCAS. 2011. "Improving kiln efficiency in the brick making industry in Bangladesh," Project Design Document Form, CDM-SSC-PDD, Version 04/03/11, pp. 31-49.

Biswas, S., Uma, R., Kumar, A. and Vasudevan, N. 2009. Energy conservation and pollution control in brick kilns, Tata Energy Research Institute, New Delhi. Retrieved from http://www.scribd.com/doc/7843305/Brick-Energy-Conservation-Pollution-Control.

BUET. 2007. Small study on air quality of impacts of the North Dhaka brickfield cluster by modeling of emissions and suggestions for mitigation measures including financing models. Prepared by the Chemical Engineering Department, vol. 11. pp. 55-61.

Butler, T.M., Gurjar, B.R., Lawrence, M.G. and Lelieveld, J. 2004. "Evaluation of emissions and air quality in megacities," Atmospheric Environment, vol. 42. pp. 59-72.

DOE, 2010. Bangladesh Government-Memo no. DOE/Enforcement/37, pp. 17-28.

Enters, T.E. 2000. Report book on carbon emission monitoring and estimation in Asia. vol.8. pp. 7-24.

Heirli, U. and Maithel, S. 2008. Brick kiln by emission: the Herculean task of cleaning up the Asian brick industry. Swiss Agency for Development and Cooperation. Report no. 14. pp. 11-19.

IPCC. 2000. Guidelines for national greenhouse gas inventories, vol. 3. Greenhouse Gas Inventor Reference Manual, pp. 2-7.

Iqbal, A. and Hasan, I. 2007. Modeling for minimizing the emitted CO_2 from brick kilning through afforestation in Bangladesh, *J. Environ. Sci.* 5 (11): 21-29.

32

COMMUNITY PERCEPTIONS TOWARDS THE ESTABLISHMENT OF AN URBAN FOREST PLANTATION: A CASE OF DZIVARESEKWA, ZIMBABWE

A. Mureva, T. Nyamugure*, C. Masona, S.M. Mudyiwa, P. Makumbe, M. Muringayi and G. Nyamadzawo

Abstract

The health of urban forest communities not only depend on the government and non-governmental organizations, but also strongly rely on local community stewardship. A study was carried out to assess community perceptions on the establishment of an urban forest plantation among urban residents in Dzivaresekwa, an urban area in Harare. Randomized systematic sampling was used to select 150 households and one resident per household was interviewed using a pretested questionnaire with both closed and open-ended questions. The objectives of the study were to determine how age and gender and employment status variables, were related to the urban residents' perceptions towards establishment of a forest plantation in an urban area. Most females (58.3%) viewed the plantation as a threat while most men (51.7%) viewed the plantation as a recreational area. The highest proportion (61.9%) of the middle age group (21-40 years) perceived the plantation as a source of employment. There was a statistically significant relationship (p = 0.040) between gender and the general perception of establishing a forest plantation in the urban area. However, there was no statistically significant relationship (p = 0.203) between age groups and the perception of establishing a forest plantation in the urban area. It is concluded that the community had diverse perceptions on urban community forestry.

Keywords: Community Perception, Urban Forestry, Plantation

Department of Environmental Science, Faculty of Agriculture and Environmental Science, Bindura University of Science Education, P. Bag 1020, Bindura, Zimbabwe

*Corresponding author's email: vamugure@gmail.com (T. Nyamugure)

Introduction

Urban and community forests play an instrumental role in the social, economic, and environmental well-being of urban residents. Trees provide a wide variety of goods and services and are one of the most important forms of vegetation in an urban environment (Lorenzo et al., 2000). Trees reduces air pollution through carbon sequestration, they enhance air quality, moderate microclimate, reduce noise level and provide a habitat for wildlife (Fraser and Kenney, 2000; Lorenzo et al., 2000). Urban trees and forests in the developing world also provide firewood for energy, timber for use in construction activities, fruits, medicines and other useful minor forest products. Wood is still the main source of energy for cooking and heating, because alternative sources, when available, are unaffordable to the majority of urban residents. Despite their importance, Zimbabwe's urban trees, woodlands and forest resources are under increasing threat from the rapid increase of urban agriculture, expansion of urbanization and the high demand of fuel wood due to prolonged electricity shortage. The loss of trees within urban areas has led to long distance marketing of firewood. This has resulted in a number of rural lands surrounding the urban centers of Harare, Chitungwiza, Bulawayo and Gweru losing vast lands of trees for firewood. (Makonese and Mushamba, 2004). Strong and responsive programs similar to other community projects in infrastructure development are required to protect urban trees and forests. In urban environments, healthy community forests require ongoing stewardship as well as cooperation at governmental, organizational, and community levels.

In the process of designing, planning, and managing urban fringe forests it is important to consider urban people's opinions regarding establishment or presence of these forests. Local resident participation in the planning, planting, and management stages of forest care is essential to urban forest sustainability. A failure to address people's requirements may generate conflicts between users, planners, and managers, as has happened historically in relation to urban forests (Konijnendijk, 2000). To encourage participation in community forestry, positive attitudes of the public towards trees and education about the benefits they provide is essential. To determine the potential for resident stewardship of urban trees in Dzivaresekwa community and the need for educational programs in the future, the study sought to gather information on residents' knowledge and perceptions of urban trees and forests.

Materials and Methods

Study area

The woodland is located in Harare, Dzivaresekwa, a high-density residential area in Zimbabwe and its coordinates are, latitude 17.8065°S; longitude 30.925°E, (Fig. 1). The study area has approximately 1500 housing units and approximately a population of 10,100 people. The area is characterized by annual precipitation between 725-974 mm/year and average maximum temperature 15.5 to 20°C (FAO, 2012).The woodland is in an area made up of the miombo and open savanna grassland. In the open grassland, there are perennial grasses that include thatching grass *(Hyparrhenia filipendula)*, spear grass *(Heteropogon contortus)*, and couch grass *(Cynodon dactylon)*. Bushes and shrubs are abundant in the forest zone with sandy loam and gravel soil supporting a wide variety of trees with *Brachystegia spiciformis* and *Jubernadia globiflora* dominating.

Data collection

A sampling intensity of 10% was used and households were sampled using a randomized systematic sampling to come up with 150 households. The first house was randomly selected and thereafter data was collected from every 10th house in the sampling frame. The respondents provided information on their homes' possible distance from the plantation area choosing from the groups (<150m, 150-300m, and >300m radius).

The household survey was carried out using a pretested questionnaire with both closed and open-ended questions to examine the level and forms of participation of local people on the establishment of an urban plantation. The information collected during the study included the perceived impacts of establishing the urban plantation on the livelihood of people in Dzivaresekwa and residents' perceptions on the urban plantation. Two focus group discussions disaggregated by sex were conducted with residents in the area, and key informant interviews were held with two representatives

from the Dzivaresekwa District office, a representative of urban farmers, the plantation supervisor, the District Councilor and a representative from the Dzikwa Trust Directors.

Analysis of data

Data were analysed using Statistical Package for Social Sciences (SPSS) version 21. To analyse the perception of residents on social, economic and environmental benefits towards plantation a multinomial regression model was used. Social, economic or environmental benefits were taken to be the dependent variable while age, gender, work status, distance from plantation and knowledge of plantation were taken to be the explanatory variables. A binary logistic model was used to test the involvement of residents in plantation activities and whether they derived any benefits from planting trees. In both cases the dependent variable was a yes or no response. Age, gender, work status, distance from plantation and knowledge of plantation were taken to be the explanatory variables. In fitting the model to the data, all explanatory variables were included in the model. These were subsequently removed if they did not significantly contributed to the model. Removal of terms was stopped when a Minimal Adequate Model (MAM) was obtained. All tests were performed at 5% level of significance

Results

Demographic information of respondents

A total of 150 residents were interviewed. The respondents constituted of 60.7 % females and 39.3 % males. Most of the respondents (45.3%) were in the 21-40 years age group followed by the 41-60 years (26%), the 61+years (18.7%), and the group with the least number of respondents was the 15-20 years (10%).

Perceived socio-economic and environmental values of the plantation on peoples' livelihoods.

Social benefits and threats

Nagelkerke pseudo r-square=0.71 which means that 71% of the variation is explained by the model.

Contribution of each parameter to the model:

Perception of residents on social benefits and threats towards plantation significantly differed based on distance from plantation ($\chi^2=50$; df=10; P<0.001), age ($\chi^2=82.8$; df=15; P<0.001), gender ($\chi^2=12.7$; df=5; P<0.027) and work status ($\chi^2=72.1$; df=10; P<0.001).

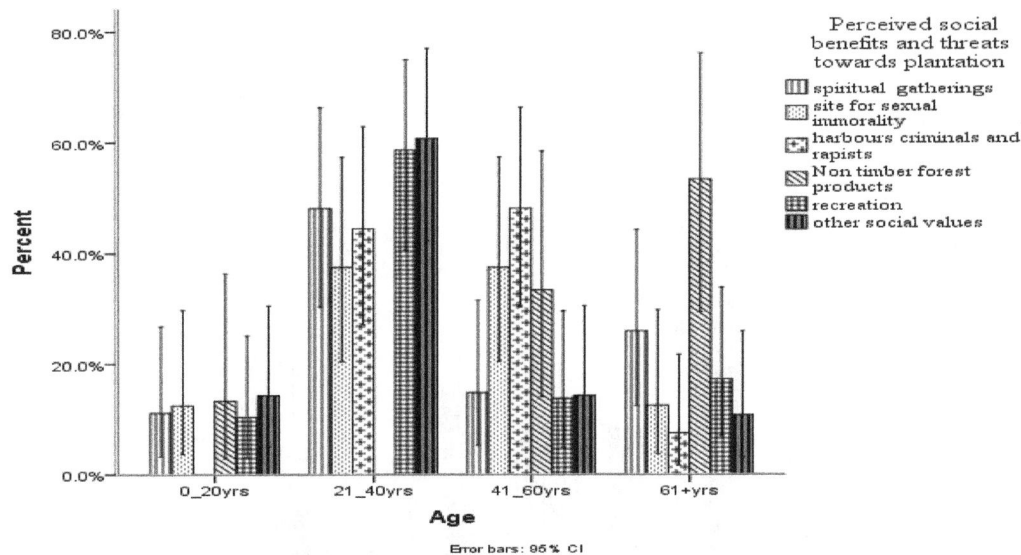

Fig. 1. Perceived social benefits and threats in relation to age

Older age (41 to 60 years and 61+ years) groups perceived plantations to be important for spiritual gatherings than young (0-20 years) and middle age (21-40 years) groups (P<0.005) (Fig. 1 and Table 1). The employed viewed the plantation to be a significantly important site for spiritual gatherings than the self- employed (P=0.002). Only the unemployed significantly perceive that

plantations are sites for sexual immorality (P=0.006). Males, middle aged people, and those staying at 150m and 300m from the plantation do not think that forests are places that significantly harbour criminals (P<0.05) while only the employed thought that plantations harbour criminals (P<0.001) [Table 1].

Table1. Perceived social benefits and threats

Category	Parameter	Level	Relative odds (coefficients)	Wald Chi sq.	P-value
Spiritual gatherings	Age	12-20 years	-2.8	4.81	0.028
		21-40 years	-3.4	8.77	0.003
	Work status	Employed	3.0	9.44	0.002
Site for sexual immorality	Work status	Unemployed	2.8	7.43	0.006
Harbours criminals	Age	21-40 years	-4.9	8.25	0.004
	Gender	Male	-2.6	4.35	0.037
	Distance from plantation	150m	-2.5	8.73	0.003
		300m	-4.2	10	0.004
	Work status	Employed	6.1	15.03	<0.001
NTFPs	Work status	Unemployed	3.5	9.65	0.002
	Age	0-20 years	-3.0	5.15	0.023
Recreation	Work status	Employed	3.1	8.17	0.004

The dependant variable, "other social benefits" was used as the reference category. For the explanatory variables: age group "61+", gender "female", distance "450+m" and work status "self-employed" were used as the reference categories. Parameters with significant negative coefficients decrease the likelihood of that response category with respect to the reference category while parameters with significant positive coefficients increase the likelihood of that response category with respect to the reference category.

Perceived economic value

Nagelkerke pseudo r-square=0.62 which means that 62% of the variation was explained by the model.

Perception of residents on economic benefits from plantations significantly differed based on age (χ^2=193.7; df=9; P<0.001), gender (χ^2=174.98; df=3; P<0.001), work status (χ^2=157.4; df=6; P=0.016) and knowledge of forest plantation (χ^2=163.1; df=3; P<0.001).

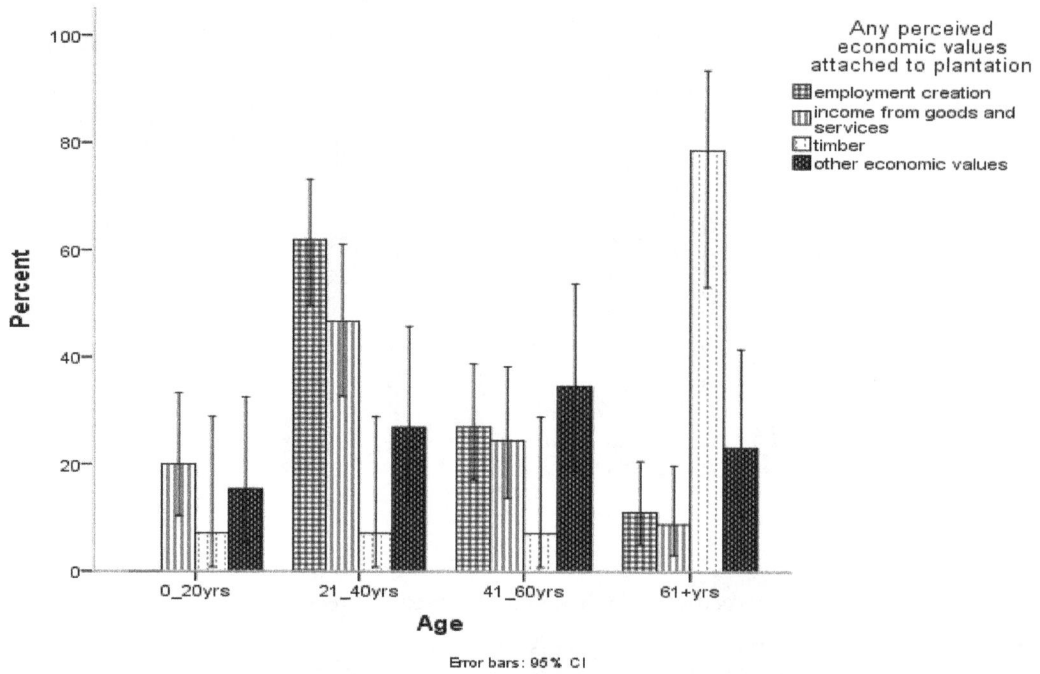

Fig. 2. Perceived economic values in relation to age

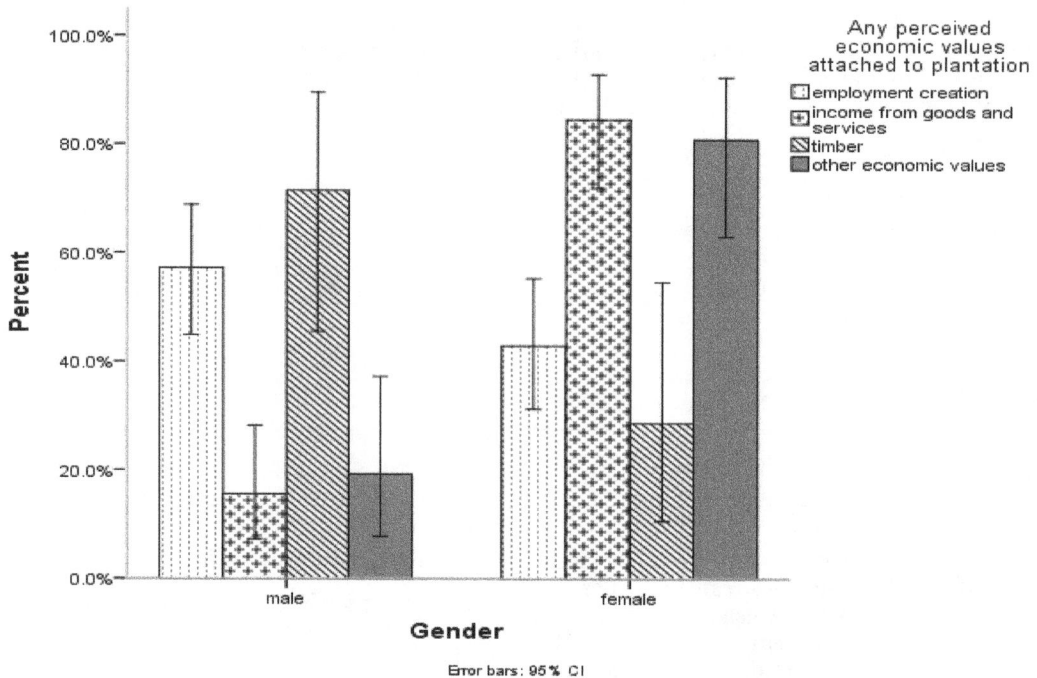

Fig. 3. Perceived economic values in relation to gender

Further investigation showed that males and age group 21-40 years perceive that the forest plantation significantly provides employment (P<0.005) (Fig. 2 and Table 2). In addition males perceive that timber is a major benefit of the plantations (P=0.004) (Fig. 6 and Table 2). Residents with knowledge of the plantation and those who were young (12-20 years old) perceived that forests are sources of income (P<0.02) [Table 2].

Table 2. Perceived economic values of the plantation

Category	Parameter	Level	Relative odds	Waldi Chi sq.	P-value
Employment benefits	Age	21-40 years	3.1	8.02	0.005
	Gender	male	2.8	11.8	0.001
Income from goods	Age	12-20 years	3.3	5.68	0.017
	Knowledge of plantation	Yes	3.4	9.08	0.003
Timber	Gender	male	3.0	8.29	0.004

The dependant variable, "other economic benefits" was used as the reference category. For the explanatory variables: age group "61+", gender "female" and knowledge of plantation "yes" were used as the reference categories. Parameters with significant positive coefficients increase the likelihood of that response category with respect to the reference category.

Perceived environmental value

Nagelkerke pseudo r-square=0.46
Perception of residents on environmental benefits from plantations significantly differed based on the following parameters: age (χ^2=90.3; 3=df; P<0.001), work status (χ^2=74.98; 2=df; P=0.007) and knowledge of forest plantation (χ^2=91.2; 1=df; P<0.001).

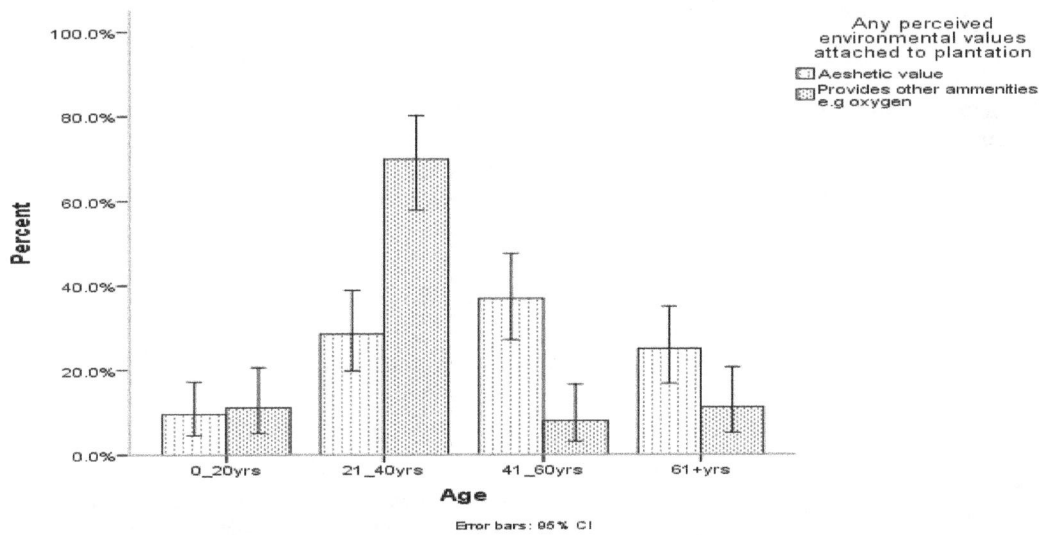

Fig. 4. Perceived environmental values in relation to age

More females appreciated the plantation as a provider of other amenities like clean air and slightly differed with males in appreciation for its aesthetic values (51.2%) (Fig. 5).

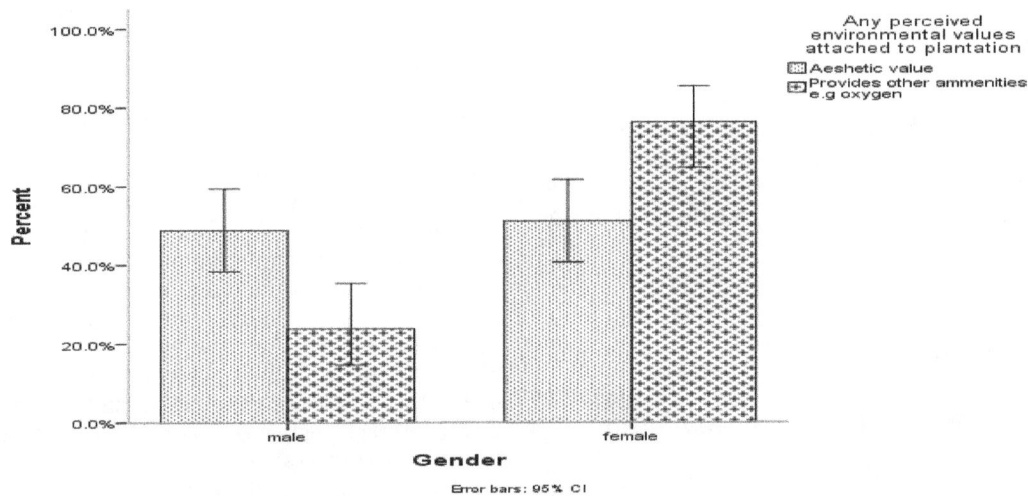

Fig. 5. Perceived environmental values in relation to gender

People aged between 21-40 years, the unemployed and those who have knowledge on plantations are about twice significantly less likely to perceive the plantations as source of aesthetic value than 64+ age group, the self-employed and those who have no knowledge of plantation (P=0.05) [Table 3].

Table 3. Perceived environmental values

Category	Parameter	Level	Relative odds	Waldi Chi sq	P-value
Aesthetic value	Age	21-40 years	-1.8	6.5	0.011
	Work status	Unemployed	-1.9	9.09	0.003
	Knowledge of plantation	Yes	-2.4	21.1	<0.001

For the dependant variable, "other environmental amenities" was used as the reference category. For the explanatory variables: age group "61+", work status "unemployed" and knowledge of plantation "yes" were used as the reference categories. Parameters with significant negative coefficients decrease the likelihood of that response category with respect to the reference category

Willingness to be part of the reforestation project

Involvement in plantation activities (y/n) using binary logistic model

Work status and distance from plantation were the two parameters that influenced significantly the involvement of residents in plantation activities [work status: Wald Chi-square=8.24; 2=df; P=0.016. Distance from plantation: Wald Chi-square=19.16; 2=df; P<0.001].

Odds of being involved in plantation activities is significantly higher when the respondent was employed (0.93x) and unemployed (1.5x) than self-employed (P<0.05). People living closer to the plantation (150m) are significantly less likely (2 times) to be involved in forest activities than people living far away (P<0.001) [Table 4].

Table 4. Willingness to be part of the reforestation project

Parameter	Level	Relative odds	Wald Chi square	P-value
Work status	Employed	0.93	4.45	0.035
	Unemployed	1.5	7.42	0.006
Distance from plantation	150m	-2.0	16.19	<0.001

The dependant variable, "yes" was used as the reference category. For the explanatory variables: work status "unemployed" and distance from plantation "450+m" were used as the reference categories. Parameters with significant negative coefficients decrease the likelihood of that response category with respect to the reference category while. Parameters with significant positive coefficients increase the likelihood of that response category with respect to the reference category

Benefits of planting trees

Perception on whether planting trees was beneficial than other activities was significantly affected by knowledge of forest plantation (Wald Chi square=4.1; 1=df; P=0.043) with those having knowledge of plantation having odds of 0.87x than those with no knowledge of plantation (Wald Chi-square=4.08; P=0.04)

Discussion

Residents' perceptions towards establishment of an urban plantation

The Dzivaresekwa people's perceptions were overall positive towards urban forestry possibly because the area is being used as a dumpsite posing health hazards to the community as compared to a forestry project, which comes as a development to the community.

Relationship of gender and age on the perceptions of establishing an urban forest plantation in Dzivaresekwa

Gender had no relationship with urban resident's perceptions of establishing a forest plantation in Dzivaresekwa. Age group differences were found to produce possible divergences in perceptions.

Age has also been a contributory factor in perceptions as stipulated by Burgess (1995); elderly people's attitudes towards urban woodland have been addressed in previous studies these have tended to concentrate on negative perceptions. The use of green space varies hugely between different social groups with underrepresented user groups including the elderly, young people, women, ethnic minorities, disabled people and the economically disadvantaged (Fairburn et al., 2005). The differences in perceptions according to age group might be as because of the generation gap. Cohen et al. (2007) strongly suggests that the proximity of (peri-) urban forests and parks to where people live has a huge impact on levels of usage and therefore the benefits that can be gained from use.

The 21-40yr age group had most respondents because it is the most active group of a society and given the high unemployment rate in Zimbabwe as a whole, the respondents in the age group were mostly found at home. The 61+ age group had few respondents because it is most likely that most of them had retired and gone to rural areas.

Perceived socio-economic and environmental values on the livelihoods of people in Dzivaresekwa

According to Meza (1992), urban forestry presents many social issues that require addressing to allow urban forestry to be seen by many as an advantage rather than a curse on their environment. Females were found to be concerned about the plantation being a site for sexual immorality and its potential to harbour criminals and rapists. Burgess (1995) indicated that people are afraid of becoming the victims of physical or sexual assault, robbery or bullying and intimidation from groups of young people in the woodland. In addition, a study carried out by Konijnendijk's (1999) showed that the concern about aspects of safety was widespread. In correlation with this study, urban forests in the Dutch cities of Haarlem and Amsterdam indicated that people do not always feel safe in urban forests. The same research also pointed out that some forests in France have gained a rather bad reputation due to safety problems. Lack of safety could mean threat by criminal elements and falling tree limbs. People generally as well as in the South Wales study were concerned by problems with urban forest abuse, such as the dumping of garbage, prostitution, homosexual activity and other ('semi') criminal or illegal activities, littering, dogs walking off their leash, illegal motorized traffic through the forest crossing, illegal fires and poaching, among others. Dzivaresekwa also has a record of high crime rates hence the varying responses both threats and benefits.

The research proved that in Dzivaresekwa females appreciated the plantation as a place with a high aesthetic value and providing other amenities. Habitat surveys, as well as studies focusing on flora and fauna, have all demonstrated the importance of tree cover for biodiversity in urban areas (Tyrvainen et al., 2005) and enhanced biodiversity. Urban plantations can often also enhance an area's recreational and aesthetic qualities (Chen and Jim, 2008).This research has proved that people are not only biased towards monetary benefits but are able to appreciate nature. Dzivaresekwa often has sewage bursts all over but the plantation can help in air purification and sewer effluent absorption.

Willingness to be part of the plantation project

The reasons for willingness seemed to be more economically biased and those whose who did not show interest in the project were mainly the ones who owned small pieces of land in the plantation area.

Conclusion

The residents of Dzivaresekwa had diverse perception towards establishment of a forest plantation in their suburb though most of them had a positive perception towards urban forestry. Gender had no significant influence on perceptions of establishing a forest plantation in Dzivaresekwa. Age group differences produce possible divergences in perceptions. After an in-depth assessment into the socio-economic and environmental values, the results indicated that people had perceived various benefits and threats.

Recommendations

Demographic information should be considered in community projects because the research clearly shows that age group differences had an effect on the people's perceptions in Dzivaresekwa. Though the people's perceptions were overall positive we do not have to ignore that there were also negatives so we have to try and conscientise them and make them realize the positivity of the forest plantation. Consideration of other factors such as education level and work status need to be considered in future projects.

To forage ahead with community forestry programmes and their implementation foresters should develop a two-way communication system with the residents. They should be involved in problem definition, the design of possible solutions, and evaluation of the proposed technological solutions. Women should be involved in community forestry programmes as decision makers and implementers as the wood collection and consumption is their responsibility, and they remain in the villages all year round. Women play a vital role in tree planting programmes as they show interest, positive attitude, and willingness to participate.

References

Burgess, J. 1995. *Growing in Confidence-Understanding People's Perceptions of Urban Fringe Woodlands.* Countryside Commission, Cheltenham.

Chen, W.Y. and Jim, C.Y. 2008. Assessment and Valuation of the Ecosystem Services Provided by Urban Forests. *In:* M.M. Carreiro, Y.C. Song and J. Wu (Eds.) *Ecology, Planning, and Management of*

Urban Forests. Springer, New York: pp. 53-83.

Cohen, D.A., McKenzie, T.L., Sehgal, A., Williamson, S., Golinelli, D. and Lurie, N. 2007. Contribution of Public Parks to Physical Activity. *American J. Public Health* 97 (3): 509-514.

Fairburn, J., Walker, G. and Smith, G. 2005. Investigating Environmental Justice in Scotland - Links Between Measures of Environmental Quality and Social Deprivation Report UE4 (03)01, Scottish and Northern Ireland Forum for Environmental Research, Edinburghh.

FAO. 2012. country profiles. http:www.fao.org/countryprofiles/maps.aspi so3=ZWE&lang=en

Fraser, E.D.G. and Kenney, W.A. 2000. Cultural background and landscape history as factors affecting perceptions of the urban forest. *J. Arboric.* 26 (2): 106-113

Konijnendijk, C.C. 1999. Urban forestry: comparative analysis of policies and concepts in Europe. Contemporary Urban Forestry Policy-Making in Selected Cities and Countries of Europe. Working Paper 20. European Forest Institute, Joensuu, Finland. pp. 266-289.

Konijnendijk, C.C. 2000. Adapting forestry to urban demands – role of communication in urban forestry in Europe. Landscape and Urban Planning: pp. 89-100.

Lorenzo, A., Blanche, C., Qi, Y. and Guidry, M. 2000. Assessing residents' willingness to pay to preserve thecommunity urban forest: A small-city case study. *J. Arboric.* 30 (1): 28-35.

Makonese, M. and Mushamba, S. 2004. The Policy, Legislative and Institutional Framework for Urban Forestry in Zimbabwe, 34 UA-Magazine: pp. 34-35.

Meza, H.M.B. 1992. Current Situation of the Urban Forest in Mexico City. *J. Arboric.* 18: 33-36.

Tyrvainen, L., Pauleit, S., Seeland, K. and de Vries, S. 2005 Benefits and Uses of Urban Forests and Trees. *In:* C.C. Konijnendijk, Nilsson, K., Randrup, T.B. and Schipperijn (Eds.) *Urban Forests and Trees.* Springer, Berlin/Hieldelberg/New York: pp. 81-144.

Permissions

List of Contributors

M. A. Hossain
Associate Professor, Department of Dairy and Poultry Science, Chittagong Veterinary and Animal Sciences University, Khulshi, Chittagong-4225, Bangladesh

J. R. Amin
School of Environmental and Rural Science, Armidale, NSW 2351, Australia

M. E. Hossain
Department of Animal Science and Animal Nutrition, Chittagong Veterinary and Animal Sciences University, Khulshi, Chittagong-4225, Bangladesh

M. K. Hasan
Senior Scientific Officer (Agricultural Economics), Spices Research Centre, Bangladesh Agricultural Research Institute, Shibgonj, Bogra, Bangladesh

M. A. A. Mahmud
Assistant Inspector General (Crime-East), Bangladesh Police, Police Headquarter, Dhaka, Bangladesh

Naima Sultana
Former MS student, Dept. of Agril. Ext. and Rural Development, BSMRAU, Gazipur, Bangladesh

M. S. I. Afrad
Professor, Dept. of Agril. Ext. and Rural Development, BSMRAU, Gazipur, Bangladesh

M. N. Uddin
Assistant Professor, Department of Agricultural Extension Education, Bangladesh Agricultural University (BAU), Mymensingh-2202, Bangladesh

N. Anjuman
Ex-MS Student, Department of Rural Sociology, BAU, Mymensingh-2202, Bangladesh

I. V. Shepelenko
Department of Agricultural Machinery Operating and Repair, National Technical University of Kirovograd, Ukraine

V. V. Cherkun
Department of Mechanic and Technology, Agrotechnological University of State Tavri, Ukraine

A. Warouma
Department of Rural Engineering & Water and Forests, University of Maradi, Niger

M. S. Alam
Department of Genetics and Plant Breeding, Bangladesh Agricultural University, Mymensingh, Bangladesh

S. N. Begum
Senior Scientific Officer, Plant Breeding Division, Bangladesh Institute of Nuclear Agriculture, Mymensingh, Bangladesh

R. Gupta
Scientific Officer, Bangladesh Institute of Nuclear Agriculture, Sub Station, Khagrachari, Bangladesh

S. N. Islam
Department of Genetics and Plant Breeding, Bangladesh Agricultural University, Mymensingh, Bangladesh

Francisco M. Iglesias
Cátedra de Cerealicultura, Dto. Producción Vegetal, Facultad de Agronomía UBA, Av San Martín 4453 (C1417DSE) Buenos Aires, Argentina

Daniel J. Miralles
Cátedra de Cerealicultura, Dto. Producción Vegetal, Facultad de Agronomía UBA, Av San Martín 4453 (C1417DSE) Buenos Aires, Argentina
CONICET and IFEVA

M. F. Hossain
Senior Scientific Officer, Regional Agricultural Research Station, BARI, Ishwardi, Pabna, Bangladesh

N. Ara
Principal Scientific Officer, Regional Agricultural Research Station, BARI, Ishwardi, Pabna, Bangladesh

M. R. Islam
Scientific Officer, Regional Agricultural Research Station, BARI, Ishwardi, Pabna, Bangladesh

J. Hossain
Scientific Officer, Regional Agricultural Research Station, BARI, Ishwardi, Pabna, Bangladesh

B. Akhter
Scientific Officer, Regional Agricultural Research Station, BARI, Ishwardi, Pabna, Bangladesh

M. K. Islam
Senior Scientific Officer, Bangladesh Agricultural Research Institute, Burirhat, Rangpur, Bangladesh

M. S. Mahfuz
Regional Farm Broadcasting Officer, Agriculture Information Service, Rangpur, Bangladesh

M. A. I. Sarker
Senior Scientific Officer, Bangladesh Agricultural Research Institute, Burirhat, Rangpur, Bangladesh

S. Ghosh
Scientific Officer, Bangladesh Agricultural Research Institute, Burirhat, Rangpur, Bangladesh

A. S. M. Y. Ali
Scientific Officer, Bangladesh Agricultural Research Institute, Burirhat, Rangpur, Bangladesh

A. Ngakou
Department of Biological Sciences, University of Ngaoundere, P.O.Box 454 Ngaoundéré, Cameroon

H. Koehler
Center for Environmental Research and Sustainable Technology (UFT), University of Bremen, Germany

H. C. Ngueliaha
Department of Biological Sciences, University of Ngaoundere, P.O.Box 454 Ngaoundéré, Cameroon

M. J. Hasan
Principal Scientific Officer, Hybrid Rice Division, Bangladesh Rice Research Institute, Joydebpur, Gazipur, Bangladesh

M. U. Kulsum
Scientific Officer, Hybrid Rice Division, Bangladesh Rice Research Institute, Joydebpur, Gazipur, Bangladesh

M. Z. Ullah
Field Coordinator (Vegetable), AVRDC, Bangladesh

M. Manzur Hossain
Assistant Information Officer (Crop Production), Agriculture Information Services, Khamarbari, Farmgate, Dhaka-1215

M. Eleyash Mahmud
PhD fellow BSMRAU and Plant Breeder, Energypac Agro Ltd. Monipur, Gazipur, Bangladesh

F. A. Prodhan
Lecturer, Department of Agricultural Extension and Rural Development, Bangabandhu Sheikh Mujibur Rahman agricultural University, Gazipur-1706, Bangladesh

M. S. I. Afrad
Professor, Department of Agricultural Extension and Rural Development, Bangabandhu Sheikh Mujibur Rahman agricultural University, Gazipur-1706, Bangladesh

M. K. Islam
Senior Scientific Officer, Bangladesh Agricultural Research Institute, Burirhat, Rangpur, Bangladesh

M. S. Mahfuz
Regional Farm Broadcasting Officer, Agriculture Information Service, Rangpur, Bangladesh

S. Ghosh
Scientific Officer, Bangladesh Agricultural Research Institute, Burirhat, Rangpur, Bangladesh

A. S. M.Y. Ali
Scientific Officer, Bangladesh Agricultural Research Institute, Burirhat, Rangpur, Bangladesh

M. Z. Hasnat
Information Officer, Agriculture Information Service, Dhaka, Bangladesh

M. U. Dimelu
Department of Agricultural Extension, University of Nigeria Nsukka, Nigeria
Department of Agricultural Extension and Economics, River State University, Port Harcout, Nigeria

F. H. Bonjoru
Department of Agricultural Extension, University of Nigeria Nsukka, Nigeria

A. I. Emodi
Department of Agricultural Extension, University of Nigeria Nsukka, Nigeria

M. C. Madukwe
Department of Agricultural Extension and Economics, River State University, Port Harcout, Nigeria

M. R. Begum
Assistant Professor, Department of Agricultural Economics and Social Sciences, Chittagong Veterinary and Animal Sciences University, Khulshi, Chittagong, Bangladesh

M. Anaruzzaman
Doctor of Veterinary Medicine, Chittagong Veterinary and Animal Sciences University, Khulshi, Chittagong, Bangladesh

M. S. I. Khan
Assistant Professor, Department of Food Microbiology, Patuakhali Science and Technology University (PSTU), Dumki, Patuakhali, Bangladesh

M. Yousuf
Director (Program), SANGRAM (Sangathita Gramunnyan Karmasuchi), Bangladesh

M. A. Islam
On Farm Research Division, Regional Agricultural Research Station, BARI, Pabna, Bangladesh

Mst. A. Begum
Department of Agronomy, Bangladesh Agricultural University, Mymensingh, Bangladesh

M. M. Jahangir
Department of Soil Science, Bangladesh Agricultural University, Mymensingh, Bangladesh

M. A. Badshah
College of Agronomy, Hunan Agriculture University, Hunan, China

Tu Nai Mei
College of Agronomy, Hunan Agriculture University, Hunan, China

Mehreen Hassan
Institute of Agricultural Sciences, University of the Punjab, Lahore, Pakistan

Sana Hanif
Institute of Agricultural Sciences, University of the Punjab, Lahore, Pakistan

N. Hamzehpour
Soil Science Department, Maragheh University, Maragheh, East Azerbaijan, Iran

M. K. Eghbal
Soil Science Department, Tarbiat Modares University, Tehran, Iran

P. Bogaert
Faculty of Bio-engineering, Agronomy and Environment, Catholic University of Leuven, Louvain-la-Neuve, Belgium

N. Toomanian
Isfahan Research Centre for Agriculture and Natural Resources, Isfahan, Iran

A. H. M. Kohinoor
Freshwater Station, Bangladesh Fisheries Research Institute, Mymensingh, Bangladesh

M. M. Rahman
Freshwater Station, Bangladesh Fisheries Research Institute, Mymensingh, Bangladesh

M. J. Alam
Department of Fisheries Management, Bangladesh Agricultural University, Mymensingh-2202, Bangladesh
Department of Fisheries, Ministry of Fisheries and Livestock, Dhaka, Bangladesh

M. Shahjahan
Department of Fisheries Management, Bangladesh Agricultural University, Mymensingh-2202, Bangladesh

M. S. Rahman
Department of Fisheries Management, Bangladesh Agricultural University, Mymensingh-2202, Bangladesh

H. Rashid
Department of Fisheries Management, Bangladesh Agricultural University, Mymensingh-2202, Bangladesh

M. A. Hosen
Department of Fisheries Management, Bangladesh Agricultural University, Mymensingh-2202, Bangladesh
Department of Fisheries, Ministry of Fisheries and Livestock, Dhaka, Bangladesh

F. Jahan
Department of Fisheries Management, Bangladesh Agricultural University, Mymensingh, Bangladesh

M. S. Rahman
Department of Fisheries Management, Bangladesh Agricultural University, Mymensingh, Bangladesh

M. A. Hossain
Department of Aquaculture, Bangabandhu Sheikh Mujibur Rahman Agricultural University, Gazipur, Bangladesh

K. U. Ahamed
Scientific Officer, PGRC, Regional Agricultural Research Station Ishurdi, Pabna, Bangladesh

B. Akhter
Scientific Officer, Plant Pathology Division, Regional Agricultural Research Station Ishurdi, Pabna, Bangladesh

M. R. Islam
Scientific Officer, Agronomy Division, Regional Agricultural Research Station Ishurdi, Pabna, Bangladesh

M. R. Humaun
Scientific Officer, Plant Pathology Division, Regional Agricultural Research Station Ishurdi, Pabna, Bangladesh

M. J. Alam
Scientific Officer, Pulses Research Centre, BARI, Ishurdi, Pabna, Bangladesh

M. M. Hernández
Instituto de Productos Naturales y Agrobiología, Consejo Superior de Investigaciones Científicas, Avenida Astrofísico Francisco Sánchez 3, 38206 La Laguna, Tenerife, Canary Islands, Spain

M. Fernández-Falcón
Instituto de Productos Naturales y Agrobiología, Consejo Superior de Investigaciones Científicas, Avenida Astrofísico Francisco Sánchez 3, 38206 La Laguna, Tenerife, Canary Islands, Spain

C. E. Álvarez
Instituto de Productos Naturales y Agrobiología, Consejo Superior de Investigaciones Científicas, Avenida Astrofísico Francisco Sánchez 3, 38206 La Laguna, Tenerife, Canary Islands, Spain

G. K. M. M. Rahman
Department of Soil Science, Bangabandhu Sheikh Mujibur Rahman Agricultural University, Salna, Gazipur, Bangladesh

M. S. I. Afrad
Department of Agricultural Extension and Rural Development, Bangabandhu Sheikh Mujibur Rahman Agricultural University, Salna, Gazipur, Bangladesh

M. M. Rahman
Department of Soil Science, Bangabandhu Sheikh Mujibur Rahman Agricultural University, Salna, Gazipur, Bangladesh

M. A. Hosen
Department of Fisheries Management, Bangladesh Agricultural University, Mymensingh-2202, BangladeshDepartment of Fisheries, Ministry of Fisheries and Livestock, Dhaka, Bangladesh

M. Shahjahan
Department of Fisheries Management, Bangladesh Agricultural University, Mymensingh-2202, Bangladesh

M. S. Rahman
Department of Fisheries Management, Bangladesh Agricultural University, Mymensingh-2202, Bangladesh

M. J. Alam
Department of Fisheries Management, Bangladesh Agricultural University, Mymensingh-2202, Bangladesh Department of Fisheries, Ministry of Fisheries and Livestock, Dhaka, Bangladesh

S. Sutharsan
Department of Crop Science, Faculty of Agriculture, Eastern University, Sri Lanka, Chenkalady- 30350, Sri Lanka

K. Sivakumar
Department of Crop Science, Faculty of Agriculture, Eastern University, Sri Lanka, Chenkalady- 30350, Sri Lanka

S. Srikrishnah
Department of Crop Science, Faculty of Agriculture, Eastern University, Sri Lanka, Chenkalady- 30350, Sri Lanka

S. Chikuta
Department of Agricultural production, Makerere University, P .O. Box 7062, Uganda

T. Odong
Department of Agricultural production, Makerere University, P .O. Box 7062, Uganda

F. Kabi
Department of Agricultural production, Makerere University, P .O. Box 7062, Uganda

M. Mwala
Department of Crop Science, University of Zambia, P. O. Box 32379, Zambia

P. Rubaihayo
Department of Agricultural production, Makerere University, P .O. Box 7062, Uganda

J. Hossain
Lecturer, Dept. of Agroforestry and Environmental Science, Exim Bank Agricultural University, Chapainawabganj, Bangladesh

T. Ahmed
Associate Professor, Dept. of Agroforestry and Environment, BSMRAU, Gazipur, Bangladesh

M. Z. Hasnat
Information Officer (Plant Protection), Agriculture Information Service, Ministry of Agriculture, Bangladesh

D. Karim
Senior Scientific Officer, Soil Resource Development Institute, Ministry of Agriculture, Bangladesh

S. Tamiru
College of Agriculture and Veterinary Sciences, Department of Plant Sciences, Ambo University, Ambo, Ethiopia

H. Ashagre
College of Agriculture and Veterinary Sciences, Department of Plant Sciences, Ambo University, Ambo, Ethiopia

M. A. Imran
Department of Environmental Science, Bangladesh Agricultural University, Mymensingh, Bangladesh

M. A. Baten
Department of Environmental Science, Bangladesh Agricultural University, Mymensingh, Bangladesh

B. S. Nahar
Principal Scientific Officer, Bangladesh Agricultural University, Mymensingh, Bangladesh

N. Morshed
Department of Environmental Science, Bangladesh Agricultural University, Mymensingh, Bangladesh

A. Mureva
Department of Environmental Science, Faculty of Agriculture and Environmental Science, Bindura University of Science Education, P. Bag 1020, Bindura, Zimbabwe

T. Nyamugure
Department of Environmental Science, Faculty of Agriculture and Environmental Science, Bindura University of Science Education, P. Bag 1020, Bindura, Zimbabwe

C. Masona
Department of Environmental Science, Faculty of Agriculture and Environmental Science, Bindura University of Science Education, P. Bag 1020, Bindura, Zimbabwe

S. M. Mudyiwa
Department of Environmental Science, Faculty of Agriculture and Environmental Science, Bindura University of Science Education, P. Bag 1020, Bindura, Zimbabwe

P. Makumbe
Department of Environmental Science, Faculty of Agriculture and Environmental Science, Bindura University of Science Education, P. Bag 1020, Bindura, Zimbabwe

M. Muringayi
Department of Environmental Science, Faculty of Agriculture and Environmental Science, Bindura University of Science Education, P. Bag 1020, Bindura, Zimbabwe

G. Nyamadzawo
Department of Environmental Science, Faculty of Agriculture and Environmental Science, Bindura University of Science Education, P. Bag 1020, Bindura, Zimbabwe

www.ingramcontent.com/pod-product-compliance
Lightning Source LLC
Chambersburg PA
CBHW080653200326
41458CB00013B/4833